Engineering Agile Big-Data Systems

RIVER PUBLISHERS SERIES IN SOFTWARE ENGINEERING

Indexing: All books published in this series are submitted to the Web of Science Book Citation Index (BkCI), to CrossRef and to Google Scholar.

The "River Publishers Series in Software Engineering" is a series of comprehensive academic and professional books which focus on the theory and applications of Computer Science in general, and more specifically Programming Languages, Software Development and Software Engineering.

Books published in the series include research monographs, edited volumes, handbooks and textbooks. The books provide professionals, researchers, educators, and advanced students in the field with an invaluable insight into the latest research and developments.

Topics covered in the series include, but are by no means restricted to the following:

- Software Engineering
- Software Development
- Programming Languages
- Computer Science
- Automation Engineering
- Research Informatics
- Information Modelling
- Software Maintenance

For a list of other books in this series, visit www.riverpublishers.com

Engineering Agile Big-Data Systems

Editors

Kevin Feeney
Trinity College Dublin, Ireland

Jim Davies
Oxford University, United Kingdom

James Welch
Oxford University, United Kingdom

Sebastian Hellmann
University of Leipzig, Germany

Christian Dirschl
Wolters Kluwer, Germany

Andreas Koller
Semantic Web Company, Austria

Pieter Francois
Oxford University, United Kingdom

Arkadiusz Marciniak
Adam Mickiewicz University, Poland

Published 2018 by River Publishers
River Publishers
Alsbjergvej 10, 9260 Gistrup, Denmark
www.riverpublishers.com

Distributed exclusively by Routledge
4 Park Square, Milton Park, Abingdon, Oxon OX14 4RN
605 Third Avenue, New York, NY 10158

First published in paperback 2024

Engineering Agile Big-Data Systems / by Kevin Feeney, Jim Davies, James Welch, Sebastian Hellmann, Christian Dirschl, Andreas Koller, Pieter Francois, Arkadiusz Marciniak.

© The Editor(s) (if applicable) and The Author(s) 2018. This book is published open access.

Open Access
This book is distributed under the terms of the Creative Commons Attribution-Non-Commercial 4.0 International License, CC-BY-NC 4.0) (http://creativecommons.org/licenses/by/4.0/), which permits use, duplication, adaptation, distribution and reproduction in any medium or format, as long as you give appropriate credit to the original author(s) and the source, a link is provided to the Creative Commons license and any changes made are indicated. The images or other third party material in this book are included in the work's Creative Commons license, unless indicated otherwise in the credit line; if such material is not included in the work's Creative Commons license and the respective action is not permitted by statutory regulation, users will need to obtain permission from the license holder to duplicate, adapt, or reproduce the material.

The use of general descriptive names, registered names, trademarks, service marks, etc. in this publication does not imply, even in the absence of a specific statement, that such names are exempt from the relevant protective laws and regulations and therefore free for general use.
The publisher, the authors and the editors are safe to assume that the advice and information in this book are believed to be true and accurate at the date of publication. Neither the publisher nor the authors or the editors give a warranty, express or implied, with respect to the material contained herein or for any errors or omissions that may have been made.

Routledge is an imprint of the Taylor & Francis Group, an informa business

Publisher's Note
The publisher has gone to great lengths to ensure the quality of this reprint but points out that some imperfections in the original copies may be apparent.

While every effort is made to provide dependable information, the publisher, authors, and editors cannot be held responsible for any errors or omissions.

ISBN: 978-87-7022-016-3 (hbk)
ISBN: 978-87-7004-381-6 (pbk)
ISBN: 978-1-003-33812-3 (ebk)

DOI: 10.1201/9781003338123

Contents

Preface xv

Acknowledgements xvii

List of Contributors xix

List of Figures xxi

List of Tables xxix

List of Abbreviations xxxi

1 Introduction 1
- 1.1 State of the Art in Engineering Data-Intensive Systems ... 2
  - 1.1.1 The Challenge 4
- 1.2 State of the Art in Semantics-Driven Software Engineering . 5
  - 1.2.1 The Challenge 8
- 1.3 State of the Art in Data Quality Engineering 8
  - 1.3.1 The Challenge 11
- 1.4 About ALIGNED 12
- 1.5 ALIGNED Partners 15
  - 1.5.1 Trinity College Dublin 15
  - 1.5.2 Oxford University – Department of Computer Science 15
  - 1.5.3 Oxford University – School of Anthropology and Museum Ethnography 15
  - 1.5.4 University of Leipzig – Agile Knowledge Engineering and Semantic Web (AKSW) 15
  - 1.5.5 Semantic Web Company 16
  - 1.5.6 Wolters Kluwer Germany 16

		1.5.7	Adam Mickiewicz University in Poznań	16
		1.5.8	Wolters Kluwer Poland	17
	1.6	Structure		17

2 ALIGNED Use Cases – Data and Software Engineering Challenges   21
Arkadiusz Marciniak and Patrycja Filipowicz

- 2.1 Introduction 21
- 2.2 The ALIGNED Use Cases 24
  - 2.2.1 Seshat: Global History Databank 24
  - 2.2.2 PoolParty Enterprise Application Demonstrator System 26
  - 2.2.3 DBpedia 27
  - 2.2.4 Jurion and Jurion IPG 29
  - 2.2.5 Health Data Management 31
- 2.3 The ALIGNED Use Cases and Data Life Cycle. Major Challenges and Offered Solutions 33
- 2.4 The ALIGNED Use Cases and Software Life Cycle. Major Challenges and Offered Solutions 36
- 2.5 Conclusions 39

3 Methodology   41
James Welch, Jim Davies, Kevin Feeney, Pieter Francois, Jeremy Gibbons and Seyyed Shah

- 3.1 Introduction 41
- 3.2 Software and Data Engineering Life Cycles 43
  - 3.2.1 Software Engineering Life Cycle 43
  - 3.2.2 Data Engineering Life Cycle 47
- 3.3 Software Development Processes 49
  - 3.3.1 Model-Driven Approaches 49
  - 3.3.2 Formal Techniques 51
  - 3.3.3 Test-Driven Development 52
- 3.4 Integration Points and Harmonisation 53
  - 3.4.1 Integration Points 54
  - 3.4.2 Barriers to Harmonisation 55
  - 3.4.3 Methodology Requirements 58
- 3.5 An ALIGNED Methodology 60
  - 3.5.1 A General Framework for Process Management . . . 60

	3.5.2	An Iterative Methodology and Illustration	63
3.6		Recommendations	65
	3.6.1	Sample Methodology	66
3.7		Sample Synchronisation Point Activities	69
	3.7.1	Model Catalogue: Analysis and Search/Browse/ Explore	70
	3.7.2	Model Catalogue: Design and Classify/Enrich	71
	3.7.3	Semantic Booster: Implementation and Store/Query	72
	3.7.4	Semantic Booster: Maintenance and Search/Browse/ Explore	72
3.8		Summary	74
	3.8.1	Related Work	74
3.9		Conclusions	76

4 ALIGNED MetaModel Overview   79
Rob Brennan, Bojan Bozic, Odhran Gavin and Monika Solanki

4.1		Generic Metamodel	80
	4.1.1	Basic Approach	80
	4.1.2	Namespaces and URIs	81
	4.1.3	Expressivity of Vocabularies	82
	4.1.4	Reference Style for External Terms	82
	4.1.5	Links with W3C PROV	82
4.2		ALIGNED Generic Metamodel	83
	4.2.1	Design Intent Ontology (DIO)	83
4.3		Software Engineering	83
	4.3.1	Software Life Cycle Ontology	83
	4.3.2	Software Implementation Process Ontology (SIP)	85
4.4		Data Engineering	86
	4.4.1	Data Life Cycle Ontology	86
4.5		DBpedia DataID (DataID)	87
4.6		Unified Quality Reports	89
	4.6.1	Reasoning Violation Ontology (RVO) Overview	89
	4.6.2	W3C SHACL Reporting Vocabulary	91
	4.6.3	Data Quality Vocabulary	93
	4.6.4	Test-Driven RDF Validation Ontology (RUT)	96
	4.6.5	Enterprise Software Development (DIOPP)	109
	4.6.6	Unified Governance Domain Ontologies	111

	4.6.7	Semantic Booster and Model Catalogue Domain Ontology	112
		4.6.7.1 Model catalogue	112
		4.6.7.2 Booster	113
	4.6.8	PROV	113
	4.6.9	SKOS	115
	4.6.10	OWL	117
	4.6.11	RDFS	119
	4.6.12	RDF	121

5 Tools   125

Kevin Feeney, Christian Dirschl, Katja Eck, Dimitris Kontokostas, Gavin Mendel-Gleason, Helmut Nagy, Christian Mader and Andreas Koller

5.1	Model Catalogue	125
	5.1.1 Introduction	125
	5.1.2 Model Catalogue	127
	5.1.2.1 Architecture	127
	5.1.2.2 Searching and browsing the catalogue . .	130
	5.1.2.3 Editing the catalogue contents	131
	5.1.2.4 Administration	134
	5.1.2.5 Eclipse integration and model-driven development	134
	5.1.2.6 Semantic reasoning	136
	5.1.2.7 Automation and search	137
	5.1.3 Semantic Booster	138
	5.1.3.1 Introduction	138
	5.1.3.2 Semantic Booster	139
5.2	RDFUnit	155
	5.2.1 RDFUnit Integration	157
	5.2.1.1 JUnit XML report-based integration . . .	158
	5.2.1.2 Custom apache maven-based integration .	158
	5.2.1.3 The shapes constraint language (SHACL)	160
	5.2.1.4 Comparison of SHACL to schema definition using RDFUnit test patterns	161
	5.2.1.5 Comparison of SHACL to auto-generated RDFUnit tests from RDFS/OWL axioms .	162

		5.2.1.6	Progress on the SHACL specification and standardisation process	163
		5.2.1.7	SHACL support in RDFUnit	163
5.3	Expert Curation Tools and Workflows			164
	5.3.1	Requirements		165
		5.3.1.1	Graduated application of semantics	165
		5.3.1.2	Graph – object mapping	165
		5.3.1.3	Object/document level state management and versioning	166
		5.3.1.4	Object-based workflow interfaces	166
		5.3.1.5	Integrated, automated, constraint validation	166
		5.3.1.6	Result interpretation	167
		5.3.1.7	Deferred updates	167
	5.3.2	Workflow/Process Models		167
		5.3.2.1	Process model 1 – linked data object creation	167
		5.3.2.2	Process model 2 object – linked data object updates	168
		5.3.2.3	Process model 3 – updates to deferred updates	168
		5.3.2.4	Process model 4 – schema updates	169
		5.3.2.5	Process model 5 – validating schema updates	170
		5.3.2.6	Process model 6 – named graph creation .	170
		5.3.2.7	Process model 7 – instance data updates and named graphs	171
5.4	Dacura Approval Queue Manager			172
5.5	Dacura Linked Data Object Viewer			172
	5.5.1	CSP Design of Seshat Workflow Use Case		173
	5.5.2	Specification		174
5.6	Dacura Quality Service			176
	5.6.1	Technical Overview of Dacura Quality Service		177
	5.6.2	Dacura Quality Service API		178
		5.6.2.1	Resource and interchange format	178
		5.6.2.2	URI	178
		5.6.2.3	Literals	178
		5.6.2.4	Literal types	178
		5.6.2.5	Quads	179
		5.6.2.6	POST variables	180

	5.6.2.7	Tests	180
	5.6.2.8	Required schema tests	180
	5.6.2.9	Schema tests	181
	5.6.2.10	Errors	182
	5.6.2.11	Endpoints	182

5.7 Linked Data Model Mapping 184
    5.7.1 Interlink Validation Tool 184
        5.7.1.1 Interlink validation 185
        5.7.1.2 Technical overview 187
        5.7.1.3 Configuration via iv_config.txt 188
        5.7.1.4 Configuration via external_datasets.txt . . 189
        5.7.1.5 Execute the interlink validator tool 190
    5.7.2 Dacura Linked Model Mapper 190
    5.7.3 Model Mapper Service 193
        5.7.3.1 Modelling tool – creating mappings 193
        5.7.3.2 Importing semi-structured data with data harvesting tool 193

5.8 Model-Driven Data Curation 195
    5.8.1 Dacura Quality Service Frame Generation 196
    5.8.2 Frames for UserInterface Design 197
    5.8.3 SemiFormal Frame Specification 197
    5.8.4 Frame API Endpoints 199

6 Use Cases   201

Kevin Feeney, Christian Dirschl, Andreas Koller, James Welch, Dimitris Kontokostas, Pieter Francois, Sabina Łobocka and Piotr Bledzki

6.1 Wolters Kluwer – Re-Engineering a Complex Relational Database Application 201
    6.1.1 Introduction 201
    6.1.2 Problem Statement 202
    6.1.3 Actors 204
    6.1.4 Implementation 206
        6.1.4.1 PoolParty notification extension 206
        6.1.4.2 rsine notification extension 206
            6.1.4.2.1 Results 206
        6.1.4.3 RDFUnit for data transformation 207
        6.1.4.4 PoolParty external link validity 211
        6.1.4.5 Statistical overview 214

	6.1.5	Evaluation	215
		6.1.5.1 Productivity	217
		6.1.5.2 Quality	217
		6.1.5.3 Agility	217
		6.1.5.4 Measuring overall value	218
		6.1.5.5 Data quality dimensions and thresholds	218
		6.1.5.6 Model agility	219
		6.1.5.7 Data agility	219
	6.1.6	JURION IPG	219
		6.1.6.1 Introduction	219
		6.1.6.2 Architecture	225
		6.1.6.3 Tools and features	227
		6.1.6.4 Implementation	228
		6.1.6.5 Evaluation	232
		6.1.6.6 Experimental evaluation	234
6.2	Seshat – Collecting and Curating High-Value Datasets with the Dacura Platform		235
	6.2.1	Use Case	237
		6.2.1.1 Problem statement	237
	6.2.2	Architecture	238
		6.2.2.1 Tools and features	240
	6.2.3	Implementation	240
		6.2.3.1 Dacura data curation platform	240
		6.2.3.2 General description	240
		6.2.3.3 Detailed process	241
	6.2.4	Overview of the Model Catalogue	246
		6.2.4.1 Model catalogue in the demonstrator system	250
	6.2.5	Seshat Trial Platform Evaluation	253
		6.2.5.1 Measuring overall value	253
		6.2.5.2 Data quality dimensions and thresholds	253
6.3	Managing Data for the NHS		259
	6.3.1	Introduction	259
	6.3.2	Use Case	260
		6.3.2.1 Quality	260
		6.3.2.2 Agility	260
	6.3.3	Architecture	261
	6.3.4	Implementation	263
		6.3.4.1 Model catalogue	263

		6.3.4.2	NIHR health informatics collaborative	263
	6.3.5	Evaluation		268
		6.3.5.1	Productivity	269
		6.3.5.2	Quality	271
		6.3.5.3	Agility	272
6.4	Integrating Semantic Datasets into Enterprise Information Systems with Poolparty			272
	6.4.1	Introduction		272
	6.4.2	Problem Statement		274
		6.4.2.1	Actors	274
	6.4.3	Architecture		274
	6.4.4	Implementation		276
		6.4.4.1	Consistency violation detector	276
		6.4.4.2	RDFUnit test generator	277
		6.4.4.3	PoolParty integration	277
		6.4.4.4	Notification adaptations	277
		6.4.4.5	RDFUnit	278
		6.4.4.6	Validation on import	278
	6.4.5	Results		284
		6.4.5.1	RDF constraints check	285
		6.4.5.2	RDF validation	286
		6.4.5.3	Improved notifications	289
		6.4.5.4	Unified governance	293
	6.4.6	Evaluation		295
		6.4.6.1	Measuring overall value	295
		6.4.6.2	Data quality dimensions and thresholds	299
		6.4.6.3	Evaluation tasks	300
6.5	Data Validation at DBpedia			302
	6.5.1	Introduction		302
	6.5.2	Problem Statement		302
		6.5.2.1	Actors	303
	6.5.3	Architecture		303
	6.5.4	Tools and Features		304
	6.5.5	Implementation		305
	6.5.6	Evaluation		309
		6.5.6.1	Productivity	309
		6.5.6.2	Quality	310
		6.5.6.3	Agility	312

7 Evaluation — 305

Pieter Francois, Stephanie Grohmann, Katja Eck, Odhran Gavin, Andreas Koller, Helmut Nagy, Christian Dirschl, Peter Turchin and Harvey Whitehouse

- 7.1 Key Metrics for Evaluation 313
  - 7.1.1 Productivity 315
  - 7.1.2 Quality 316
  - 7.1.3 Agility 316
  - 7.1.4 Usability 317
- 7.2 ALIGNED Ethics Processes 318
- 7.3 Common Evaluation Framework 320
  - 7.3.1 Productivity 320
  - 7.3.2 Quality 320
  - 7.3.3 Agility 321
- 7.4 ALIGNED Evaluation Ontology 323

Appendix A – Requirements — 325

Index — 395

About the Editors — 399

Preface

As digital processes become more embedded in all facets of life, the ability to deal with big data has become not just an advantage, but a necessity. The massive increase in scale of computer systems has led to new challenges for the builders of software and data systems. The data these systems consume is heterogeneous and unstructured, requiring innovative approaches in how to deal with its volume, variety, and velocity. Manual management of data becomes impossible when dealing with billions or trillions of data points, necessitating the development of software systems which can automatically handle this magnitude of information. These systems must capable of automatic reconfiguration to deal with the changes in data required by the business needs of users and consumers.

This book outlines a suite of approaches which can be used to deal with the continuing growth of scale in software and data engineering. By utilising a lightweight alignment methodology and a variety of semantic web tools, users can ensure that software and data remain synchronised throughout multiple development cycles. This approach allows users to give structure to massive and diverse data collections, giving legibility to data which would otherwise be practicably unworkable.

The methodology and software developments covered in this book arose out of the ALIGNED project. ALIGNED – Aligned, Quality-centric Software and Data Engineering, was a European Union Horizon 2020 project that ran from February 2015 to January 2018. Partners from five European Union member states, all with deep interests in the semantic web and its opportunities, came together to develop new ways of dealing with large scale big data and semantic web approaches.

The research outlined in this book been incorporated into the Semantic Web Company's PoolParty Semantic Suite, the Seshat Global History Databank, the release process for DBpedia, and two of Wolters Kluwer's systems – JURION, a legal portal for German and EU law, and JURION IPG, a legal-commercial information system. It also led to the spin out of a start-up from TCD, backed by Atlantic Bridge, called DataChemist.

Building on lessons learned and tools developed during the project, DataChemist provides companies with a way to build clean consistent datasets at large scale from messy, unstructured data. After devising a schema incorporating semantic intelligence for all the organisation's data, artificial intelligence approaches map relationships usually undiscovered by conventional approaches. DataChemist enables the identification and visualisation of relationships between entities, at depths unmatched by any other competing approach, and enforces compliance with previously specified rules concerning those relationships.

We would like to thank all of the people who were involved in the ALIGNED project during its inception and when it was running for their hard work. This book would not have been possible without all of their contributions. We thank the members of the following groups who have participated in the project: the School of Computer Science and Statistics, Trinity College Dublin; the Department of Computer Science and the School of Anthropology and Museum Ethnography, University of Oxford; the Agile Knowledge Engineering and Semantic Web Research Group, University of Leipzig; the Institute of Archaeology, Adam Mickiewicz University in Poznań; Wolters Kluwer Germany; Wolters Kluwer Poland; and Semantic Web Company. We are grateful to the European Union for funding the ALIGNED project under the Horizon 2020 Programme.

Editors:

Dr. Kevin Feeney, Trinity College Dublin, Ireland & DataChemist

Prof. Jim Davies, Oxford University, United Kingdom

James Welch, Oxford University, United Kingdom

Dr.-Ing. Sebastian Hellmann, University of Leipzig, Germany

Christian Dirschl, Wolters Kluwer, Germany

Andreas Koller, Semantic Web Company, Austria

Dr. Pieter Francois, Oxford University, United Kingdom

Prof. Arkadiusz Marciniak, Adam Mickiewicz University, Poland

July 2018

Acknowledgements

In addition to the contributors listed, the editors would like to thank the ALIGNED team members for their work on the project: Markus Ackermann, Sunduz Akkus-Keles, Marta Bartkowiak, Martin Brummer, Robert David, Diego Esteves, Ruth Fiddy, Markus Freudenberg, Robbie Gallagher, Mahek Hanfi, Katja Harms, Steve Harris, Sebastian Hellmann, Nadine Janicke, Martin Kaltenbock, Jens Lehmann, Marie Lemon, Michael Leuthold, Alan Meehan, Declan O'Sullivan, Sandra Prator, Thomas Thurner, Andre Valdestilhas, Katharina Weissenberg, Simon Westhues, Ornella Zampieri, and Anrapali Zaveri. The editors would also like to thank the ALIGNED External Advisory Board: Eelco Visser, Heimo Hanninen, Gerard Kuys, Christoph Goller, Peter Turchin, and Gabriel Hogan.

The ALIGNED project received funding from the European Union's Horizon 2020 re-search and innovation programme under grant agreement No 644055, the ALIGNED project (www.aligned-project.eu) and from the ADAPT Centre for Digital Content Technology, funded under the SFI Research Centres Programme (Grant 13/RC/2106) and co-funded by the European Regional Development Fund.

List of Contributors

Piotr Bledzki, *Wolters Kluwer Poland, Poland*
Bojan Bozic, *Trinity College Dublin, Ireland*
Rob Brennan, *Trinity College Dublin, Ireland*
Jim Davies, *University of Oxford, UK*
Christian Dirschl, *Wolters Kluwer Germany, Germany*
Katja Eck, *Wolters Kluwer Germany, Germany*
Kevin Feeney, *Trinity College Dublin, Ireland*
Patrycja Filipowicz, *Adam Mickiewicz University, Poland*
Pieter Francois, *University of Oxford, UK*
Odhran Gavin, *Trinity College Dublin, Ireland*
Jeremy Gibbons, *University of Oxford, UK*
Stephanie Grohmann, *University of Oxford, UK*
Andreas Koller, *Semantic Web Company, Austria*
Dimitris Kontokostas, *University of Leipzig, Germany*
Sabina Łobocka, *Wolters Kluwer Poland, Poland*
Christian Mader, *Semantic Web Company, Austria*
Arkadiusz Marciniak, *Adam Mickiewicz University, Poland*
Gavin Mendel-Gleason, *Trinity College Dublin, Ireland*
Helmut Nagy, *Semantic Web Company, Austria*
Seyyed Shah, *University of Oxford, UK*
Monika Solanki, *University of Oxford, UK*
Peter Turchin, *University of Connecticut, USA*
James Welch, *University of Oxford, UK*
Harvey Whitehouse, *University of Oxford, UK*

List of Figures

Figure 2.1	Seshat World Sample 30.	25
Figure 2.2	PoolParty Application Suite.	26
Figure 2.3	DBpedia Extraction Pipeline.	28
Figure 2.4	Jurion IPG.	30
Figure 3.1	The waterfall process for software development.	43
Figure 3.2	A modified waterfall process.	44
Figure 3.3	An iterative software development process.	45
Figure 3.4	A data engineering life cycle.	48
Figure 3.5	Comparison of terminology in software and data engineering.	56
Figure 3.6	An incomplete grid for analysing integration points.	62
Figure 3.7	A parallel life cycle with synchronisation.	64
Figure 3.8	Model catalogue interface: browsing the SESHAT code book.	70
Figure 3.9	Example semantic booster system with annotations.	73
Figure 4.1	The ALIGNED metamodel layers.	80
Figure 4.2	The Design Intent Ontology (DIO).	84
Figure 4.3	The Software Life cycle Ontology.	85
Figure 4.4	Core Concepts of the Software Implementation Process (SIP) Ontology.	86
Figure 4.5	Generic data life cycle metamodel (DLO).	87
Figure 4.6	The DataID Ontology.	88
Figure 4.7	Reasoning Violation Ontology (RVO) Base Classes.	90
Figure 4.8	RVO Instance and Schema Violation Classes.	91
Figure 4.9	Resulting RDF Graph after Validation.	91
Figure 4.10	Data model showing the main relevant classes and their relations.	94
Figure 4.11	Using the property prov:wasDerivedFrom to interrelate quality metrics and other quality statements.	95

Figure 4.12	The ALIGNED domain-specific ontology for E-research in the Social Sciences and Humanities.	100
Figure 4.13	The Seshat ontology.	100
Figure 4.14	The Dacura ontology.	102
Figure 4.15	Dacura console usage example.	103
Figure 4.16	New candidate example part 1.	104
Figure 4.17	New candidate example part 2.	105
Figure 4.18	The Crowd-sourced Public Datasets ontology.	106
Figure 4.19	The Organisation of PROV.	115
Figure 4.20	An RDF Graph Describing Eric Miller.	122
Figure 5.1	The layered architecture of the Model Catalogue.	128
Figure 5.2	Core concepts – data model components – within the Model Catalogue.	128
Figure 5.3	A model showing the datatypes represented in the Model Catalogue.	129
Figure 5.4	Model Catalogue interface: browsing the Seshat code book.	131
Figure 5.5	Model Catalogue interface: data element view.	132
Figure 5.6	Model Catalogue interface: keyword search for a data item.	132
Figure 5.7	Model Catalogue interface: editing a data item.	133
Figure 5.8	Model Catalogue Eclipse Integration.	135
Figure 5.9	Screenshot showing RDF representation of catalogue contents.	137
Figure 5.10	A Booster specification edited with the Eclipse IDE.	140
Figure 5.11	The Booster generation pipeline.	141
Figure 5.12	The architecture of a Booster information system.	142
Figure 5.13	The Booster Web-based user interface.	142
Figure 5.14	Generating Booster systems from Model Catalogue models.	143
Figure 5.15	Excerpt from the Booster system generated from Prov-DM Core.	145
Figure 5.16	Model Catalogue information in the Booster interface.	146
Figure 5.17	Semantic Booster – generation menu in the Eclipse IDE.	147
Figure 5.18	Semantic Booster – generated R2RML file.	148
Figure 5.19	Booster specification with semantic annotations.	149

Figure 5.20	Booster user interface showing semantic annotations from the Model Catalogue.	150
Figure 5.21	Semantic Booster Web-based editor.	152
Figure 5.22	Default Booster data explorer.	153
Figure 5.23	D2RQ and SNORQL for exploration of Semantic Booster data.	154
Figure 5.24	d3sparql for visual exploration of Semantic Booster data.	155
Figure 5.25	RDFUnit Web interface.	156
Figure 5.26	RDFUnit architecture.	157
Figure 5.27	RDFUnit report from the IntelliJ IDE.	159
Figure 5.28	Example Bamboo overview from an RDFUnit JUnit XML report.	159
Figure 5.29	Custom JUnit integration with RDFUnit as a library for JURION Use Case in ALIGNED.	160
Figure 5.30	Overview for Fundamental Concepts of SHACL. .	162
Figure 5.31	Excerpt of an EARL test report for the SHACL test suite.	164
Figure 5.32	Key to workflow/process models.	168
Figure 5.33	Process Model 1 – Object Creation.	168
Figure 5.34	Process Model 2 – Object Update.	169
Figure 5.35	Process Model 3 – Updates to deferred update. . . .	169
Figure 5.36	Process model 4 – Schema Updates.	170
Figure 5.37	Process model 5 – Validating schema updates. . . .	171
Figure 5.38	Process Model 6 – Named Graph Creation.	171
Figure 5.39	Process model 7 – instance data updates in named graphs.	172
Figure 5.40	Screenshot of Dacura Linked Data Approval Queue Manager Tool.	172
Figure 5.41	Screenshot of Dacura Linked Data Object Viewer Tool showing version browsing toolbar.	173
Figure 5.42	Automatically generated workflow diagram from CSPm specification.	175
Figure 5.43	Dacura platform Quality Test Interface that calls the DQS.	177
Figure 5.44	Interlink Validation Process.	186
Figure 5.45	Operation of the Interlink Validation Tool. The arrows indicate the flow of information/data among the different components.	187

Figure 5.46	Example of seshat code book page.	192
Figure 5.47	Importing a model from semi-structured HTML source.	193
Figure 5.48	Process for associating property definitions in a model with a pattern within a semi-structured HTML page.	194
Figure 5.49	Process for using patterns to extract data from semi-structured html pages.	194
Figure 5.50	Screenshot showing results of automated importing of semi-structured HTML data into structured model.	195
Figure 5.51	Graphical Representation of ontology fragment.	199
Figure 6.1	JURION Content Pipeline and Semantic Search.	203
Figure 6.2	Distribution of the Linked Data stack components w.r.t. Linked Data Publishing cycle.	204
Figure 6.3	ALIGNED Use Cases.	205
Figure 6.4	Notification message.	207
Figure 6.5	Transformation process with RDFUnit.	209
Figure 6.6	RDFUnit results.	210
Figure 6.7	Jenkins-CI Test Report.	211
Figure 6.8	Validation Data stored for Analysis.	212
Figure 6.9	Example defect: the Image file of the external source does not exist anymore.	213
Figure 6.10	Validation Results.	214
Figure 6.11	Statistical checks.	215
Figure 6.12	JURION: Overview.	216
Figure 6.13	JURION Content Pipeline, showing ALIGNED tools integrated with existing functionality and datasets.	220
Figure 6.14	IPG problem statement.	221
Figure 6.15	Screenshot of a subset of the IPG model in the Model Catalogue.	223
Figure 6.16	The Eclipse-based Booster tool.	224
Figure 6.17	Screenshot of the Booster administrator interface for the JURION IPG system.	224
Figure 6.18	Results of using the RDFUnit tool against data from a Semantic Booster database.	225
Figure 6.19	Jurion IPG unsolvable issues.	226

List of Figures xxv

Figure 6.20	Jurion IPG use-case architecture showing integration across all major project tools and partners.	227
Figure 6.21	Integration Paradigms and vocabularies supported by ALIGNED tools and platforms.	228
Figure 6.22	Complexity of the Jurion IPG use case.	229
Figure 6.23	Integrating Semantic Booster and the Model Catalogue.	229
Figure 6.24	IPG Data Error detection and correction using Dacura.	230
Figure 6.25	Ontology generated from IPG SQL database by Dacura's Model Mapper Tool.	231
Figure 6.26	Using Dacura's curation tools to analyse the IPG data model.	231
Figure 6.27	Seshat Use Case Trial System Architecture, showing the tools provided to different Seshat users, the use of ALIGNED integration standards and interoperation paradigms.	238
Figure 6.28	Features of the ALIGNED tools used to support the Seshat trials.	240
Figure 6.29	The Dacura platform in the context of the ALIGNED Seshat use case.	241
Figure 6.30	Screenshot of TCD's Seshat Data Entry/Validation tool in Demonstrator System.	242
Figure 6.31	Modifying Seshat Schema.	242
Figure 6.32	Screenshot of TCD's Schema Management component using the prototype integrity enforcement framework in the Demonstrator System.	243
Figure 6.33	Screenshot of TCD's Schema Validation Service in Demonstrator System.	244
Figure 6.34	Screenshot of TCD's Wiki Export Component.	245
Figure 6.35	Seshat Errors per variable.	246
Figure 6.36	Managing Complex Workflows.	247
Figure 6.37	Importing data to Seshat from DBpedia with Unified Views.	248
Figure 6.38	Publication.	248
Figure 6.39	Services to support software engineering.	249
Figure 6.40	The Model Catalogue user interface showing a section of the code book.	249

Figure 6.41	Screenshot of the Model Catalogue Web interface, showing the 'tree view' and a section of the Seshat code book.	251
Figure 6.42	Screenshot of the Model Catalogue Web interface showing the comparison between two versions of the Seshat code book.	252
Figure 6.43	Seshat: Comparison.	258
Figure 6.44	Health Informatics Collaborative system architecture.	262
Figure 6.45	The front page of the catalogue interface.	264
Figure 6.46	Data comparison in the Health Informatics Collaborative.	265
Figure 6.47	Data elements in the UK 100,000 Genomes Project catalogue.	266
Figure 6.48	An example shopping cart in the Health Data Finder.	267
Figure 6.49	The model catalogue in the Health Data Finder.	267
Figure 6.50	Dataset metadata in the NIHR Health Data Finder.	268
Figure 6.51	Screenshot from the NIHR HIC Model Catalogue.	271
Figure 6.52	PoolParty Architecture.	275
Figure 6.53	Import dialogue.	284
Figure 6.54	Consistency constraint violations as reported by RDFUnit.	284
Figure 6.55	High level technical overview.	285
Figure 6.56	RDF validation conformance checks.	286
Figure 6.57	Repair strategy for the constraint check.	287
Figure 6.58	RDF Validation Screenshot.	288
Figure 6.59	Improved notification system.	290
Figure 6.60	UnifiedViews pipeline for PoolParty use case.	296
Figure 6.61	Unified Governance Search.	297
Figure 6.62	Issue Integration reporting dialogue.	297
Figure 6.63	Issue Integration created dialogue.	297
Figure 6.64	Semantic search over development artefact – Graph Search.	298
Figure 6.65	Details view of specific issue with the option to select similarity algorithm – PP Recommender.	298
Figure 6.66	DBpedia Use Case Trial System Architecture, showing the ALIGNED tools used in different stages of the DBpedia data workflow.	304

Figure 6.67	ALIGNED Tools and Features used in the DBpedia trial platform.	305
Figure 6.68	Instance data validation report with RDFUnit. . . .	305
Figure 6.69	Mapping validation report with RDFUnit and RML.	306
Figure 6.70	The new Mappings UI (using RDFUnit for validating mappings).	307
Figure 6.71	DBpedia Link Viz tool.	307
Figure 6.72	SUMMR Mapping tool.	308
Figure 6.73	Active extraction monitoring (here: extraction summaries forwarded to Slack).	308
Figure 6.74	DBpedia download page through DataID.	309
Figure 6.75	Dockerised DBpedia.	309
Figure 7.1	The ALIGNED Evaluation Framework.	314
Figure 7.2	ALIGNED metrics ontology – classes.	324
Figure A.1	Seshat Architecture for Month 19 Demo.	358

List of Tables

Table 3.1	A usecase-oriented synchronisation table for the ALIGNED project	66
Table 4.1	Generic metamodel namespace declarations	81
Table 4.2	Domain-specific metamodel namespace declarations	97
Table 4.3	JURION actors	98
Table 4.4	JURION entities	98
Table 4.5	JURION activities	99
Table 4.6	DBpedia actors	106
Table 4.7	DBpedia entities	107
Table 4.8	DBpedia activities	107
Table 4.9	PoolParty actors	110
Table 4.10	PoolParty entities	110
Table 4.11	PoolParty activities	111
Table 5.1	CSPm specification of workflow	174
Table 5.2	Dacura Quality Service Frame Grammar	198

List of Abbreviations

ACID	Atomic, Consistent, Isolated, Durable
AKSW	Agile Knowledge Engineering and Semantic Web group, University of Leipzig
CI	Continuous integration
CJDE	PoolParty Confluence/JIRA Data Extractor
CMS	content management system
CSP	Communicating Sequential Processes
CSPDO	Crowd-sourced Public Datasets
CWA	Closed World Assumption
D2RQ	D2RQ Platform, a system for accessing relational databases as virtual, read-only RDF graphs
DC	Dublin Core
DCAT	W3C Data Catalog Vocabulary
DIEF	DBpedia Information Extraction Framework
DIO	Design Intent Ontology
DIOPP	Enterprise Software Development Ontology
DLO	Data Life Cycle Ontology
DQS	Data Quality Service
DQV	Data Quality Vocabulary
DSL	domain specific language
DTD	Document Type Definition
EBNF	extended Backus-Naur form
EIPDM	Enterprise Information Processing Domain-specific Meta-Model
ELV	External Link Validation
EMF	Eclipse Modeling Framework
ETL	Extract Transform Load
FOAF	Friend Of A Friend
FOL	First Order Logic
GIS	geographic information system
GUID	globally unique identifier

HIC	Health Informatics Collaborative
IBIS	Interactive Intent-Based Illustration
ICV	Integrity Constraint Validator
IRI	International Resource Identifier
JSON	JavaScript Object Notation
KPI	key performance indicator
LD	Linked Data
LOD	Linked Open Data
MDA	Model-driven architecture
MDE	Model-driven engineering
MDSE	Model-driven software engineering
MIREOT	Minimum information to reference an external ontology term
MUTO	Modular and Unified Tagging Ontology
NESSI	Networked European Software and Services Initiative
NIHR	National Institute for Health Research
OMG	Object Management Grou
ORE	Ontology Repair and Enrichment
ORM	Object-relational mapping
OWL	Web Ontology Language
PCI	Portal Content Interface
PP	PoolParty
PPT	PoolParty Thesaurus Manager
PPX	PoolParty Extractor
PROV	W3C PROV Ontology
R2RML	RDB to RDF Mapping Language
RDF	Resource Description Framework
RDFS	RDF Schema
REST	Representational State Transfer
RF	Repair Framework and Notification
RUT	Test-Driven RDF Validation Ontology
RVO	Reasoning Violations Ontology
SE	Software engineering
SEON	Software Evolution Ontologies
SHACL	Shapes Constraint Language
SIP	Software Implementation Process Ontology
SKOS	Simple Knowledge Organisation System
SLO	Software Lifecycle Ontology

SME	Small and medium-sized enterprises
SNORQL	front-end for exploring RDF SPARQL endpoints
SOAP	Simple Object Access Protocol
SPARQL	SPARQL Protocol and RDF Query Language
SPIN	SPARQL Inferencing Notation
SUS	System Usability Scale
SWC	Semantic Web Company
SWO	Software Ontology
UL	University of Leipzig
UML	Unified Modeling Language
UNA	Unique Name Assumption
URI	Uniform Resource Identifier
WKD	Wolter Kluwer
XLST	Extensible Stylesheet Language Transformations
XML	eXtensible Markup Language
XSD	XML Schema Definition

1

Introduction

To be effective, data-intensive systems require extensive ongoing customisation to reflect changing user requirements, organisational policies, and the structure and interpretation of the data they hold. Manual customisation is expensive, time-consuming, and error-prone. In large complex systems, the value of the data can be such that exhaustive testing is necessary before any new feature can be added to the existing design. In most cases, precise details of requirements, policies and data will change during the lifetime of the system, forcing a choice between expensive modification and continued operation with an inefficient design.

In 2013, the Networked European Software and Services Initiative (NESSI) identified "Collaborative Service Engineering based on convergence of software and data" as an EU research priority. Information systems are composed of software and data components that must co-evolve as requirements change. In existing development methodologies, software and data engineering are considered as separate concerns.[1] New techniques and tools are required to support the development of effective solutions in the presence of changing requirements, policies, schemas, and data. NESSI also identified "Integration of Big Data Analytics into Business processes" as a research priority, emphasising the importance of data-centric or "Big Data" approaches. This serves only to emphasise the relative value of the data and the need for agility. Big Data approaches involve the imposition of multiple, changing models upon unstructured heterogeneous Linked Data. A single static data model will not suffice, and the manual development of customised code against multiple changing models is unsustainably expensive. Automatic support for customisation, driven by domain models of knowledge and requirements, is an essential component of effective, sustainable Big Data solutions, building on underlying technology from both domains.

---

[1]A. Cleve, T. Mens, J-L. Hainaut, Data-Intensive System Evolution, IEEE Computer, August 2010.

In software engineering, there are meta-modelling frameworks of the kind that support the Unified Modeling Language (UML), allowing engineers to describe and design features that work for whole classes or families of data models, rather than for a specific instance. There is widespread language support for higher-order programming, in which programs are managed as data. There are mature formal program specification approaches and languages that enable programs to be described mathematically and to be provably correct. We have model-driven, product-line, and generative programming techniques, in which a single set of validated transformations is used to produce or customise many different applications or many different versions of the same application. However, evidence is lacking for the effectiveness of these techniques except in narrow domains.[2]

In data engineering, we have meta-formats such as eXtensible Markup Language (XML), allowing us to describe and design data formats and representations. We have languages such as the Resource Description Framework (RDF) for recording and communicating relationships between different data items; Resource Description Framework Schema (RDFS) for detailing relationships between classes of entities; the Web Ontology Language (OWL) for describing domain knowledge, axioms, and inference rules; and powerful, scalable tools for applying knowledge and rules to large collections of data and metadata. These tools overlap with the expressivity of UML, but in practice, the tractability of code or transformation generation and the ability to reuse data from these syntax-focussed expressions are much weaker than those of native semantic models. More important is perhaps the skills and engineering culture gaps that divide the software and data engineering communities. Common tools that bridge this gap will lead to a deeper shared understanding.

The challenge is to bring these aspects together in a practical, proven methodology, which can be instantiated in software, and which enables the effective, sustainable development of large, complex, and data-intensive systems.

1.1 State of the Art in Engineering Data-Intensive Systems

While the topic of co-evolution between software artefacts and other artefacts produced during software development is an active area of research, its

---

[2]J. Hutchinson et al. "Model-driven engineering practices in industry," Software Engineering (ICSE), pp. 633,642, 21–28, 2011.

application to data-intensive software systems is not trivial.[3] Although the research focus had been fixed firmly on software interacting with traditional data environments of relational databases[4] and data warehousing,[5] recently, a more technology-independent approach has emerged. Mori and Cleve[6] introduced the notion of data-intensive self-adaptive systems as data-intensive systems able to perform context-dependent data access. They proposed adoption of a framework that supports feature-based data tailoring by means of a filtering design process and a run-time filtering process. Manousis et al.[7] introduced a method for the adaptation of data-intensive ecosystems based on three algorithms that (i) assess the impact of a change, (ii) compute the need of different variants of an ecosystem's components, depending on policy conflicts, and (iii) rewrite the modules to adapt to the change.

Naturally, a prerequisite to assessing impact is the ability to represent the interdependency of the artefacts in a machine-processable manner. Terwilliger et al.[8] stated that "bi-directional mappings" are emerging as a mechanism in the software engineering domain to represent such interdependency. They also identify, characterise, and compare a representative set of tools implementing the approach. Compatible with the concepts, but emerging from the data community, are semantic mappings, where progress has been made in representing and characterising complex mappings through correspondence patterns.[9]

---

[3] A. Serebrenik & T. Mens. Emerging trends in software evolution. In Evolving software systems, pp. 329–332, Berlin: Springer, 2014.

[4] A. Cleve, T. Mens, and J.-L. Hainaut, Data-intensive system evolution, IEEE Computer, vol. 43, no. 8, pp. 110–112, 2010.

[5] A. Abelló, J. Darmont, L. Etcheverry, M. Golfarelli, J. Mazón, F. Naumann, T. Pedersen et al. "Fusion cubes: Towards self-service business intelligence." International Journal of Data Warehousing and Mining (IJDWM) 9, no. 2, pp. 66–88, 2013.

[6] M. Mori, A. Cleve, Towards Highly Adaptive Data-Intensive Systems: A Research Agenda, Advanced Information Systems Engineering Workshops, Lecture Notes in Business Information Processing Volume 148, pp. 386–401, 2013.

[7] P. Manousis, P. Vassiliadis, G. Papastefanatos, Automating the Adaptation of Evolving Data-Intensive Ecosystems, Conceptual Modelling, Lecture Notes in Computer Science Volume 8217, pp. 182–196, 2013.

[8] J. F. Terwilliger, A. Cleve, C. A. Curino, How Clean Is Your Sandbox?, Theory and Practice of Model Transformations, Lecture Notes in Computer Science Volume 7307, pp. 1–23, 2012.

[9] J. Keeney, A. Boran, I. Bedini, C. Matheus and P. Patel-Schneider, "Approaches to Relating and Integrating Semantic Data from Heterogeneous Sources." In Proc. 2011 IEEE/WIC/ACM International Conferences on Web Intelligence and Intelligent Agent Technology-Vol 01, pp. 170–177. IEEE Computer Society, 2011.

Empirical studies and research that help to motivate the need for strongly integrated system co-evolution are also emerging. Goeminne et al.[10] reported on early results obtained in the empirical analysis of the co-evolution between code-related and database-related activities of contributors in a large open source data-intensive system. Their study investigated questions such as: what is the effect of introducing a new database technology? And how do developers divide their effort between the activity types involved in evolving a data-intensive system? Papastefanatos et al.[11] proposed a set of graph-theoretic metrics for the prediction of impact of schema evolution upon ETL software and evaluated them over seven systems. Meurice and Cleve,[12] in a short study, described the type of schema evolution that emerged in four systems over a period of months and the utility of having a tool to aid the analysis. Sen and Gotlieb[13] proposed a methodology for testing data-intensive systems and present results achieved when applied to a case study in the Norwegian Customs and Excise governmental department.

1.1.1 The Challenge

There is a body of research studying data-intensive systems, from a unified point of view, but the focus to date has been largely on relational data models. These are, of course, important for current enterprise systems. However, the Web is currently undergoing a data revolution, where machine-to-machine communication will eventually dominate over human-centric, document-oriented Web traffic. A key driver of this data revolution is graph-based data, whether in the form of the Facebook Graph API[14] for searching their social graph, Google, Bing, Yandex and Yahoo's schema.org for annotating Web pages with graph-based metadata or the W3C's Linked Open Data (LOD)

---

[10] M. Goeminne, A. Decan, T. Mens, (2014, February). Co-evolving code-related and database-related changes in a data-intensive software system. In Proceedings of the IEEE CSMR-WCRE 2014 Software Evolution Week.

[11] G. Papastefanatos, P. Vassiliadis, A. Simitsis, Y. Vassiliou, Metrics for the Prediction of Evolution Impact in ETL Ecosystems: A Case Study, Journal on Data Semantics, Volume 1, Issue 2, pp. 75–97, August 2012.

[12] L. Meurice & A. Cleve, DAHLIA: A Visual Analyzer of Database Schema Evolution, CSMR-WCRE 2014, Belgium, 2014.

[13] S. Sen and A. Gotlieb, Testing a Data-intensive System with Generated Data Interactions: The Norwegian Customs and Excise Case Study, 25th International Conference on Advanced Information Systems Engineering (CAISE'13) (2013).

[14] https://developers.facebook.com/docs/graph-api/

community[15] that builds on over a decade of semantic Web research. For the next generation of Web-scale data-intensive systems, it is not enough to transfer legacy data models to the cloud. Instead, the research on controlled co-evolution of software and data must be extended to deal natively with Linked Data-based systems.

Many of the techniques developed for traditional data-intensive systems, such as data transformation generation, are still relevant, but Linked Data versions must be developed. The richer models of semantic, RDF-based methods offer new opportunities: for leveraging domain knowledge expressed as ontologies; applying semantic mapping techniques for correspondence classification to schema evolution evaluation (to drive controlled transformations for programs, queries, and data); and modelling the software and data life cycles in a machine-computable way, enabling heterogeneous tools to collaborate in combined software and data engineering tool chains.

1.2 State of the Art in Semantics-Driven Software Engineering

Model-driven software engineering is the automatic production of software artefacts from abstract models of structure and functionality. This approach can reduce the costs of development and maintenance and increase the quality and reliability of the software produced. It has been adopted for the development of control and embedded systems,[16] for aspects of data warehousing,[17] and for service implementations.[18] It has yet to achieve any widespread adoption outside these domains. Multiple reasons are suggested by Den Haan,[19] but the two most common explanations are a lack of adequate

---

[15] http://www.w3.org/standards/semanticweb/data

[16] D. Hästbacka, T. Vepsäläinen, S. Kuikka, Model-driven development of industrial process control applications, Journal of Systems and Software, Volume 84, Issue 7, pp. 1100–1113, July 2011.

[17] J. Mazón, J. Trujillo, M. Serrano, and M. Piattini. "Applying MDA to the development of data warehouses." In Proceedings of the 8th ACM international workshop on Data warehousing and OLAP, pp. 57–66. ACM, 2005.

[18] J. Bezivin, S. Hammoudi, D. Lopes, and F. Jouault. "Applying MDA approach for web service platform." In Enterprise Distributed Object Computing Conference, 2004. EDOC 2004. pp. 58–70. IEEE, 2004.

[19] J. Den Haan, "8 Reasons Why Model-Driven Approaches (will) Fail". http://www.infoq.com/articles/8reasons-why-MDE-fails, July 2008.

tool support[20] and, as a consequence, a lack of any proven, empirically tested methodology.

Existing tools are focussed on the production of structural, static components of an implementation. Beyond a handful of tightly constrained domains, these tools lack any means to model and generate anything beyond the most basic aspects of functionality.

Technology platforms are available to support more general model transformation and code production. Many of these have been implemented in the widely used Eclipse environment and address the Object Management Group's (OMG) Model-Driven Architecture (MDA) proposal,[21] with tools for domain-specific modelling,[22] developing model transformations,[23] and performing model edits and manipulations.[24]

The Atlas Transformation Language, in particular, is based on the Query View Transformation proposal[25] for transformation languages and acts on models written in UML: the de facto industry standard for software systems modelling. Techniques have been developed that support genericity and bi-directional transformation,[26] with the aim of facilitating round-trip engineering and iterative development. Specialised tools, such as Stratego,[27] have been developed for program transformation or meta-programming.

---

[20] J. Whittle, J. Hutchinson, M. Rouncefield, B. Håkan, and R. Heldal. "Industrial Adoption of Model-Driven Engineering: Are the Tools Really the Problem?" In Model-Driven Engineering Languages and Systems, pp. 1–17. Springer, 2013.

[21] A. Kleppe, J. Warmer, W. Bast, "M.D.A. Explained. The model driven architecture: practice and promise", 2003.

[22] F. Jouault, J. Bézivin, and I. Kurtev, "TCS: a DSL for the Specification of Textual Concrete Syntaxes in Model Engineering," in Procs of the 5th Int. Conf. on Generative programming and Component Engineering (GPCE '06). New York, NY, USA: ACM, pp. 249–254, 2006.

[23] F. Jouault, F. Allilaire, J. Bézivin, I. Kurtev, ATL: A model transformation tool, Science of Computer Programming

[24] M. Del Fabro, J. Bézivin, and P. Valduriez. "Weaving Models with the Eclipse AMW plugin." In Eclipse Modelling Symposium, Eclipse Summit Europe (2006).

[25] MG, Meta Object Facility (MOF) 2.0 Query/View/Transformation Specification, OMG Document formal/2011-01-01, Object Management Group, http://www.omg.org/spec/QVT/1.1/ (2011).

[26] J. Cuadrado, E. Guerra, and J. De Lara. "Generic model transformations: write once, reuse everywhere." In Theory and Practice of Model Transformations, pp. 62–77, Springer Berlin Heidelberg, 2011.

[27] E. Visser, Program transformation with Stratego/XT, in: Domain-Specific Program Generation, Vol. 3016 of Lecture Notes in Computer Science, pp. 216–238, Springer Berlin Heidelberg, 2004.

1.2 State of the Art in Semantics-Driven Software Engineering

There has been work on mappings between the ISO/IEC 11179 metadata registry standard and description logics, such as OWL,[28] but this has focussed purely on the representation of modelling constructs, with no consideration of the implications for software and data engineering. Similarly, within the OMG, efforts have focussed on how to enable the use of UML notation and tools for ontology modelling.[29] There has been related work on representing systems specifications as ontologies for project planning that draws on the OMG MDA specification as inspiration.[30]

The most significant effort to date on the incorporation of semantic models into software engineering has been the FP7 MOST project (2007–2011), which investigated the utilisation of ontologies in an MDA approach.[31] Their work developed new techniques for applying semantic reasoners to MDA tasks,[32] such as model checking, specification validation, or supporting domain specific languages (DSLs) with strong semantics. Much effort was focussed on model translation or bridging[33] between non-mainstream UML variants such as grUML and OWL ontologies. In a 2013 update,[34] one of the project's principal investigators laid out a vision for Ontology-Driven Software Engineering that targets 2030 as the year when this technology will be mature. This timescale indicates the difficulty of building formal ontologies into the heart of software engineering. It also distinguishes this work from the approach of ALIGNED, which is based on a more lightweight Linked Data methodology that aims to enable reuse of rich dataset and metadata descriptions by software engineering tools while supporting co-evolution

---

[28] C. Tao, G. Jiang, W. Wei, H. R. Solbrig, and C. G. Chute. "Towards semantic-web based representation and harmonization of standard meta-data models for clinical studies." AMIA Summits on Translational Science Proceedings: 59 (2011).

[29] S. Brockmans, R. M. Colomb, P. Haase, E. F. Kendall, E. K. Wallace, C. Welty, G. Tong Xie. A Model Driven Approach for Building OWL DL and OWL Full Ontologies, ISWC 2006.

[30] M. Líška and P. Navrat, An Approach to Project Planning Employing Software and Systems Engineering Meta Model Represented by an Ontology, ComSIS Vol.v7, No. 4, December 2010.

[31] http://www.slideshare.net/malgorzatasiwiec/ontologies-and-software-technologies-the-most-project.

[32] http://www.slideshare.net/fparreiras/filling-the-gap-between-semantic-web-owl-ontology-technology-andmodel-driven-engineering-mde-mdsd-mda.

[33] T. Walter, Bridging Technological Spaces: Towards the Combination of Model-Driven Engineering and Ontology Technologies, PhD thesis, Universite Koblenz-Landau, 2011.

[34] U. Assmann, Current Trends and Perspectives in Ontology-Driven Software Development, August 2013, available at http://www.computational-logic.org/content/events/iccl-ss-2013/download/assmann-1-odsd.pdf.

of software and data assets. In 2012, Katasonov[35] pointed the way forward, "beyond model checking and transformations", with a call to apply semantics in software engineering for its known capabilities in describing software and data assets, as well as semantic search and multi-layered modelling of systems.

1.2.1 The Challenge

There is a large body of research on model-driven engineering (MDE), and, in principle, its benefits are clear, especially for evolvable systems. Despite this and the high-profile OMG MDA initiative of the early 2000s, it has not succeeded in proliferating to the mainstream of software engineering practice other than in embedded systems and certain niches. Modern data-intensive systems are characterised by the need to meet changing application requirements and to integrate multiple data sources whose ownership may lie outside the authority of the application developers. The goal of the ALIGNED project was to change this by collecting quantitative evidence of the benefits of deploying model-driven technology in enterprise information processing systems. The basis of this was aggregating formal system specifications for both data and software, based on a common set of metamodels or vocabularies.

There is already evidence that ontologies or semantic models can provide benefit as input domain models for model-driven development. Despite this, semantic data engineering is a marginal activity at the periphery of software engineering. There is an opportunity to create a more holistic view of the data-intensive system engineering process. By modelling design intents, life cycles, and inter-life cycle communication, it was possible to better integrate the tools and methods used in the software and data engineering processes, in order to enable loosely coupled co-evolution of systems and external Web data resources.

1.3 State of the Art in Data Quality Engineering

Data quality engineering is an issue that exists independently of data representation and technology and arises wherever data are stored for incorporation into business processes. However, in general, the older and more

---

[35] A. Katasonov, Ontology-driven software engineering: Beyond model checking and transformations, International Journal of Semantic Computing, Vol. 6 (2012) No. 2, pp. 205–242, 2012.

1.3 State of the Art in Data Quality Engineering

established a language and technology, the more mature the tools, standards, and processes are for dealing with data quality engineering issues. For example, where XML is concerned, Schematron[36] is an ISO standard for validation and quality control of XML documents based on XPath and XSLT. Similarly, in database research, there are related approaches to formulate common integrity constraints[37] using First Order Logic (FOL). The work of Fan,[38] for example, uses FOL to describe data dependencies for quality assessment and suggests repairing strategies. The development of similar mechanisms for RDF is of crucial importance to provide solutions to allow the use of RDF in settings that require either high-quality data or at least an accurate assessment of its quality.

Several approaches for assessing the quality of Linked Data have been proposed, which can be broadly classified into (i) automated;[39] (ii) semi-automated;[40] and (iii) manual[41] methodologies. These approaches introduce systematic methodologies for assessing the quality of an RDF dataset at the process level. Additionally, there have been efforts to assess the quality of large-scale Web data,[42] which included the analysis of 14.1 billion HTML tables from Google's general-purpose Web crawl in order to retrieve tables with high-quality relations. Similarly, Hogan et al.[43] assessed the quality of published RDF data. This study described the errors characteristically associated with publishing RDF data, catalogued the available techniques to improve the quality of structured data on the Web, and analysed each technique's effectiveness. In a recent study, 4 million RDF/XML documents were analysed, which provided insights into the level of conformance these

---

[36] http://www.schematron.com/

[37] A. Deutsch. Fol modelling of integrity constraints (dependencies). In L. LIU and M. ÖZSU, editors, Encyclopedia of Database Systems, pp. 1155–1161, Springer US, 2009.

[38] W. Fan. Dependencies revisited for improving data quality. In Proceedings of the Twenty-seventh ACM SIGMOD-SIGACT-SIGART Symposium on Principles of Database Systems, ACM, pp. 159–170, New York, NY, USA, 2008.

[39] C. Guéret, P. T. Groth, C. Stadler, and J. Lehmann. Assessing linked data mappings using network measures. In Proceedings of the 9th Extended Semantic Web Conference, volume 7295 of LNCS, pp. 87–102. Springer, 2012.

[40] A. Flemming. Quality characteristics of linked data publishing datasources. MSc thesis, Humboldt-Universität Berlin, 2010.

[41] C. Bizer and R. Cyganiak. Quality-driven information filtering using the WIQA policy framework. Web Semantics, 7(1), pp. 1–10, January 2009.

[42] M. J. Cafarella, A. Y. Halevy, D. Z. Wang, E. Wu, and Y. Zhang. Webtables: exploring the power of tables on the web, PVLDB, 1(1), pp. 538–549, 2008.

[43] A. Hogan, A. Harth, A. Passant, S. Decker, and A. Polleres. Weaving the pedantic web. In LDOW, 2010.

documents had with the Linked Data guidelines. This effort assessed a vast amount of Web and RDF/XML data; however, most of the analysis was performed automatically, thereby overlooking the problems arising due to contextual discrepancies. In earlier work, similar ideas were used for describing knowledge base evolution.[44]

The approach described in Fürber and Hepp[45] advocates the use of SPARQL and SPARQL Inferencing Notation (SPIN) for RDF data quality assessment. However, their approach requires a domain expert for the instantiation of test case patterns. SPIN[46] is a W3C submission aimed at representing rules and constraints on Semantic Web models. SPIN also allows users to define SPARQL functions and reuse SPARQL queries. In a similar way, Fürber et al. also defined a set of generic SPARQL queries to identify missing or illegal literal values and datatypes and functional dependency violations. Another related approach is the Pellet Integrity Constraint Validator (ICV).[47] Pellet ICV translates OWL integrity constraints into SPARQL queries. The execution of those SPARQL queries identifies violations. An implication of the integrity constraint semantics of Pellet ICV is that a partial unique names assumption (all resources are considered to be different unless equality is explicitly stated) and a closed world assumption are adopted. qSKOS defines rules to detect potential quality problems in datasets using the Simple Knowledge Organisation System (SKOS) schema. The rules are based on existing thesaurus construction guidelines and are evaluated using SPARQL queries and graph algorithms (e.g., to find weakly connected components). Finally, Lausen et al.[48] suggested extensions to RDF by constraints akin to RDBMS in order to validate data using SPARQL as a constraint language. This is achieved by providing an RDF view on top of the data.

---

[44]C. Rieß, N. Heino, S. Tramp, and S. Auer. EvoPat – Pattern-Based Evolution and Refactoring of RDF Knowledge Bases. In Proceedings of the 9th International Semantic Web Conference (ISWC2010), LNCS, Berlin/Heidelberg, Springer 2010.

[45]C. Fürber and M. Hepp. Using SPARQL and SPIN for data quality management on the semantic web. In W. Abramowicz and R. Tolksdorf, editors, BIS, volume 47 of Lecture Notes in Business Information Processing, pp. 35–46, Springer, 2010.

[46]H. Knublauch, J. A. Hendler, and K. Idehen. SPIN – overview and motivation. W3C Member Submission, February 2011.

[47]E. Sirin and J. Tao. Towards integrity constraints in OWL. In Proceedings of the Workshop on OWL: Experiences and Directions, OWLED, 2009.

[48]G. Lausen, M. Meier, and M. Schmidt. SPARQLing constraints for RDF. In Proceedings of the 11th International Conference on Extending Database Technology: Advances in Database Technology, EDBT '08, ACM, pp. 499–509, New York, NY, USA, 2008.

While there has been considerable research into quality assessment of Linked Data sets, work that attempts to incorporate such efforts into quality engineering frameworks, which operate to improve data quality over time, is only starting to emerge. Feeney et al.[49] described a semi-automated methodology, framework, and process, which integrate RDF quality assessment mechanisms with human workflows for achieving quality control of published RDF datasets.

1.3.1 The Challenge

The challenge that ALIGNED faced in data quality engineering was twofold. First, the data quality engineering processes that the partners developed for Linked Data required further development, validation, and standardisation. Secondly, mechanisms were required to allow quality control actions of software and data teams, which have generally been developed in isolation, to be aligned and synchronised. For example, if a customer bug report arrives, then it can often be solved by modifications in either the applications or the data. How is this responsibility allocated in diverse teams and what solution will have the best outcome in terms of both the short- and long-term agility and integrity of the combined system?

When data quality is vital, the ultimate resource to deploy is human expertise. In some cases, it may be necessary to deploy human experts to annotate and interpret datasets in order to elevate the raw data to useful information or knowledge for the planned application tasks. However, this is very expensive in terms of both time and the limited resource of domain expertise. Fully automated solutions are popular in research applications, but in enterprise, the deployment of human talent dominates. This is because of the persistent gulf in quality between human-curated content and automated approaches. Thus, the challenge for pragmatic systems is to define semi-automated methods and tools that involve human expert curators in the loop while minimising their workload. By partitioning curation tasks into different levels of required expertise, it is possible to lower the expertise required for participation in the data processing pipeline and thus broaden the base of contributors, hence lowering costs and increasing the productivity of the highest-value experts. Curation workflow tools that provide this functionality

---

[49] K. Feeney, D. O'Sullivan, W., Tai, R. Brennan, Improving curated web-data quality with structured harvesting and assessment (2014), International Journal on Semantic Web and Information Systems.

based on an explicit data life cycle model will result in higher-quality systems at lower cost.

One of the attractions of Linked Data, from an enterprise point of view, is the widespread availability of compatible datasets with which to enrich or annotate an application-specific dataset. However, in practice, this is often seen as an advantage that is still to be realised, since the quality of datasets published on the Web varies widely and it is only recently that mature Linked Data quality frameworks have appeared. Importing low-quality datasets often results in a large clean-up exercise for the application owners. Given that system integrity depends directly on the quality of data input, there is an opportunity to control dataset integrity by limiting updates to datasets based on a strong, semantic specification of the system, the application and schema needs, and design intents. A repository integrity gateway could utilise both data quality frameworks and the system specification to limit the data input, referring offending data to human administrator-based intervention or to other automated checks.

Just as unit testing has entered the mainstream of software development, it is possible to create automated data testing based on rich models of domains, application data needs and design intents and to integrate these into semi-automated processes, which maximise the utilisation of new technologies without dispensing with the ability to use human expertise to provide the highest-quality data. Developing and validating processes that successfully integrate these processes was the challenge tackled by ALIGNED.

1.4 About ALIGNED

ALIGNED is an EU research project, which ran from February 2015 to January 2018. It brought together world-class researchers, representing stakeholders from across the value-chain. It combined model-driven software engineering (Oxford are leading the development of the next generation of UK National Health Service systems), Linked Data quality (Leipzig and Trinity College have published foundational papers) with innovative enterprises (Wolters Kluwer has pioneered the use of Linked Data in complex mission critical systems; the Semantic Web Company (SWC) leads the world in enterprise Linked Data), and expert-driven data curation (Oxford Anthropology and Poznań) to work on high-impact use cases such as DBpedia (Leipzig are co-creators). The project's ambition was to develop the foundations for the next generation of Big Data systems by enabling model-driven creation of Linked Data applications that can effectively deal with the dynamism, complexity, scale, and data quality challenges (e.g., inconsistency and incompleteness) of Web data while retaining the reliability, security, and robustness that come with model-driven software engineering.

The objective of the ALIGNED project was to align semantics-based model-driven software engineering with full life cycle Linked Data engineering to produce powerful and flexible service engineering systems and enable rapid development cycles based on reuse and extension of heterogeneous data sources. This approach supports an aligned engineering process spanning the full service life cycle, based on rich, semantic Linked Data representations, which enable expressive models to be specified for open extensible systems in such a way that flexibility and reusability are prioritised. This will facilitate a step change in the development[50] of Web-scale data-intensive systems. Successfully attaining this objective requires innovations in three distinct technical areas:

- Model-driven software engineering is a maturing research field with well-developed tools and methods like UML, XML, and DSL creation, code, and transformation generation tools like Stratego/Spoofax.[51] The ALIGNED project evolved this research with more expressive and shareable data models based on the modern Web of data.
- Enterprise Linked Data-based systems are starting to appear,[52] and while Linked Data quality engineering processes have started to emerge,[53] they suffer from inadequate tool support. Most Linked Data life cycle management tools also suffer from being oriented towards knowledge engineers, specialising in semantics, rather than the domain experts or software engineers that build and administer enterprise data-intensive systems. ALIGNED addressed this shortcoming by developing, testing, and validating collaborative Linked Data engineering tools and integrating them into user-friendly data curation services and platforms.

---

[50] http://www.uml.org/ & http://www.w3.org/XML/

[51] http://strategoxt.org/view/Spoofax/WebHome

[52] C. Dirschl, K. Eck, and J. Lehmann, "Supporting the Data Lifecycle at a Global Publisher using the Linked Data Stack", ERCIM News, 96, January 2014.

[53] A. Zaveri, A. Rula, A. Maurino, R. Pietrobon, J. Lehmann, S. Auer, "Quality assessment methodologies for linked open data", under review, Semantic Web Journal, IOS Press.

- Linked Data schemas, expressed in standardised languages such as RDFS[54] and OWL,[55] enable self-describing data structures with rich semantics included within the data itself. Aspects of program logic previously encapsulated in software are now embedded in data models, meaning that the software engineering life cycle of data-intensive systems needs to be aligned with the data engineering life cycle. For example, changes to data schemas may require updates to the software that consumes it, and vice versa. ALIGNED addressed this challenge by identifying common phases and signalling between the parallel processes and tools to support alignment at higher levels. This supports both integrated, model-driven unified processes and loosely coupled, co-evolving systems through the specification of common vocabularies and domain-specific metamodels.

ALIGNED leveraged Linked Data as the common technical platform to support integration at three levels: first, by applying semantics and Linked Data to model-driven software engineering to develop rich domain and application-specific specification models; second, as a means to integrate tools for combined software and data engineering; and third, as the basis for exemplar data-intensive systems that combine software and data to manage, publish, process, and consume data.

NESSI has identified "Collaborative Service Engineering based on the convergence of software and data" and "Integration of Big Data Analytics into Business processes" as EU research priorities.[56] This is a response to the parallel trends which see increasingly complex and dynamic service-delivery collaborations alongside the ongoing explosive growth of data available via the Web. The increasing prevalence of rich and flexible standardised semantic languages[57] has created opportunities for service providers to add value to their services with readily available machine-processable knowledge.[58] To take advantage of these opportunities, service and software engineering organisations must integrate data engineering and service engineering processes.

---

[54] http://www.w3.org/RDF/ & http://www.w3.org/TR/rdf-schema/

[55] http://www.w3.org/TR/owl2-overview/

[56] Strategic Research and Innovation Agenda Version 2.0, NESSI Position Paper, April 2013.

[57] C. Bizer, K. Eckert, R. Meusel, H. Mühleisen, M. Schuhmacher, and J. Völker: Deployment of RDFa, Microdata, and Microformats on the Web – A Quantitative Analysis In: 12th International Semantic Web Conference, 21–25 October 2013.

[58] P. Hitzler, K. Janowicz, Linked Data, Big Data, and the 4th Paradigm, Semantic Web Journal, IOS Press, 2013.

1.5 ALIGNED Partners

1.5.1 Trinity College Dublin

Trinity College Dublin is Ireland's leading university. TCD, founded in 1592, hosts over 15,500 students. It enjoys an esteemed reputation in research and innovation with an outstanding record of publications in high-impact journals and a track record in winning research funding. Its research impact is currently ranked 44th in the world by the Times Higher Education Ranking of World Universities and 9th in Europe by the 2013 Leiden University Ranking of World Universities' research performance.

1.5.2 Oxford University – Department of Computer Science

The Department of Computer Science, ranked first in Europe in the Shanghai tables, has particular strengths in software engineering, programming languages, and information systems. The Software Engineering Group works across all three areas and has a strong track record of interdisciplinary collaboration in medical and scientific research, humanities, and social sciences. It has also a strong track record of effective engagement with industry, delivering a substantial programme of advanced education aimed at full-time professionals: designers, developers, managers, and users.

1.5.3 Oxford University – School of Anthropology and Museum Ethnography

The School of Anthropology and Museum Ethnography is one of the oldest and most distinguished anthropology departments in the world. It is also one of the broadest, comprising five units that cover a wide range of subfields of anthropology (social and cultural, cognitive and evolutionary, visual and material, medical and biological) as well as a range of specialised foci (e.g., migration, science and technology) with long-established field projects all around the globe. Of particular importance for ALIGNED, it is home to the Institute of Cognitive and Evolutionary Anthropology, which employs staff with expertise in database construction and analysis.

1.5.4 University of Leipzig – Agile Knowledge Engineering and Semantic Web (AKSW)

The Institute for Applied Computer Science (InfAI, http://infai.org) at Universität Leipzig hosts world-class research groups in service and Web science.

The approximately 40 researchers of the Agile Knowledge Engineering and Semantic Web research group (http://aksw.org) at InfAI are establishing theoretical results and scalable implementations for realising the Semantic Data Web. Particular emphasis is given to areas such as ontology creation and manipulation, knowledge extraction, ontology learning and information, and data integration on the Linked Data Web. The scientific publications of the group, founded in 2006, have already attracted more than 6,000 citations (according to Google Scholar).

1.5.5 Semantic Web Company

SWC is an SME, based in Vienna, Austria, founded in 2001, which offers ICT consulting services and solutions in semantic information management. This includes data and metadata management, knowledge and information management systems, LOD, enterprise search, and social software. SWC is the vendor of the PoolParty Semantic Suite (http://poolparty.biz) for enterprise-ready solutions in taxonomy management and data integration. SWC's work is always based on open semantic Web standards to ensure interoperability and sustainability for solutions.

1.5.6 Wolters Kluwer Germany

Wolters Kluwer Germany is an information services company specialising in the legal, business, and tax sectors. Wolters Kluwer provides pertinent information to professionals in the form of literature, software, and services. Headquartered in Cologne, it has over 1,200 employees located at over 20 offices throughout Germany, conducting business on the German market for over 25 years. Wolters Kluwer Germany is part of the leading international information services company, Wolters Kluwer n.v., located in Alphen aan den Rijn (the Netherlands). The core market segments, targeting an audience of professional users, are legal, business, tax, accounting, corporate and finance services, and healthcare.

1.5.7 Adam Mickiewicz University in Poznań

Adam Mickiewicz University in Poznań is the major academic institution in Poznań and one of the top Polish universities. Its reputation is founded on tradition, the outstanding achievements of the faculty, and the attractive curriculum offered to students. It is a centre of academic excellence, where research and teaching are mutually sustaining, and where the context within which research is conducted and knowledge is sought and applied is international as much as regional and national. The University was founded in 1919 and its current student population is nearly 49,000. The University currently employs nearly 3,000 teaching staff, including 264 tenured professors, 439 associate professors, and 1,617 adjunct professors and senior lecturers.

1.5.8 Wolters Kluwer Poland

Wolters Kluwer Poland the largest publisher of legal and business information in Poland. It provides a large database of legal and business information under the IPG brand. Wolters Kluwer Poland is part of the leading international information services company, Wolters Kluwer n.v., located in Alphen aan den Rijn (the Netherlands).

1.6 Structure

The remainder of the book is organised as follows. Chapter 2, Use Cases, briefly describes the five use cases undertaken in the book. It focusses on the data engineering and software engineering challenges, where they are the same and where they differ across the use cases. Chapter 3, Methodology, describes a general methodology for understanding Big Data systems, their requirements, the different families of modelling approaches that are suitable for different systems, and the integration of software and data engineering life cycles by way of signalling points and common vocabularies. Chapter 4, Vocabularies and Ontologies, describes the use of layered common taxonomies, vocabularies, and ontologies as a basis for semantic integration. These include foundational schemas such as RDF, RDFS, and OWL; common widely used standards such as PROV and SKOS; new general-purpose ontologies to describe validation errors and dataset identities such as RVO and DataID; and high-level custom ontologies to describe processes (DLO and SLO). Chapter 5, Tools, describes the software tools used to solve the problems of the use cases, which include RDFUnit, DataID, the Model Catalogue, Semantic Booster, the PoolParty Semantic Suite, and the Dacura semantic curation platform. It focusses on describing the vocabularies and APIs supported by each tool with a little bit on implementation for each.

Chapter 6, Integrated Systems, describes the integrated systems that were developed to solve the problems of the use cases introduced in Chapter 2.

It is split into five parts:

- Wolters Kluwer – Re-engineering a complex relational database application: In every enterprise environment, relational databases are used for a long time to process critical data. It is a common situation that the database schema has heavily evolved over time and no one in the company understands the impact of any change in its entirety anymore. Therefore, companies continue to use these databases without touching them anymore, reducing its overall value over time. Sooner or later, a complete re-engineering or even complete new development is required, which means a significant investment and a high risk of failure. In this presentation, we will show that it is possible to reduce this risk by using semantic technologies when replacing the old application and which also better prepares the company for any re-engineering effort in the future.
- Seshat – collecting and curating high-value datasets with the Dacura platform: This section uses the Seshat project as a case study – a huge distributed effort by social scientists to compile an authoritative databank describing the evolution of all human societies that have existed since 10,000 BCE. We show how the system uses semantic models both to provide strong data consistency assurances and to generate user interfaces for crowd-sourcing and human expert approval. Although this use case is an academic endeavour, the technology is entirely agnostic to the application and can be applied in any scenario where an organisation wishes to collect and curate high-quality datasets.
- Managing data for the NHS: This section examines the ALIGNED Data Catalogue system: a set of tools for automating aspects of data management at scale. At the heart of the system is the metadata catalogue, a tool for capturing and linking key information about data: information that can be used to determine, automatically, how data are to be processed, transformed, and accessed. Other tools support the processes of metadata capture and curation, as well as system configuration and generation. We explore the application of the Data Catalogue system to the management of health data in the United Kingdom. The Oxford ALIGNED partners have deployed the metadata catalogue and other tools in support of several, large health data projects in collaboration with the NHS. One of these, the 100,000 Genomes Project, required the coordination of data specifications, form designs, database schemas, and messages, for a wide range of diseases, across 70 hospitals.

- Integrating semantic datasets into Enterprise Information Systems with PoolParty: The Linked Data movement has seen increasingly large semantic datasets published on the Web, as part of the web of data. This creates opportunities for integrating public sources of data with enterprise information sources to create enriched high-quality semantic knowledge bases. ALIGNED is developing tools and processes to integrate with PoolParty, SWC's semantic technology suite. PoolParty Thesaurus Server is a Thesaurus and Taxonomy Management Tool to build and maintain information architectures. In this section, we showcase how we use SHACL and the RDFUnit test framework as a basis for the import assistant to run automatically and manually generated test cases for validating data consistency constraints.
- Data Validation at DBpedia: Data validation is a crucial part of data integration – integrated data must meet a minimum validation criterion before it can be considered integrated. Reducing the manual time and effort required to validate data is a critical enabler of dealing with the volume and velocity of Big Data. In this section, we show how DBpedia has used ALIGNED tools including RDFunit to develop a high-quality curated dataset offering.

Finally, Chapter 7, Evaluation, describes a suite of evaluation techniques and measures focussed on agility, productivity, and quality in big-data systems and presents an ontology in which the various types of measures are related to one another and an abstract framework for evaluating such systems.

2

ALIGNED Use Cases – Data and Software Engineering Challenges

Arkadiusz Marciniak and Patrycja Filipowicz

Adam Mickiewicz University, Poland

2.1 Introduction

The ALIGNED project developed an aligned methodology for parallel software and data engineering of Web-scale information systems with Linked Data as a unifying technical foundation for system specification and process and tool integration. This methodology (see Chapter 3) is based on a metamodel describing the complete software and data life cycles, domain models, and design intentions. This metamodel specifies tools to produce software development models, including transformations that generate or configure software applications as well as data development models, incorporating data quality and integrity constraints, data curation workflows, and data transformations.

Software and data engineering are different disciplines, with different practices and processes. Significant differences between these fields mean that a single prescriptive approach could not work. Instead, the project has identified a matrix of synchronisation points between different stages of the software and data life cycles. Each point represents a key area where software and data engineers may need to interact and define formats and processes for working together. This approach is flexible enough to accommodate many different workflows, while still identifying key areas where alignment of the two life cycles can lead to significant savings in effort. The approach adopted endeavoured to make it possible to improve the overall quality, productivity, and agility in a variety of different use cases. In order to achieve these objectives, the project sought to develop Linked Data schemata for alignment that enabled the software engineering life cycle of data-intensive systems to be integrated with the data engineering life cycle, by identifying common phases and signalling between the parallel processes and tools to support alignment at higher levels.

The decision to adopt data-model-driven approaches in the project had far-reaching consequences. In particular, it required that every step in the process be directly driven by the model, rather than independently configured. Harvested datatypes also could not be consumed directly, but through a model, which dictated the shape and structure that the data must take. Accordingly, a model-driven approach led to the creation of explicit models at each stage of the development process.

MDE describes a development process in which the components of the final software artefact are derived – either manually or automatically – from models that typically form part or all the specifications or requirements of the system. The software needs to be written in such a way that it understands the modelling language and is capable of handling updates to the model. Such software can be reused in different applications within a similar domain, minimising the time spent on the implementation phase, and capturing common repeating patterns that would otherwise have to be repeated in each cycle of an iterative development. In the MDE world, it is required that the data model is provided in the form of the ontologies available at a well-known URL, which is typically achieved by providing a metadata registry.

In order to achieve the postulated goals, a number of tools from both domains were developed and used in order to make the advocated integration of both life cycles efficient, particularly in relation to challenges posed by the different use cases. These comprise Booster, the Model Catalogue, RDFUnit, Repair Framework and Notification (RF), Ontology Repair and Enrichment (ORE), Dacura, the PoolParty Confluence/JIRA Data Extractor (CJDE), External Link Validation (ELV), and the Unified Governance Plugins.[1] Similarly, a set of open, public ontologies and vocabularies were adopted and used wherever possible by all tools to support integration (for details, see Chapter 4). These include foundational schemas, such as RDF, RDFS, and OWL, and common widely used standards such as PROV and SKOS. Where ontologies did not exist to cover the advocated integration needs, new models were created and made publicly available (RVO, RUT, DataID). This collection of common, project-wide ontologies gave the ability

---

[1] Shah, Seyyed M., James Welsh, Jim Davies, and Jeremy Gibbons. 2017. In Mahmood, Z (ed.), Software Project Management for Distributed Computing: Life-Cycle Methods for Developing Scalable and Reliable Tools, 367–385. Springer: Cham.

to exchange rich, structured information covering the most significant entities within research focus.

The ultimate objective of the project, however, was to produce tools, methods, and standards, which lead to real improvements in productivity, quality, and agility of different types of data. The rapidly increasing size and complexity of Web and Big Data often makes their management virtually impossible, where even specialists struggle to harness them. Hence, five use cases representing different domains from legal to health and complex archaeological and historical datasets were chosen to adopt a broad bottom-up approach to system development and integration. Accordingly, the project tackled problems in a wide range of areas with the intention to show how the latest semantic technologies can help create means of managing and using these datasets. The selection of uses cases was also driven by a need of testing interoperability between the tools, particularly those who support both data and software engineering that were developed in the project. The chosen use cases were: (i) Seshat: Global History Databank, (ii) PoolParty Enterprise Application Demonstrator System, (iii) DBpedia, (iv) Jurion and Jurion IPG, and (v) Health Data Management.

Each use case is a large-scale, real-world project with large user communities and complex sets of data. The project's research has thus had a practical focus, which has seen the application of innovative tools and solutions in real life. The use cases represent diverse domains, both commercial and non-commercial, which have their own requirements and data characteristics. They also represent a significantly different level of advancements in both the data and software engineering tools and procedures. Each use case has its own problems with quality, agility, and productivity. The project has built tools and processes that improve software and data engineering for each of these use cases. Every tool appears in more than one use case, and every use case involves tools developed by different partners. In each case, trial platforms were constructed in multiple phases, which integrate research outputs from multiple work packages and partners, served to offer the greatest potential for real improvements to the existing processes employed within these use cases.

The objectives of this chapter are thus threefold: (1) to present the five case studies used in the ALIGNED project, (2) to analyse the major challenges identified by these use cases in the data engineering life cycle as well as present their proposed solutions, and (3) to analyse the major challenges identified by the use cases in the software life cycle and propose solutions to these challenges.

2.2 The ALIGNED Use Cases

2.2.1 Seshat: Global History Databank

The Seshat: Global History Databank[2] is an international initiative of humanities and social science scholars to build an open repository of expert-curated historical time-series data.[3] The Seshat project began by selecting a sample of 30 areas from around the world. For each area, all societies that had controlled it throughout history were recorded. This made it possible to answer a wide range of questions about each of them – describing its population, technology, religion, infrastructure, and so on. The Seshat has been designed to test theories about the evolution of social complexity, from the point of view of historians and anthropologists.[4] The databank extracts data from a combination of databases, Linked Data sources, websites, academic publications, and human experts. Figure 2.1 shows the initial sample of 30 geographical areas chosen for the databank.

A special code book defined the full list of questions, and researchers added data to the system by creating a copy of the code book page for each society and adding data points using a special syntax that encoded uncertainty, disagreement, and temporal scope, along with comments and citations in relation to domain-specific provenance information. In the initial stages of the Seshat project, a wiki was used to collect the data. The system amassed over 200,000 data points on hundreds of civilisations, but whilst the unstructured wiki data store allowed great flexibility at the start of the project,

---

[2] http://seshatdatabank.info

[3] Turchin, Peter, Thomas E. Currie, Kevin C. Feeney, Pieter Franois, Daniel Hoyer, J.G. Manning, Arkadiusz Marciniak, Daniel Mullins, Alessio Palmisano, Peter Peregrine, Edward A.L. Turner and Harvey Whitehouse Harvey. Seshat, The Global History Databank, Cliodynamics. The Journal of Quantitative History and Cultural Evolution 6(1), pp. 77–107.

[4] Turchin, Peter, Thomas E. Currie, Harvey Whitehouse, Pieter Franois, Kevin Feeney, Daniel Mullins, Daniel Hoyer, Christina Collins, Stephanie Grohmann, Patrick Savage, Gavin Mendel-Gleason, Edward Turner, Agathe Dupeyron, Enrico Cioni, Jenny Reddish, Jill Levine, Greine Jordan, Eva Brandl, Alice Williams, Rudolf Cesaretti, Marta Krueger, Alessandro Ceccarelli, Joe Figliulo-Rosswurm, Po-Ju Tuan, Peter Peregrine, Arkadiusz Marciniak, Johannes Preiser-Kapeller, Nikolay Kradin, Andrey Korotayev, Alessio Palmisano, David Baker, Julye Bidmead, Peter Bol, David Christian, Connie Cook, Alan Covey, Gary Feinman, Árni Daníel Júlíusson, Axel Kristinsson, John Miksic, Ruth Mostern, Cameron Petrie, Peter Rudiak-Gould, Barend ter Haar, Vesna Wallace, Victor Mair, Liye Xie, John Baines, Elizabeth Bridges, Joseph Manning, Bruce Lockhart, Amy Bogaard and Charles Spencer. Single dimension of complexity in human societies. Proceedings of the National Academy of Sciences, 115 (2) E144-E151; DOI:10.1073/pnas.1708800115.

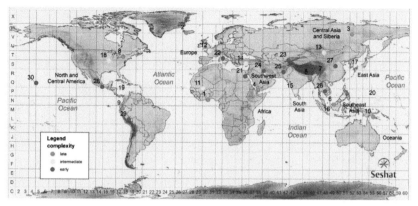

Figure 2.1 Seshat World Sample 30.

it did not scale to the number of contributors, data users, data points, or the complexity of the data.

Seshat also evolved to encompass new areas that were not originally anticipated. In particular, this involved recording societies from the prehistoric past, which required a collection of archaeological data. It soon became obvious that many Seshat variables were unsuitable for capturing this part of human past. There was also a lack of relevant proxies that would allow translation of archaeological evidence into coding templates. Accordingly, the Archaeological Seshat code book was designed and developed in order to fill in the gap, and the data were collected independently.

A wiki-based approach, used in Seshat for the data collection task, posed numerous problems, in particular for the verification of data correctness, and the extraction of data in usable forms. As the dataset grew and the focus moved from collection to integration and analysis, several other significant problems emerged. The fundamental problem is that a wiki is designed for human presentation and editing of data. To a machine, it is semi-structured, lacks any type information, and the meaning of the elements depends on their context within a jumble of HTML. Without any support for validation, errors proliferated.

The limitations of the wiki also impacted agility. As the Seshat code book was rapidly evolving, any changes needed to be manually copied to all existing data pages. This was a costly and error-prone task. There was also no easy way to express spatial data through the wiki, so these data were stored in a separate geographic information system (GIS). The wiki-based system offered no support for publication. Furthermore, while the scraping tool could extract raw datapoints, important citations and comments were encoded in totally unstructured HTML.

Productivity suffered as increasing resources had to be devoted to curation and cleaning. Some of the corrections were not copied back to the wiki and spreadsheets became the authoritative source for some sections of the data. Moreover, there was no way of incorporating third-party data into Seshat dataset.

2.2.2 PoolParty Enterprise Application Demonstrator System

The PoolParty Semantic Suite[5] is the SWC's platform for enterprise information integration based on Linked Data principles. The PoolParty semantic technology suite comprises a number of tools based on the extraction, curation, and management of linked open datasets. These tools are split into three categories: data portals and collaboration platforms, tools for knowledge engineering and graph management, and functionality for content enrichment and data integration. Any data is transformed into RDF graphs and can be queried with SPARQL (SPARQL Protocol and RDF Query Language). Since it was created, the product has evolved to include entity extraction from unstructured information. PoolParty's API provides a rich set of methods for text mining and entity extraction. Figure 2.2 shows the tools of the PoolParty Application Suite.

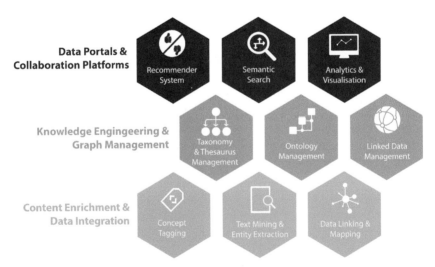

Figure 2.2 PoolParty Application Suite.

---

[5]http://www.poolparty.biz

As a loosely coupled collection of tools, additional functionality has been enabled through the integration of third-party tools. An example of this is the use of Atlassian Confluence (a team collaboration tool), Atlassian Jira (a tool for issue tracking and project management), and Media Sonar (a Web-mining tool), for a general-purpose requirements engineering system. However, the systems concerned are typically document-oriented and require extensive human interaction in order to link their data to development tasks recorded in PoolParty against standard ontologies. The system lacks the required integration and alignment of data management issues with the software development life cycle, so that each supports the other.

2.2.3 DBpedia

DBpedia[6] publishes authoritative RDF-based datasets that are used as a common point of reference for interlinking and enriching most of the structured data on the Web today. It relies on an automated data extraction framework to generate open RDF data from Wikipedia documents, published in the form of file dumps, Linked Data, and SPARQL hosting on the Linked Data Stack. This structured information resembles an open knowledge graph, which is a kind of database, which stores knowledge in a machine-readable form and provides a means for information to be collected, organised, shared, searched, and utilised. DBpedia passes all published data through RDFUnit, validating it against an up-to-date version of the DBpedia ontology. The validated outputs generate consistent data termed DBpedia+, whereas the wider, more exhaustive data are published as the standard DBpedia datasets.

To create high-quality data, a validation method for DBpedia instance data has to provide sufficient metadata to distinguish between three different possible sources of a violation: (i) the Wikipedia editor (entering erroneous values), (ii) incorrect mappings between source and DBpedia ontology, and (iii) a software issue in the DBpedia Extraction Framework. Accordingly, RDFUnit provides the necessary metadata for any violation found and creates links between a software issue and the violating instance. The resulting violations and associated metadata provide the exact coordinates of a violation, the grounds for this violation, and the possible source. Thus, violations recorded in such a manner are used as a feedback medium, relating possible mistakes to Wikipedia editors, to the mapping community, or to software developers. In addition to validating the resulting instance data, DBpedia started to validate

---

[6]http://wiki.dbpedia.org

the mappings between DBpedia ontology and the Wikimedia data sources on a regular basis with RDFUnit. Thus, most of the mapping-related violations can be caught before ever starting the data extraction, preventing possible reruns of whole extraction steps and increasing productivity.

The DBpedia Links repository maintains linksets between DBpedia and other LOD datasets. A system for maintenance, updates, and quality checks, which validates various aspects of the link submission, is in place and is integrated with common continuous integration services, such as Travis CI. It offers a way to publish linksets between DBpedia and any given dataset, which are published alongside the DBpedia dataset files.

The major productivity issues identified for DBpedia involve code maintenance, release management, ontology editing, release documentation creation, and dealing with user queries. Further complications involved dealing with the increasing number of published datasets that tend to increase over time when incorporating new extraction methods and algorithms.

To ensure quality regarding the extraction workflow, DBpedia extended the Extraction Framework to produce metadata for any extraction process, extensive logging of progress and exceptions, as well as high-level summaries of extractions. These efforts support extensive monitoring, metadata propagation and logging (on both the triple and dataset level), and the deployment of ETL frameworks and Workflow Management Systems to further decrease the time needed for extraction and to automate this process completely. Figure 2.3 shows this pipeline.

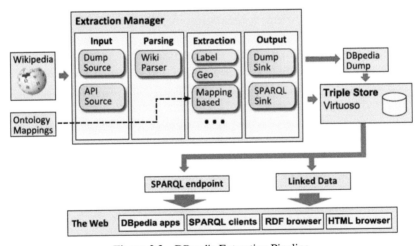

Figure 2.3   DBpedia Extraction Pipeline.

The greatest need for agility in DBpedia is the ability to rapidly respond to changes in source datasets like Wikipedia. These may involve, among others, the introduction of new pages that represent new concepts and the introduction of new infobox templates that represent additional instance data in DBpedia and changes in infobox structures. Adapting to those changes in a (semi-) automated way will prevent the loss of data (due to changes to Wikipedia templates) and incorporate new instance data automatically.

2.2.4 Jurion and Jurion IPG

The Wolters Kluwer[7] use case within ALIGNED is twofold. On the one hand, the project worked with a legal research database application called Jurion (www.jurion.de) from Wolters Kluwer Germany. In this use case, it mainly focussed on addressing data quality issues. Second, the project re-engineered the IPG system from Wolters Kluwer Poland, which is a commercial intelligence system, based on huge amounts of data in a relational database system.

Jurion merges and interlinks over one million documents of content and data from diverse sources such as national and European legislation and court judgements, extensive internally authored content and local customer data, as well as social media and Web data (e.g., from DBpedia). In collecting and managing this data, all stages of the Data Life cycle are present – extraction, storage, authoring, interlinking, enrichment, quality analysis, repair, and publication. Wolters Kluwer concentrated mainly on the enhancement of data quality and repair processes. Based on the requirements, it started to work on data transformation issues and the improvement of data quality processes in PoolParty in parallel to some tasks within the PoolParty use case. Based on large amounts of XML data, governed by a DTD, continuous transformation from XML to RDF, based on XSLT scripts, needs to take place. This process is complicated and error-prone, especially when it comes to schema changes. The second major data quality challenge is around domain thesauri and controlled vocabularies. Very often, these data are initially created and stored in XLS files and when it comes to a systematic usage of more powerful tools like PoolParty, the import process of this data needs to be optimised, so that errors and inconsistencies can be detected very early in the process.

---

[7] www.wolterskluwer.com

The Jurion IPG system is a commercial intelligence system, providing a means for business contractors to perform due-diligence queries, serving historical data about companies and their relationships with other companies, responsible individuals, and business documents. It has been developed by Wolters Kluwer Poland and it contains data on 450,000 companies, 1.1 million people, and 3.5 million documents. The existing data are currently stored in a relational format. The complexity of the system stems from huge amounts of daily processed data originating from pdf sources and their maintenance through a proprietary, obsolete CMS. In order to remain a reliable provider of credibility and financial information for over five million entities, the integrity and consistency of the data is of vital importance, and increasingly hard to manage at scale. Business value of the system is dependent on the maintenance and evolution of a large, semantically consistent dataset. The overall goal is to ensure the quality of the system used to enter and maintain the data and to improve the value by linking to external datasets. The major requirements involve deploying new tools to find problems in the existing data, improving the integrity of data submitted in the future as well as help increasing the scope of the data by enabling the linking of data stored within the system to external related datasets. Figure 2.4 shows the JURION IPG workflow.

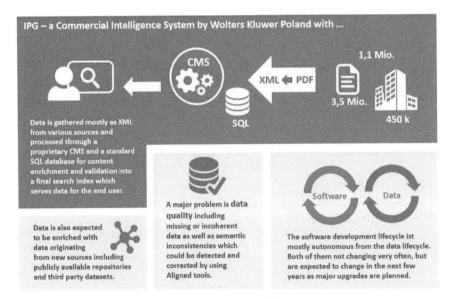

Figure 2.4 Jurion IPG.

2.2.5 Health Data Management

The Health Research Data use case involved four separate projects related to health research data in the United Kingdom:

- the Health Data Finder[8] – an online tool for discovering national healthcare datasets commissioned from the National Institute for Health Research (NIHR). They primarily contain routine hospital data for audit and economic reasons, but may be made available to researchers in academia and industry with appropriate governance approval. The datasets are maintained by a number of separate organisations, and so data users wishing to discover data and request access may have to make a number of requests, often with inconsistent results.
- the NIHR Health Informatics Collaborative[9] – routine clinical data in five therapeutic areas provided by the largest teaching and research hospital trusts. These include critical care, ovarian cancer, acute coronary syndromes, hepatitis, and renal transplantation. Each trust maintains data to differing standards and semantics, and rather than unifying data to a lowest common denominator, sites are asked to build their own data warehouses for a federated data store. Users of the data can make a request to the hospitals, and data can be linked and unified on a per-usage basis, taking into account the research purpose.
- the UK 100,000 Genomes Project[10] – a UK Government project aimed at sequencing whole genomes from National Health Service patients. It is focussed on rare diseases, major types of cancer, and infectious diseases. The patients give consent for the genome data to be linked to information about their medical condition and health data. The ultimate goal of the project is to improve knowledge of the causes, treatment, and care of these diseases.
- the construction of a data warehouse for Oxford University Hospitals Foundation Trust[11] – this is a detailed asset register for the hospital, detailing field-level metadata about databases and spreadsheets of patient data around the hospital, as well as describing dataflows and message-passing between systems, and specifications for audit and research datasets.

---

[8] http://www.hdf.nihr.ac.uk

[9] https://www.nihr.ac.uk/about-us/how-we-are-managed/our-structure/infrastructure/health-informatics-collaborative.htm

[10] https://www.genomicsengland.co.uk/the-100000-genomes-project

[11] http://www.ouh.nhs.uk

In all four applications, reuse of existing data without detailed documentation causes major problems, particularly in relation to poorly developed semantics. Furthermore, linkage between datasets may be inaccurate, transformation of data into different formats may be incorrect, and interpretation of statistical results is error-prone. In the Health Data Finder, such data reuse is minimal. Researchers do not know what data may be available to them, different providers may return inconsistent results on data governance, and data must be re-interpreted each time, which may result in costly errors. In similar projects preceding the Health Informatics Collaborative and 100,000 Genomes projects, collecting comparable data from multiple hospitals has proven difficult. Precise specifications have been hard to produce, mechanisms for data capture and transfer have been manually programmed, often by non-technical domain experts, and inconsistencies have resulted in data that is often incomplete, incomparable, or completely unusable.

The quality and accuracy of data documentation is difficult to maintain during an iterative process. In all the health data research projects, datasets are continually evolving and data specifications are continually being improved. Without careful version management and automation, it is very easy for the documentation to get left behind. Similarly, software artefacts must keep pace with the changes in requirements: changes to the data or the software specifications must invoke updates to the XML schema, database schema, or Case Report Forms. Manual coding slows the iteration process, which in turn can result in outdated or inaccurate specifications.

Furthermore, domain experts find it difficult to provide documentation or simple modelling because of the technicalities involved. XML schema and Case Report Forms require specialist technical knowledge. Implementing efficient database structures requires a lot of repetitive work such as implementation of a domain class will involve a familiar pattern of tables, association tables, keys, and indexes. Such work is time-consuming and error-prone, yet ripe for automation. Data scientists looking to reuse health data currently spend a lot of time searching for usable datasets, often requiring long periods of interaction where inventories and documentation are not available online. Applying for governance, asking technical questions, and retrieving data in a suitable format often require further time and energy. Interpretation and curation of the data is a typically manual task, which may be repeated and reproduced by every scientist receiving a data extract.

2.3 The ALIGNED Use Cases and Data Life Cycle. Major Challenges and Offered Solutions

The LOD life cycle consists of eight stages for data engineering.[12] These are: (i) extract – taking information in unstructured form or conforming to other structured or semi-structured formalisms and mapping it to the RDF data model, (ii) storage and querying – retrieving and persisting information in triple form to be included as part of the dataset; (iii) manual revision/authoring – processes for manual creation, modification, and extension of the structured data; (iv) interlinking/fusion – creating and maintaining links between datasets; (v) classification/enrichment – creating and maintaining links between data, and models of data (which themselves may be linked and part of the dataset); (vi) quality analysis – testing for data completeness and correctness; (vii) evolution/repair – correcting invalid data resulting from a quality analysis phase, via either manual or automated processes; and finally, (viii) search/browse/exploration – making data artefacts available to domain experts or to users beyond the original authors.

Different stages of data engineering in the ALIGNED project have been identified primarily for building tool support and integrated frameworks as well as encouraging compatibility of independent tools within a particular framework. Feedback from one phase is to be fed into another. For example, the models linked during the classification or enrichment stage will determine the scope of the quality analysis stage, or any errors found during quality analysis may need to be resolved in the evolution/repair phase.

As the dataset grew and the focus moved from collection to analysis, several significant problems with agility, quality, and productivity emerged. First, the fundamental problem was that a wiki is designed for human presentation and not machine-readable. Second, the limitations of the wiki impacted agility: manual data harvesting has been very time-consuming. Finally, productivity suffered as increasing resources had to be devoted to curation and cleaning.

In each use case, ALIGNED technologies are being used in slightly different ways. In case of Seshat, these tools are automatically generated from the Seshat ontology. These comprise the Model Mapping Tool, Real-time Instance Data Validation, and curation workflows, all deployed as Dacura services. Dacura is a data curation platform developed by Trinity

---
[12]Shah et al. 2017: 370.

College Dublin, which incorporates several techniques. The adopted solutions improved the process of data collection.

The model catalogue tool is used in the analysis phase of model-driven software engineering to explore and gather metadata related to the system under construction. It is also used in its search browse and phase life cycle. In the project, it is primarily the Model Catalogue that is used along with components of Semantic Booster, both developed by Software Engineering at Oxford University. In the data engineering context, tools generated by Booster can be used to provide a well-defined API as well as to search and gather data into the data store. Booster-generated systems provide, create, read, update, and delete functionality for data in a data store, as well as implement any user-specified action, which can then be accessed as triples via an API (Shah et al. 2017: 381).

In the Jurion use case, an enhanced data quality and repair pipeline was established with the help of RDFUnit and PoolParty, so that data life cycle process was suffering from less data errors and schema inconsistencies and the overall process was accelerated, especially when data or schema changed over time.

In the Jurion IPG use case, the Model Catalogue is used to provide accurate descriptions of data fields, including those from linked external data sources. Such descriptions can aid correct data entry and permit additional reuse of data within the organisation. The Model Catalogue also aims at serving as a provider of models to the generated tools and as an environment where new versions of the data model can be created and evolved. Dacura was instantiated as an alternative approach, covering the overall process from model storing, mapping, and a complete automatic generation of the final future data schema, accompanied by automatic data testing with RDFUnit.

In the NIHR Health Data Finder, the Model Catalogue is the central resource, holding the master copy of models and documentation. In the NIHR Health Informatics Collaborative, each site hosts its own instance of the Model Catalogue, documenting their own data landscape including a data warehouse, source patient record systems, research systems, and local data flows. A central installation of the catalogue contains the shared data specifications, along with local variations, and relevant national specification. Local catalogue installations can automatically import the latest version of the central models, and the central catalogue is used to generate XML schema for use by all partners. In the UK 100,000 Genomes Project, the architecture of the pilot is of particular interest: information is provided by the hospitals in the form of XML, matching a schema generated by the Model Catalogue, or manually through online Case Report Forms, hosted in a system called OpenClinica. Information is extracted via an ETL (extract, transform, load) process from OpenClinica, combined with a shredded form of XML, and stored in a matching relational database, generated by a component of Semantic Booster. Finally, the architecture of the OUH data warehouse follows a similar pattern. Almost 100 local databases and data specifications are modelled within the catalogue, along with the design for the main data warehouse. The catalogue is used to document field-level metadata, summary metadata, and dataflows, and this information is to be used in the construction of research data extracts and for generating hospital auditing and service improvement metrics.

One of the major steps in data engineering life cycle was the development of new approaches to data validation. In particular, it comprised a new tool developed for the PoolParty semantic suite. The process involved importing RDF data in PoolParty and using the integrated validation checks to identify problems, which are reported to the user as constraint checks. The user is then given options to repair the data consistency. After fixing the inconsistencies, the user can then import the data without the risk of application failure.

RDFUnit is integrated in PoolParty RDF Validation for performing constraint checks. The checks are defined as RDFUnit test cases using RDF. These test cases can also be run by RDFUnit independently of PoolParty on external data. For each of the constraint checks, there is an RDFUnit test case, which is based on a SHACL constraint or a SPARQL query that identifies resources that cause violations. They together formed the basis for the Data Quality Framework and the Automated Data Testing and Verification Framework.

In most cases, constraint violations only become apparent after the import has been done. In the worst case, this may even cause issues displaying the data or errors displayed to the user. In other cases, issues could pass through unnoticed or may only become apparent at a later stage. This means that users interpret data issues as software issues and report those as bugs in the SWC support space. Import validation has the potential to provide major improvements of productivity and data quality in the data development life cycle. The prototype import validation implemented in PoolParty using RDFUnit enabled the users to get direct user feedback on violations of data constraints. The feature provides direct feedback on data consistency constraint violation before data are imported. Being able to detect violations of consistency constraints on data import increases data quality, since problems are not imported into the system in the first place. The import validation features provide increased agility, empowering users to import data without quality issues. That means, users can react to issues themselves and fix the data before it gets imported. This improves the connection between the data development and the software development life cycle.

Notification can improve the usability of the PoolParty software, actively providing notifications to users based on activity in projects. Currently, staying informed about activities in projects can only be achieved by reviewing the project history regularly. The ORE tool suggests new ontology axioms (enrichment) and recommends semi-automatic fixes (for resolving violations).

Another important contribution of the ALIGNED project was in the domain of search, browsing, and exploration. Of particular importance is Dacura, which is in a position to produce data quality tolerance requirements to constrain the data to be harvested.[13] The CJDE tool is also responsible for extracting relevant requirement information and hence, tickets and creates RDF data.

The Dacura approval queue allows also dataset administrators to monitor added data for quality and completeness. Administrators can approve, deny, publish, and unpublish the Linked Data objects submitted by Seshat researchers. From a Dacura point of view, it is possible to import large volumes of IPG data into a structured, rich semantic format according to a predefined model that is amenable to statistical analysis and offers automated quality control. Dacura ensures consistency requirements and allows users to monitor newly added data with respect to quality and completeness conditions according to defined constraints.

The Unified Views tool allows data to be imported via SPARQL from third-party datasets; in this case, DBpedia is used as a source of data. The Unified Views tool also allows the establishment of processing workflows to automate the importation of such data.

2.4 The ALIGNED Use Cases and Software Life Cycle. Major Challenges and Offered Solutions

The LOD life cycle consists of five stages for software engineering, including: (i) planning – assessment of the feasibility of software to fulfil the requirements of the user; (ii) analysis – identifying potential problems;

---

[13] Shah et al. 2017: 381.

(iii) design – specification of software intended to achieve the specified goals, including recognition of necessary components and existing constraints; (iv) implementation – installation of the software on user machines; and finally, (v) maintenance – controlling and checking the performance of the software.

The simplest form of the software development is the waterfall model. It is assumed that each element in the life cycle is completed in an unproblematic fashion and there is no need to refer to the previous stage in implementing the process. However, the major problem with this model is that the execution of one phase of design may influence the previous stage. This is particularly apparent in the verification stage when issues in implementation and verification will require further effort in design, which means that design may be said to be unfinished until verification is complete. This may also hold true in the case of planning and specification while the process of producing a clear, precise specification may uncover ambiguities or inconsistencies in the requirements provided.

The integration of both life cycles is only possible when the data engineering systems, such as Dacura, provide several services to software engineers, developing software that utilises the data curated by the system. These include reliable access to data models, change notifications, and the automatic production of simpler formats, which are more familiar to traditional Web developers. For example, a GeoJSON stream is automatically made available describing all the features in the dataset that have a geographical location associated with them.

The data model developed by the Semantic Web community was made available to software engineers by providing a metadata registry. The Model Catalogue discussed above is such a registry. It can also be defined as a toolkit for creating and managing data models. The Model Catalogue tool was used to help develop and manage the ontologies used by the system – it supports OWL models and provides a RESTful API to support easy integration with third-party tools and incorporation into complex workflows. It was also integrated into the Eclipse Modelling Framework, allowing existing tools to more easily use the catalogue for development. Plugin capabilities were added, facilitating the extension of the catalogue to allow it to interact with more data sources. Semantic reasoning and search were also added, allowing the more efficient reuse of ontologies and concepts. The Model Catalogue was used for Seshat, Jurion IPG and Health Data use cases. In case of Seshat, the Model Catalogue tool allowed the creation of complex ontology to capture the complex historical data the project is collecting.

The Unified Views tool is an ETL tool for RDF data developed as part of the PoolParty semantic suite. It was used to manage the integration of datasets from third-party datasets. The development artefacts are imported into the triple store using a UnifiedViews pipeline. This pipeline runs daily to keep the data up to the date. The pipeline also calculates similarities between the issues and requirements. This solution was adopted to Seshat, PoolParty, DBpedia, Jurion, and Jurion IPG. In the latter use case, it was used to ensure that the results of the validation processes carried out by Dacura and Semantic Booster be evaluated, manage this mapping and transformation, and save the transformed data to a triple store.

As regards the design phase in software engineering life cycle, the evidence for this benefit can be seen particularly strongly in the automated harvesting and curation interface generation tools developed in the project. This is particularly evident in case of Dacura that informs the software engineering analysis phase by defining what data is to be harvested (Shah et al. 2017: 381). The Dacura Linked Data Model Mapping Service tool creates rich ontological models from semi-structured HTML and automates harvesting of data conforming to this model and was heavily tested within the Jurion IPG use case.

For the implementation phase, the University of Leipzig developed a set of tools around RDFUnit and DataID, which together formed the basis for the Data Quality Framework and the Automated Data Testing and Verification Framework. Dacura makes it possible to define statistical data quality measures to be met to support software engineering and suggest UI refinements to eliminate errors. Repair Framework and Notification tool is used in both implementation and maintenance phase as the defined data constraints influence the implementation of algorithms and as taxonomies are changes, the constraints need to be satisfied.[14]

Semantic Booster tool allows the automatic generation of software systems from formally specified system specifications. Hence, it supports both semantic domain models and models of the software and data engineering life cycles. In particular, Semantic Booster has its strengths in the automatic model and software code creation process. It has also strong quality constraints, so that no invalid data gets into the transformation process. This approach was augmented by using RDFUnit for further data quality checks and which is the prerequisite to connect external open datasets to the IPG application in an easy and sustainable way. A Booster specification is

---

[14] Shah et al. 2017: 380.

designed, which creates a model from the SQL database, along with formal constraints, which ensure that the data remains correct by construction. The Model Catalogue tool is then used to manage this data model. Semantic Booster is used to make this data available as RDF via an API.

This system was deployed in Jurion IPG and Health Data. The use of Semantic Booster in Jurion IPG allows the introduction of a wider range of semantic integrity constraints and business rules, to be applied on the data upon entry – ensuring availability of high-quality data. The automatic data migration tools provided with Semantic Booster minimise the impact of upgrading and evolving the underlying data model whilst still maintaining data consistency. Whilst Semantic Booster can already help enforce a range of integrity constraints, there are some consistency checks, which would be more reliably performed using RDF and reasoning. Hence, it was decided to use the existing D2RQ tool to convert data stored within a Booster database into RDF format, making it available to the RDFUnit testing tool. The additional testing and monitoring also provides insight into productivity and quality gains through the use of the ALIGNED tool stack.

The effective maintenance can be achieved in two alternative ways. The first approach is provided by a configuration of the Oxford MDE approach, while the second is by Dacura. A Booster specification is created, which (i) generates SQL statements to extract the data from the legacy SQL DB and saves it in a format that can be managed by the Model Catalogue tool and (ii) the Booster specification should ensure that this extracted data are correct by construction according to the Booster specification. Then, this extracted data are made available as RDF via Semantic Booster. In the approach offered by Dacura services, the Model mapping tool transforms the SQL schema of the legacy DB into an OWL ontology, which is then used by the schema checking tool to ensure that all data conforms to the model. The curation and workflow tools allow data managers to change the model and migrate the data and manage the process. This ontology is deployed as the schema for the graph into which the instance data are imported.

2.5 Conclusions

All five use cases in the ALIGNED project were thoroughly analysed to achieve its major goal, namely to create effective methods and tools for integrating software and data engineering processes and develop full life cycle workflows for combined software and data engineering. The deployment of the project designed and produced software and tools led to significant enhancement of all case studies and significant improvements in data productivity, quality, and agility and eventually user satisfaction and customer support. In Jurion IPG, of particular significance turned out to be Semantic Booster, showing significant improvements in agility, with the addition of new attributes being up to 45 times faster. Also, Dacura significantly improved the management of re-engineering from the old relational database schema to the new one. In addition, Wolters Kluwer's Jurion and Jurion IPG business information database was enhanced with ALIGNED tools, significantly improving their ability to correct errors and change data schemas over their previous tools.

The introduction of import validation in the PoolParty use case improved data quality and reduced customer support time as well as significantly contributed to the ability to fix a number of violations. Overall improvements in data curation, data agility, model agility, and software development processes were also achieved. The major achievement in DBpedia was the error rate improved.

The rebuilding of the Seshat data and tools used the full suite of Dacura tools to import the data, ensure it met consistency requirements, automatically produce user interfaces and curation tools, and finally publish the data. It resulted in a quantifiable reduction in the number of errors in data entry, while the amount of data entered dramatically increased. The new format of the dataset enabled the ability to link to external datasets to enrich the Seshat data. The data generators and users reported an increase in usability and productivity, and the technical users reported an increase in agility: the speed in which tools and data can adapt to changes in the model. The shared model for data validation and software generation involves integration points with the planning phase of the software engineering life cycle, and the quality analysis, manual revision/authoring, and search/browse/explore phases of the data engineering life cycle. In addition, the collaborative consensus required for updating the model brings additional dependencies on the interlinking and extraction phases of data engineering.

3

Methodology

James Welch[1], Jim Davies[1], Kevin Feeney[2], Pieter Francois[1], Jeremy Gibbons[1] and Seyyed Shah[1]

[1]University of Oxford, UK
[2]Trinity College Dublin, Ireland

3.1 Introduction

Software engineering is concerned with the development of reliable computer applications using a systematic methodology. Data engineering involves the collation, organisation, and maintenance of a dataset, or data product, and may be seen as the dual of software engineering. The two processes are typically treated as separate concerns – largely as a result of different skill sets. However, there is often a great deal of overlap: dependable software is reliant on consistent, semantically correct data; processing data at scale requires high-quality tools and applications.

For most enterprises, the data they hold may well be their most valuable asset. Day-to-day operations will be dependent on data concerning customers, payments, and stock. It is vital that this data is of high quality: any loss of integrity or inconsistencies with operating practices or business processes, may be costly, and in many cases irreparable. Furthermore, the ongoing success of the business is increasingly reliant on analysis of the data: historical reporting, predictive analytics, and business intelligence. These latter processes, along with decreasing costs for storing and managing data, drive an increase in scale: minimising human effort is vital, and new Big Data tools and techniques are required to manage ever-larger datasets.

For some organisations, the data may be the primary artefact or the product in itself. From research enterprises to social networks, the value of the data stems from its quality, coverage, and completeness. These curated datasets may be the product of many smaller ones, perhaps different in structure or domain, and linked to create new, richer datasets. For these combined datasets, the ability to version and update individual components is critical: users of the data require up-to-date input, new features, and access to corrections and clarifications. Tool support must be sympathetic to changes in requirements and the acquisition of new data, and must scale accordingly.

It therefore follows that Software Engineering and Data Engineering are closely related. Mission-critical software is reliant on high-quality data, and the construction and maintenance of large datasets is dependent on secure, reliable software. Many of the key challenges are common to both disciplines: correctness, scale, and agility; tools and techniques for improving software quality may also result in improved data quality and vice versa.

The increase in popularity of "Big Data" analytics means that solutions to these challenges are required more than ever. The rise in data-intensive applications – those systems that deal with data that is large in scale, complex, or frequently changing[1] – has brought about a requirement to abandon traditional methodologies and explore new processes and techniques. A broader range of software applications for processing data, including visualisation, natural language processing, and machine learning, have provided new areas for innovation, and the integration of a range of software components around an underpinning data corpus has become a typical system architecture.

Engineering processes for both data and software are also required to be sympathetic to the so-called "Five V's of Big Data": velocity, volume, value, variety, and veracity. The speed at which data can be acquired – manually through the efforts of large groups, or automatically through complex applications – can impact the processes of data curation, enriching, and analysis. The ever-increasing amount of data collected – which can include static "historical data" and changing contemporaneous data – can reach scales challenging existing software scalability. The perceived value of data captured requires precision software, and rigorous data engineering processes, to ensure continuing accuracy and integrity. The ever-greater heterogeneity of data to be handled creates semantic issues, which must be resolved when linking and analysing data. Finally, the quality or trustworthiness creates further semantic issues – understanding the meaning, provenance, and accuracy of data is vital to realising its worth, and all phases of both software and data engineering processes need to take this into account.

Modern approaches to software engineering consider automation for agility and correctness, formal techniques for reliability and iterative approaches to improve delivery time and adapt to requirements. Data

---

[1]M. Kleppmann, Designing Data-Intensive Applications: The Big Ideas behind Reliable, Scalable, and Maintainable Systems, O'Reilly Media, 2016.

engineering as a discipline is less mature, although certain phases of an iterative process have been identified, and dependencies between phases can infer a natural development life cycle. However, both life cycles remain independent, and finding an integrated process, which considers both software and data in parallel, remains a considerable challenge.

The content in this chapter is adapted from a paper submitted (in January 2018) to Elsevier's Journal of Information Sciences.

3.2 Software and Data Engineering Life Cycles

3.2.1 Software Engineering Life Cycle

Modern software development methodologies can be seen as refinements to the original waterfall process for hardware systems development. First conceived as a "stagewise" model,[2] an instantiation targeting software development is typically summarised by the diagram in Figure 3.1.

In this most basic process, progress flows one way, through each of the stages, and one phase cannot begin until the previous phase has been completed. Each of the stages can be "signed off" by either the customer or the developer in such a way that completion of a phase can be recognised and made final. For example, the requirements for the system determine the scope

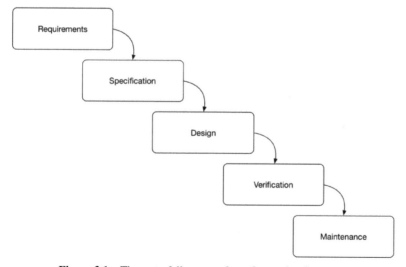

Figure 3.1 The waterfall process for software development.

---

[2]H. D. Benington, Production of large computer programs, IEEE Annals of the History of Computing 5 (1983), pp. 350–361.

of the specification; the completed specification document may be seen as a contract for the design work.

The first major problem with the waterfall model is that the execution of one phase of design may influence the previous stage. This is particularly apparent in the verification stage: issues in verification will require further effort in design; design may be said to be unfinished until verification is complete. This may also hold true in the case of specification: the process of producing a clear, precise specification may uncover ambiguities or inconsistencies in the requirements provided.

One solution to this problem is to allow feedback from one phase to modify earlier decisions. This leads to a modified version as proposed by Boehm,[3] in which backward arrows lead from one phase to the preceding one (see Figure 3.2). Although this allows for some notion of iteration in development, allowing decisions made in each phase to be revisited, it suffers from another flaw, that is, estimating delivery time (and therefore cost) can be very difficult. Without specific bounds on revisiting decisions, overall implementation can take unspecified amounts of time, leading to frustration for both developer and customer.

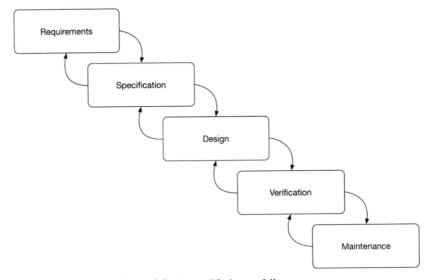

Figure 3.2 A modified waterfall process.

---

[3]B. W. Boehm, Software Engineering, IEEE Transactions on Computers 25 (12), pp. 1226–1241, 1976.

3.2 Software and Data Engineering Life Cycles

This uncertainty can be exacerbated by another common problem in software development: customers often do not know, or understand, precisely what they want until they have had a chance to see it, or interact with it. Business rules that may seem fixed at the time of requirements and specification may need revising in light of constraints in subsequent design or implementation stages. A good software engineering process must be sympathetic to revisiting even the earliest requirements decisions after design and implementation are underway, but still be amenable to stable project management in order to allow predictable costs and timescales.

More modern approaches to these problems can take two forms. The first of these is more technical, and directed at the actual design and implementation process: by reducing the length of time taken to get from requirements to implementation, decisions can be revisited quickly, and with less development effort. Prototyping allows the customer or user to get a feel for the solution earlier, permitting the requirements or specification to be revisited sooner in the overall implementation process. Automation in the implementation phase can reduce the effort involved in updating implementations to match updated requirements.

The second approach is another update to the software engineering life cycle, allowing multiple iterations of the traditional model, typically reducing the retrograde steps in the previous model in favour of completing an implementation and starting a new requirements and specification iteration sooner. Figure 3.3 shows a typical iterative software development life cycle.

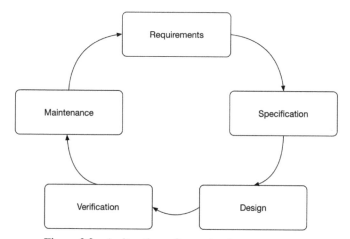

Figure 3.3 An iterative software development process.

The iterative model allows a more flexible approach to contracts and timelines: short cycles of the entire process allow prioritisation of features; early implementations can be used as prototypes and complex details can be saved for future iterations when there may be more clarity. Cycles are typically kept to a predefined length; at the start of each cycle, the scope of each phase is determined, managing time and cost expectations. Although system-wide requirements will be gathered throughout the whole cycle, some analysis will be performed at the start of each cycle in order to confirm the scope for the next cycle. Overall, time and cost estimation can be managed more effectively[4] and revised at the end of each cycle.

Another advantage to the iterative approach is that it changes the nature of the maintenance phase. Typically, during the life of the software, functionality will need adjusting to match evolving business requirements. With the standard waterfall model, the final phase of maintenance is often insufficient to deal with updated requirements, and the whole process needs to begin again; an iterative approach takes this into account, and maintenance can be merged in as part of the overall development and re-evaluation cycle.

The "Manifesto for Agile Software Development"[5] proposes 12 principles for such a development process, including to "satisfy the customer through early and continuous delivery of valuable software", to "welcome changing requirements, even late in development", and to "deliver working software frequently, from a couple of weeks to a couple of months, with a preference to the shorter timescale". The iterative approach is typically referred to as an "agile" approach, although the principles as set out for an agile process extend beyond the software life cycle itself and provide guidance for the way in which developers work as a team and interact with their customers.

Managing an iterative process effectively can still be difficult: although individual cycles can be fixed in duration, and development effort within the cycle may be reasonably estimated, it can still be difficult to manage priorities and overall development direction. A number of variations on the iterative, "agile" process have been proposed, and frameworks built around

---

[4]A. Begel, N. Nagappan, Usage and perceptions of agile software development in an industrial context: An exploratory study, in: First International Symposium on Empirical Software Engineering and Measurement, pp. 255–264, IEEE, 2007.

[5]K. Beck, Manifesto for agile software development, http://agilemanifesto.org, accessed: November 2017 (2001).

them, for example, Scrum',[6] Kanban,[7] and Extreme Programming,[8] all of which can help with cost estimation, reducing the time spent on verification and enhancing code quality.

An agile approach can also be counter-productive for building certain types of software where solutions are complex and irreducible. Such solutions require a high degree of planning and design and architectural decision-making in advance. An iterative development methodology can restrict the solution space to one in which development time may be reasonably estimated, where progress may be demonstrated at the end of each iteration and where prioritisation stays consistent.

3.2.2 Data Engineering Life Cycle

As an emerging field of research, the processes of data engineering used in industrial applications are still relatively immature. The LOD stack LOD2[9] is a collection of integrated tools supporting a life cycle for creating and managing Linked Data. Auer et al.[10] proposed an iterative process for developing linked open datasets. Eight core activities of Linked Data management are identified and managed as phases in an iterative life cycle, consistent with the principles of Linked Data:

- storage/querying: retrieving and persisting information to be included as part of the dataset;
- manual revision/authoring: processes for manual curation of content;
- interlinking/fusing: creating and maintaining links between datasets;

---

[6]K. Schwaber, M. Beedle, Agile Software Development with Scrum, Vol. 1, Prentice Hall, 2002.

[7]M. O. Ahmad, J. Markkula, M. Oivo, Kanban in software development: A systematic literature review, in: Software Engineering and Advanced Applications (SEAA), 2013 39th EUROMICRO Conference on, IEEE, pp. 9–16, 2013.

[8]K. Beck, Embracing change with extreme programming, Computer 32 (10), pp. 70–77, 1999.

[9]S. Auer, V. Bryl, S. Tramp, Linked Open Data–Creating Knowledge out of Interlinked Data: Results of the LOD2 Project, Vol. 8661, Springer, 2014.

[10]S. Auer, L. Bühmann, C. Dirschl, O. Erling, M. Hausenblas, R. Isele, J. Lehmann, M. Martin, P. N. Mendes, B. van Nuffelen, C. Stadler, S. Tramp, H. Williams, Managing the life-cycle of linked data with the LOD2 stack, in: P. Cudre-Mauroux, J. Heflin, E. Sirin, T. Tudorache, J. Euzenat, M. Hauswirth, J. X. Parreira, J. Hendler, G. Schreiber, A. Bernstein, E. Blomqvist (Eds.), International Semantic Web Conference, pp. 1–16, Springer Berlin Heidelberg, Berlin, Heidelberg, 2012.

48   *Methodology*

- classification/enrichment: creating and maintaining links between data and models of data (which themselves may be linked and part of the dataset);
- quality analysis: testing for data completeness and correctness;
- evolution/repair: correcting invalid data resulting from a quality analysis phase via either manual or automated processes;
- search/browsing/exploration: making data artefacts available to domain experts or to users beyond the original authors;
- extraction: producing or publishing profiles or projections of data to be used in other applications.

Figure 3.4 shows an iterative life cycle incorporating these stages.

These stages of data engineering have been identified primarily for building tool support and integrated frameworks, encouraging compatibility of independent tools within a particular framework. As with the software engineering process, feedback from one phase may be fed into another. For example, the models linked during the classification or enrichment stage will determine the scope of the quality analysis stage; any errors found during quality analysis may need to be resolved in the evolution/repair phase.

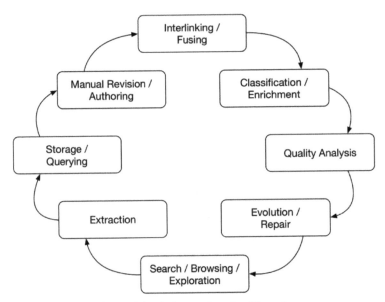

Figure 3.4   A data engineering life cycle.

3.3 Software Development Processes

In order to design a combined software and data engineering process, we will first consider some modern approaches to software engineering at scale and how phases of the data engineering life cycle might be integrated or merged. As explained in Section 3.2, automation is often seen as key to improving both the speed of software delivery and the correctness of the delivered product. In this section, we will consider three cross-cutting techniques of software engineering and discuss their advantages and disadvantages. We first consider model-driven approaches to software development, and the trade-off between automation and customisation. We then look at formal techniques, in particular formal specification, validation, and verification. Finally, we discuss test-driven development and its role in an iterative development process.

3.3.1 Model-Driven Approaches

MDE describes a development process in which the components of the final software artefact are derived – either manually or automatically – from models that typically form part or all the specifications or requirements of the system. Rather than writing software that understands the data itself, software is written that understands the modelling language and is capable of handling updates to the model. Such software can be reused in different applications within a similar domain, minimising the time spent on the implementation phase and capturing common repeating patterns that would otherwise have to be repeated on each cycle of an iterative development.

MDE is a promising starting point for our combined methodology: by choosing well-suited models that fit the application domain, updating software to match evolving data can be achieved by simpler updates to a model. With suitable tool support, this methodology may also allow ordinary business users to manipulate these models and help bridge the gap between requirements and specification.

As implied above, MDE approaches fall on a sliding scale between a fully automated generation process, and something much more manual, allowing greater flexibility and customisation. An overview of some of the possible approaches and discussion of their practicality follows below.

The first MDE approach can be described as "full automation": everything is modelled – including future-proofing – and machine learning is used to learn how to change the model from the flows of data as their format changes over time. This approach has significant advantages in terms of maintenance cost: once the system is deployed and operational, minimal further intervention is required. However, although learning how to self-adapt a model is theoretically possible, it remains impractical for real-world applications. Another problem is the generation of training datasets for the machine-learning component: this is currently beyond the scope of most data engineering projects.

A second approach that appears more practical is where a full model of system behaviour is manually produced, but a fully functional software implementation can be generated automatically from the model. The modelling language should be designed in such a way that a broad range of likely future developments and feature requests can be handled without any custom code needing to be written. If such tools are written with evolution and upgrade in mind, they may be used for rapid prototyping, as part of an iterative agile process or as a technique to manage and enable software change beyond initial deployment. If the tools for editing and managing models are good enough, such changes may even be carried out by business users and deployed instantly, rendering the whole process cost-free from a technical resources point of view.

Although feasible within particular domains of application, this approach does not work universally: there can be no theoretical basis for automatically implementing arbitrary behaviours and functionalities. However, subsets of the overall problem are tractable, and such modelling languages – also referred to as DSLs – exist with usable tool support. The UML[11] is the most significant attempt to create a complete modelling language. It does not have a formal semantics itself, but can be given one for a specific purpose, and there are many tools based on subsets of the language. UML has been successfully deployed for building large, complex model-driven systems. However, in practice, the development and testing of the models takes considerable amount of time and effort to get right. Such systems are most appropriate for domains in which a lot of effort is spent moving data through highly stereotyped workflows that do not change rapidly over time and where significant resources can be allocated to testing and managing model updates.

The remaining modelling approaches do not attempt to model behaviours, limiting their scope to data. The third approach is where a complete data model, containing a full specification of all the classes and properties that

---

[11] J. Rumbaugh, I. Jacobson, G. Booch, Unified Modelling Language Reference Manual, Pearson Higher Education, 2004.

are present in the data, is used to constrain or guide the manipulation of data corresponding to that model. This approach has the advantage that constraints on data are easier to define and use than those upon behaviours. A complete data model can be used to generate a large proportion of software components in an information system – for example, the data storage mechanism and user interfaces.

The disadvantages to this third approach are that although generation processes have been formally solved and public standards such as OWL are available, in practice, automated software generation from such models is still very hard and requires tools to be built from scratch. Most importantly, the conceptual framework and the assumptions underlying the logic of OWL need to be changed. Existing tools for model management are typically focussed on knowledge engineers with specific goals and as such are not really suited to business users.

A fourth approach is that of partial data modelling: where a subset of the information domain is specified – limited to ad-hoc or incomplete positive or negative constraints on the data. Here the assumption is made that the model is not exhaustive, that there are states of the data that are not addressed in the model. This technique has a particular advantage in processing large datasets: where data are messy and do not necessarily conform to any model, we can identify and filter out the most important problems caused by the lack of structure. A model may be incrementally built, adding rules to specifically address any issues with the data as they are encountered.

A disadvantage with this approach is that the incompleteness of the model prevents most automation techniques. Another is that the models are built up by accumulation of ad-hoc rules and become difficult to manage over time, invariably becoming a barrier to agility, and may become inconsistent. Changes to the model may result in large changes to the data – or worse, required changes to the data may go unnoticed or their calculation or derivation may be infeasible from the model.

3.3.2 Formal Techniques

The use of formal methods in the development of programs has been the traditional practice for those systems that may be seen as safety-critical: typically those systems whose failure could endanger human life. Such formal techniques include the mathematical derivation of program code from precise specifications, the logical proof that code exactly implements specifications in the form of contracts, or the exhaustive verification of software to show that unwanted behaviours are precluded. Each suffers from the same problems: that formal techniques are slow and expensive, and do not scale to large complex software systems. A rigorous, mathematical approach will require developers with very specialised skills and experience.

However, there have been some successful applications of formal techniques in practical software development. Automation can solve problems of scalability, but a completely automatic process is impossible in the general case. One solution is to restrict the problem domain: pattern matching can be applied to the specification and particular refinements applied; proof libraries and verification results can be stored for reuse. Another solution is to focus automation on part of a stepwise process; for example, automatically generating method stubs or proof obligations for manual completion.

In many cases, formal techniques are associated with a more traditional waterfall method development. This can be because there is a need for a detailed, comprehensive specification before the mathematical process can begin – requiring that much of the solution is explored before any programming starts. Hall[12] described the development life cycle of the specification itself: from Learning through Production and Simplification. These stages are necessary within any development method, but in a formal code derivation process, these must typically happen before any code has been written. This may result in an overall speed increase, but does not incorporate the fundamental component of an iterative process: feedback – the user's response to an initial implementation.

However, the construction of a complete, precise specification is not without merit. The explication and analysis of the problem space is invaluable when developing code, most importantly when a team of developers require a shared understanding. Human-readable documentation is also important for giving context and addressing subtleties not obvious from the plain mathematical statement. By addressing both specification and requirements in this way, developers have a clearer sense of direction, customers can make better judgements on the suitability of a solution, and managers can better manage expectations of time and cost.

3.3.3 Test-Driven Development

A test-driven (or "test-first") software development proceeds in an iterative fashion, but relies on a short development cycle, focussed on building

---

[12] A. Hall, Seven myths of formal methods, IEEE Software 7 (5), pp. 11–19, 1990.

functionality to meet requirements, rather than specification. At the start of each iteration, acceptance tests are written to validate the implementation of the next round of features: the expectation is that these new tests will initially fail. Minimal changes to the code are made in order to get the test suite completely passing; once all tests pass, the feature development is complete. An optional refactoring phase can be used to tidy the code, whilst maintaining a full suite of passing tests.

As well as measuring the suitability of the latest iteration of development, tests also provide a valuable restraint on regressions: that previously correct functionality is not broken by the latest updates. This can give users confidence in the stability of the software and reduce the burden for developers.

An agile test-first approach can lead to high-quality, timely software. However, some of the caveats about agile, iterative development also apply: maintaining long-term objectives whilst focussing on short-term goals can be difficult. Finding appropriate levels of code coverage requires experience: total coverage is often impossible; tests covering trivial or non-realistic cases can waste developer time, but too few tests may lead to a reduction in quality.

The test-driven approach to software has obvious parallels in the development of large datasets: the quality analysis phase of development can be used to measure the correctness of the other phases – in particular those of manual revision, interlinking, and enrichment. Tools for finding inconsistencies in data – and highlighting areas of concern – are readily available and well understood by data engineers.

3.4 Integration Points and Harmonisation

Although the processes for software engineering and data engineering discussed so far are complementary, it is more than likely that in the development of a data-intensive system, there will be dependencies between the two processes. In general, an integration point corresponds to any pair of points in the software and data engineering life cycles where specific artefacts and processes should be shared. In this section we enumerate three different forms of integration point: overlaps, synchronisation points and dependencies; we discuss the importance of each, and consider the difficulties in spotting them. We conclude the section by examining potential barriers to harmonising the two processes, in terms of terminologies, development roles, models, and tool support.

3.4.1 Integration Points

The first type of integration point between the data and software engineering processes is that of a natural overlap. This will be particularly prominent at the start of the project: for example, where the initial implementation of the software may run in parallel with a manual curation of the initial dataset. Similarly, in some projects, a phase of testing the software for correctness may coincide with a phase of quality analysis for the data: bugs in the software may be a cause of inconsistencies in the data; errors in the data may uncover issues in the software. In general, overlapping phases such as these can indicate a requirement for software engineers and data engineers to work together to ensure successful conclusions in both life cycles.

More generally, we can consider synchronisation points: where phases in both cycles are required to start, or finish, at the same time. This could be due to a release of software coinciding with the linking of a new dataset. It may be due to external pressures: the implementation of software and manual update of data to match new business processes; the completion of a cross-cutting software and data concern before a member of staff leaves the organisation.

More generally still, it is important to consider dependencies between phases in cycles. Typically this can mean that a phase in one cycle must finish before another starts, but may simply be that one phase must reach a certain level of completion. One example where a software engineering phase might depend on a data engineering phase would be when data quality analysis must be completed before the requirements for the next iteration of software development can be signed off. An example where a data engineering phase may depend on a software engineering phase might be where a particular software feature must be tested and deployed before some manual data curation may start.

Such integration points may happen regularly with every iteration – for example the requirement to migrate data to match the deployment of new software, or may happen irregularly, for example in response to changes in business processes, the implementation of new features, or updates to external data sources. Thus it becomes important to regularly review known integration points and assess the potential for new integration points in the future. As this requires insight into both data and software engineering development plans, along with an understanding of overall roadmaps and business direction, the integration analysis will involve many stakeholders across a range of disciplines or technical competencies.

As with any project management activity, care should be taken to ensure that dependencies can be appropriately managed. It is conceivable that in rare cases, cyclic dependencies appear: this may indicate that data and software engineering phases need more carefully defining – split up or merged – or that requirements and design need revising. Generic tool support for such project management is readily available, but specialist tooling – as discussed in Section 7 – is really only available for software development processes.

The nature of each integration point needs investigation to explore the best way of addressing it. For example, although some straightforward dependencies may be seen to be sufficiently addressed by a simple sign-off process, the criteria for completion must be agreed beforehand. More complicated dependencies, especially where an overlap in phases is concerned, may require more substantial collaboration between data engineers and software engineers, perhaps with intermediate checkpoints and combined requirements.

3.4.2 Barriers to Harmonisation

There are a number of barriers to the easy combination of software and data engineering processes. Although both processes have foundations in computer science and information engineering, the two disciplines have different terminology, and different reference or metamodels. The participants in each will also vary: roles may not have obvious counterparts in the other discipline, and the people carrying out each role will have different backgrounds and skills. Highlighting barriers and potential pitfalls is important so that they can be anticipated and worked around.

An integration point will usually indicate some shared resource between software and data engineering: typically a requirement, a model or a metamodel. It can be important to recognise this shared resource and ensure that both data engineers and software engineers share a collective understanding. A common barrier is that of terminology: although engineers may typically share a common language in the domain of application, with differing skills and backgrounds, software and data engineers may have different technical terminology. An example of this is shown in Figure 3.5 – showing a standard equivalence between terms of abstraction in different domains: data engineering, model-driven software engineering and more general programming.

Based on the scope of the project, however, the equivalence may not be as direct as those shown. For example, in a particular project, one specific

Meta-level	Data engineering	Software engineering	Programming
M3	Schema, Ontology Language	Meta-metamodel	Grammar notation
M2	Upper Ontology	Metamodel	Language Grammar
M1	Domain Ontology, Schema	Model	Program definition
M0	Triple, Dataset	Instance, Object	Program runtime

Figure 3.5 Comparison of terminology in software and data engineering.

Upper Ontology may be used as a Model in software engineering, which may be represented at Program Runtime in practice. The abstraction level at which each artefact is expected to be used when shared between software and data engineering processes should be documented as part of the process, and any changes in notation – for example, a process used to turn UML software models into an OWL ontology – should be automated if possible.

As well as the differing terminologies, the models themselves may differ. In order to facilitate interlinking, data engineers typically make good reuse of existing models – for example Dublin Core (DC)[13] for generic metadata, Friend Of A Friend (FOAF)[14] for social relationships, or PROV[15] for data provenance information, are all commonly reused or extended. This extension is an essential part of the data engineering process, allowing dataset linking. In software engineering, however, reuse of such pure data models is less common: reuse happens in terms of libraries of functionality. While there are some libraries that do implement standard data models,[16] most are typically restricted to the most trivial – for example hash maps – or the domain-specific – for example models of Microsoft Word documents.[17] Without common models for software and data, harmonising the two development processes will prove difficult.

Enumerating the participants involved in each of the two processes can also highlight potential hurdles. A wide variety of roles may be involved: in software engineering, these might be systems or software analysts, developers and testers; in data engineering these might be data architects, data

---

[13]S. L. Weibel, T. Koch, The Dublin Core metadata initiative, D-lib Magazine 6 (12), pp. 1082–9873, 2000.

[14]D. Brickley, L. Miller, FOAF vocabulary specification 0.91 (2007).

[15]P. Groth, L. Moreau, PROV-overview. an overview of the PROV family of documents, project Report, April 2013.

[16]C. Mattmann, J. Zitting, Tika in Action, Manning Publications Co., 2011.

[17]The Apache Software Foundation, Apache POI, http://poi.apache.org, accessed: November 2017 (2017).

harvesters and data consumers. There may be roles which can, or should, be shared across the two processes: requirements engineers, system administrators, technical or development managers. Users may be technical or domain experts; they may be users of the software, the data, or both. It is important that interaction between roles is between both sides of the process: software developers should understand the concerns of data quality analysts, for example, and the data architects should collaborate with the software architects.

Another area where software and data engineers can be divided is on the use of tools for managing the development process. In software development, the usual practice is to use an issue-tracking or defect-tracking tool, such as Atlassian Jira,[18] or JetBrains YouTrack.[19] Such tools can help orchestrate an iterative process: plugins are available to manage agile variants such as Kanban or Scrum. Technical problems can be managed through this process too: issues can be raised directly by users, taken through a workflow from prioritisation through development to testing by the developers, and "signed off" as complete by management or the original users. Customisable workflows allow this process to be adapted according to particular development processes or business culture.

Typically, such tool support for data engineering processes does not exist, in part due to the relative immaturity of formalised processes, and in part due to the wide variety of workflows for data curation, some of which will be specific to particular domains. In some cases, customisable tools such as Jira can be re-purposed, and plugins developed, but data engineers – especially the domain experts, who may be non-technical – can often be reluctant to use such tools aimed at software developers. Processes can often be managed in a more ad-hoc fashion without tool support or building additional bespoke support into data curation tools.

Having identified a number of potential barriers to integrating two different engineering processes, we can consider approaches to success. Collaboration and harmonisation between two typically distinct teams in an organisation requires a detailed understanding of the other process and those

---

[18] J. Fisher, D. Koning, A. Ludwigsen, Utilizing Atlassian JIRA for large-scale software development management, Tech. rep., Lawrence Livermore National Laboratory (LLNL), Livermore, CA (2013).

[19] JetBrains, JetBrains YouTrack, https://www.jetbrains.com/youtrack/documentation/, accessed: November 2017 (2017).

participating in it; of compromise in terms of terminology and modelling; a sympathy for those solving orthogonal problems within the same space; and shared sets of resources and tools for collaboration.

3.4.3 Methodology Requirements

Data-intensive systems require careful alignment between data engineering and software engineering life cycles to ensure the quality and integrity of the data. Data stored in such systems typically persist longer than, and may be more valuable than, the software itself, and so it is key that software development is sympathetic to the aims of "Big Data": scalability to large volumes of data; distributed, large-scale research across multiple disciplines; and complex algorithms and analysis. These are normally described in the literature as the Five V's of Big Data: velocity, volume, value, variety, and veracity.

In existing development methodologies, software and data engineering are considered as separate concerns.[20] Integrating these will introduce a number of new challenges: software engineering aims of software quality, agility and development productivity may conflict with data engineering aims of data quality, data usability, and researcher productivity. Further challenges include federation of separate data sources, dynamic and automated schema evolution, multi-source data harvesting, continuous data curation and revision, data reuse and the move towards unstructured/loosely structured data.

Auer et al. identified challenges within the domain of life cycles for Linked Data.[21] These include extraction, authoring, natural-language queries, automatic management of resources for linking, and Linked Data visualisation. Typically seen as concerns for data life cycles, they all have a major impact upon software development: the authors mentioned component integration, the management of provenance information, abstraction to hide complexity, and artefact generation from vocabularies or semantic representations.

---

[20] M. Kleppmann, Designing Data-Intensive Applications: The Big Ideas behind Reliable, Scalable, and Maintainable Systems, O'Reilly Media, 2016.

[21] S. Auer, J. Lehmann, A.-C. N. Ngomo, A. Zaveri, Introduction to linked data and its lifecycle on the web, in: Reasoning Web. Semantic Technologies for Intelligent Data Access, pp. 1–90, Springer, 2013.

3.4 Integration Points and Harmonisation

Mattmann et al.[22] used their experience of data-intensive software systems across a range of scientific disciplines to identify seven key challenges which may be summarised as:

- data volume: scalability issues that apply not just to the hardware of the system, but may affect the tractability and usability of the data;
- data dissemination: distributed systems bring challenges of interoperability and can lead to complex system architectures;
- data curation: supporting workflows and tools for improving the quality of data, in a way that allows subsequent inspection or analysis;
- use of open source: complex technologies will depend upon reliable, reusable components supporting generic functionality;
- search: making the data collected available in a usable fashion to users, including access to related metadata;
- data processing and analysis: boiling down to workflows, tasks, workflow management systems, and resource management components;
- information modelling: the authors state that "the metadata should be considered as significant as the data".

The authors split these challenges into further subcategories and pointed out many interdependencies between these problems. Zaveri et al.[23] took a broader view, highlighting inadequate tool support for Linked Data quality engineering processes. Where tool support does exist, these tools are aimed at knowledge engineers rather than domain experts or software engineers.

Anderson agreed with this issue,[24] describing a more wide-ranging lack of support for developers of data-intensive systems. He also identified "the necessity of a multidisciplinary team that provides expertise on a diverse set of skills and topics" as a non-technical issue that can be addressed by projects dealing with large, distributed datasets. A technical equivalent to this issue is to understand notions of iteration with respect to the data modelling – he argued that domain knowledge is required in order to understand data collection and curation. Subsequently, he also argued for technical knowledge

---

[22] C. A. Mattmann, D. J. Crichton, A. F. Hart, C. Goodale, J. S. Hughes, S. Kelly, L. Cinquini, T. H. Painter, J. Lazio, D. Waliser, et al., Architecting data-intensive software systems, in: Handbook of Data Intensive Computing, pp. 25–57, Springer, 2011.

[23] A. Zaveri, A. Rula, A. Maurino, R. Pietrobon, J. Lehmann, S. Auer, Quality assessment for linked data: A survey, Semantic Web 7 (1), pp. 63–93, 2016.

[24] K. M. Anderson, Embrace the challenges: Software engineering in a big data world, in: Proceedings of the First International Workshop on BIG Data Software Engineering, pp. 19–25, IEEE Press, 2015.

in order to match frameworks with requirements, emphasising the need for a multi-disciplinary team.

Some solutions to these challenges have been identified – most notably in the area of model-driven software engineering, DSLs, and generative programming. These approaches, in combination with Linked Data languages and schemas, enable self-describing data structures with rich semantics included within the data itself. Aspects of program logic previously encapsulated in software are now embedded in data models, meaning that the alignment between data and software engineering becomes even more important. But these approaches can lead to further problems: Qiu et al.[25] identified two issues: firstly the interaction between domain experts and application developers, and secondly that change to schema code may not always impact application code in a straightforward manner.

3.5 An ALIGNED Methodology

This section outlines the proposed methodology for combined software and data engineering. We describe it as "lightweight", because the technique requires some initial setup and maintenance, and its exact form can be heavily determined by the exact software and data engineering processes, by the tools available and the technical members of the team. However, in this methodology, we propose a general framework for process management, an iterative methodology, and a number of guidelines or recommendations for successful integration. We conclude the section by considering tool support for such a process.

3.5.1 A General Framework for Process Management

In Section 5, we outlined a number of potential barriers to harmonising the data and software engineering processes. Our general framework is concerned with reducing the effect of these issues, as well as providing an iterative methodology that is suitably adaptive in response to changes in context. The framework is split into two phases: the first, a "setup" phase, involves some analysis of the preferred engineering processes, the shared resources and integration points, and the impact of any tools, project roles or terminology where managing integration points will prove problematic. The second phase

---

[25]D. Qiu, B. Li, Z. Su, An empirical analysis of the co-evolution of schema and code in database applications, in: Proceedings of the 2013 9th Joint Meeting on Foundations of Software Engineering, pp. 125–135, ACM, 2013.

is the iterative development, where the outputs of the setup phase are under a process of continuous revision, such that problems can be foreseen at the start of each cycle.

The setup phase is broken into four consecutive steps – the first of which is to perform some basic analysis on the preferred software and data engineering processes. This will be most greatly influenced by the skills of the technical collaborators, the preferred management style, and the requirements laid down by the users. As part of the guidelines later in the section, we strongly recommend iterative development approaches to both software and data, and for the remainder of the section assume processes similar to those outlined in Figures 3.3 and 3.4 – generic iterative approaches corresponding with an agile approach. However, specific projects may choose, for example, a specific software testing phase apart from the more general software maintenance; or a detailed requirements phase within the data engineering life cycle.

At this point, we can assume that there is some shared understanding of the requirements – not necessarily a full detailed consensus, but a general appreciation for the tools and techniques required to produce a satisfactory solution. This is not an unreasonable assumption, as in most cases some contractual negotiations will have preceded a team starting on a development, or the new development will be part of a rolling series of features given to an in-house team of engineers working on a particular project.

The second step of the setup phase is to consider the resources that should be shared between software and data engineers. Typically, this will include requirements or specification in the form of models, or perhaps metamodels, that can be shared rather than creating two incompatible versions. Unifying terminology and semantics is important here: if software and data engineers have differing interpretations of the same model, any potential advantage may be lost. Creativity in this part of the process may result in gains later on: other potentially sharable resources may include test suites and other quality analysis tools, technical and user-facing documentation, and project management tools or support. As with all analysis carried out in this setup phase, it can be revised in later iterations, and so any decision taken here need not be final.

The third step is to consider the integration points for this particular project, in the context of the decisions made in the previous two analyses. Given iterative approaches to software and data processes, and a list of shared resources, it is possible to build a grid, similar to that shown in Figure 3.6. The software engineering life cycle steps are enumerated along the top, and the

62 Methodology

Data Engineering	Software Engineering				
	Requirements	Specification	Design	Verification	Maintenance
Manual revision/Author					
Inter-link/fuse					
Classify/Enrich					
Quality Analysis					
Evolve/Repair					
Search/Browse/Explore					
Extract					
Store/Query					

Figure 3.6 An incomplete grid for analysing integration points.

data engineering life cycle steps are enumerated on the left-hand side. Each box in the main part of the grid therefore corresponds to a potential integration point – for example, the first column in the first row represents a potential synchronisation between the requirements phase in software engineering with the manual revision/authoring phase in data engineering.

The grid can now be populated with two pieces of information. The first is to highlight any squares in which a potential integration point is possible – this will be based on the shared resources analysed in the previous step. For example, if a data schema is to be shared, then any changes made as part of the specification phase could impact some or all the data engineering phases. Similarly, any shared test cases which are updated as part of the quality analysis phase in data engineering, will affect the verification phase of the software development process. The second piece of information is the tooling that can be used to facilitate the integration at each point in the grid. In Section 7, we outline some of the tools built by the ALIGNED project that can be used to support and manage these integration points, but appropriate tools may be found off-the-shelf, repurposed from software or data engineering, developed in-house, or built for this specific development. As the need for data-intensive systems development increases, it is expected that such tools will be more widely available.

It should be obvious at this point that any identified integration point without specific tool support may need addressing. In many cases, simple awareness could be sufficient: highlighting such unsupported integration points and ensuring greater effort on collaboration at these points in the process. Alternatively, new tools could be sourced, or processes adjusted to minimise potential integration. The fourth and final step in this setup phase is to consider the other barriers to harmonisation, in the context of each integration point. Software and data engineers involved in the project should come together to consider how their terminology, standard models, developer roles and tools can be made compatible in order to ensure maximum integration at each feasible point.

3.5.2 An Iterative Methodology and Illustration

Once the setup phase is complete, a more traditional iterative development can begin. In the setup phase, an iterative process for each of the software and data engineering components was selected. In our methodology, these may now continue independently in parallel, but constrained by the integration points previously discussed: overlap, synchronisations, and dependencies. To ensure that these integration points may be sufficiently addressed, it is our recommendation that the cycles are aligned, or are coincident at a particular phase in each cycle – this will be determined by the integration points, and the shared resources.

To illustrate, we consider a typical scenario encountered by our ALIGNED project use cases. In this scenario, the software engineering process is approximately equivalent to the iterative methodology in Figure 3.3, and the data engineering process can be seen as similar to that defined in Figure 3.4. The key shared resource is a complex data model, used as a reference by the data engineers, but also forming part of the software model: data modification functionality, business rules, and additional internal data points are added to the external-facing data model, and used as a specification document for the software engineers.

In such a process, updates to the data model can occur as part of the storage/querying phase of the data engineering activity, where new data are added to the existing data corpus, or as part of the specification phase of the software engineering activity, where new requirements give rise to updates in the intended functionality of the system. This forms the key integration point: there is an overlap in process here, as both software and data engineers should agree on any updates to the data model, and neither may continue until the updates made are complete and consistent. It is important that such a key integration point is well managed: problems here could result in wasted time and effort in curating a dataset against an incorrect model, developing software against an invalid or inconsistent schema, or managing a difficult merge operation between two parallel versions of the same data model. However, managed properly, having a shared data model is worth the effort: a reduction in duplication can save time and money; automation based on this model can be shared; a common understanding can lead to a more coherent, better designed solution.

In this scenario, we insist that iterative processes in software and data engineering may now continue independently, but must synchronise on this overlapping event: storage/querying and specification. Figure 3.7 shows an example of such a parallel, synchronising process. In theory, this means that the iterations of each process should be the same length, and while in some projects this may be feasible, in others, where a particular phase may be more burdensome, this may prove to be overly restrictive. In such situations, it may be possible to relax this guideline, by simply insisting that the iterations synchronise whenever a change affecting both processes is made to the data model. For example, after a major release of a combined software and data product, minor, or patch releases may be made to the software if no changes are made to the data model, or any changes made do not affect the current iteration of data engineering. This will allow the software engineers to iterate a few times within a single iteration of the data engineering process, ensuring that data engineers have time to satisfactorily complete their iteration, and that software engineers are not kept waiting before beginning a new iteration.

Such synchronous iterations must be managed with care – those managing the projects must be made aware of any potential delays, since a delay to one process will impact the other. In software engineering, developers may be used to working within time-bounded "sprints" – in which the scope of a release may be reduced in order to ensure that completion is not delayed.

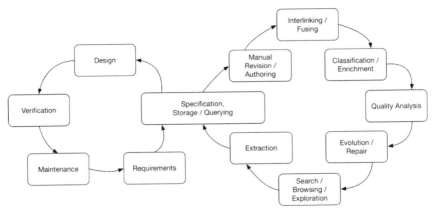

Figure 3.7 A parallel life cycle with synchronisation.

In data engineering, such practices are less common, and so some training may be required to ensure all technical staff understand the restrictions. In developments where software and data iterations coincide, but are of differing lengths, care must be taken to ensure that any additional iterations do not impact the shared resources. For example, in the scenario outlined above, any additional software iterations for a minor or patch release must not update the shared part of the data model, for otherwise the current data engineering iteration may be inconsistent with the software that will next be deployed.

3.6 Recommendations

The iterative approach outlined above can provide a framework for combining software and data engineering processes, in such a way that a certain amount of autonomy can be maintained in two quite separate disciplines, but also in a way that can improve consistency and efficiency in the delivery of a solution made up of two closely coupled components. We now give some recommendations, based on our experience on a number of use-case projects, for ensuring that integration points are managed efficiently, and to maximise collaboration between software and data engineers.

Our first recommendation is that models are shared between software and data specifications, wherever possible. As previously discussed, this increases the opportunities for reuse and helps ensure that software and data remain consistent. We further recommend that these models are formalised in such a way that removes ambiguities, reducing the chance of inconsistent assumptions being made by software and data engineers.

Second, we recommend that development is driven by these shared models, in an automated fashion wherever possible. This reduces the chance of error in development and can ensure consistency such that developers can rely on the solutions produced in a parallel iteration.

Third, any solutions for either software or data should be rigorously tested, where tests are also developed – automatically if possible – directly from the model. Sharing or reusing test components can prove efficient, as well as ensuring consistency between data and software.

Fourth, tool support should be used to effectively manage the iterative process on both software and data sides. As discussed in Section 5, software engineers are used to using project management software to coordinate and administer an agile process, but such tools are not commonly used in data engineering applications. Such tools would need specialist support for managing the integration points, and a wider range of developer roles and responsibilities.

Our final recommendation is that whenever meetings are held to discuss the iterative process – in particular the planning and feedback stages – these meetings should be attended by representatives of all solution stakeholders. The purpose for this is twofold: so that integration points and shared resources can be carefully managed; and so that the overall roadmap and architecture can be maintained whilst engineers focus on small iterations addressing short-term goals.

These five recommendations are derived from the combined experience of the project use cases, but in every project, their priorities differed, according to the experience of the development and project management teams, the tools available, and the particular iterative steps used in each development.

3.6.1 Sample Methodology

As an illustration, in this section, we look at the synchronisation points required for the ALIGNED use cases.

Table 3.1 outlines the usecase-oriented view of the synchronisation between Data and Software Engineering life cycles. Each entry of the table represents a synchronisation point within in the project. The use cases will be

Table 3.1 A usecase-oriented synchronisation table for the ALIGNED project

Data Engineering	Software Engineering				
	Requirements	Specification	Design	Verification	Maintenance
Manual revision/ Author		PS5.1, PS5.2, JURION [WKD1]	PS5.1, PS5.2 JURION [WKD2]	JURION [WKD3] PS1.4, [Seshat1]	
Inter-link/ fuse		PS4.1, PS4.2, DBpedia [DBP1]	PS4.1, PS4.2, PS4.4		DBpedia [DBP2]
Classify/ Enrich					[Seshat2]
Quality Analysis		JURION [WKD4] DBpedia [DBP3]		PS1.1, PS1.2, PS1.3, PS2.3, PS3.1	PS1.1, PS1.2, PS1.3, PS2.3, DBpedia [DBP4]
Evolve/Repair				PS5.3	PS3.1, PS3.2
Search/Browse/ Explore		PS5.1, PS5.2	PS5.1, PS5.2	DBpedia [DBP5]	
Extract		PS4.1, PS4.2	PS4.1, PS4.2	DBpedia [DBP6]	
Store/Query		PS5, JURION [WKD5]		JURION [WKD6] DBpedia [DBP7]	

3.6 Recommendations   67

used to enact the methodology with the tools in Section 7. The following summary describes the high-level features of each intersection point, in terms of use cases:

- Manual Revision/Author
    - A2: [WKD1] In the schema change use case (JS7), it is reflecting the situation that when a schema change is introduced and forwarded to the software manager in the SE life cycle, which initiates a process of validating the suitability of the model for use in SE. [WKD2] In the bug reporting governance use case (JS8), when a bug is reported and the software analyst finds that the bug is caused by a data error, he informs the DE expert to fix the data error via manual revision. [PS5.1] Develop plugins for Confluence and JIRA [PS5.2] Make use of collected process-related data.
    - A3: [WKD2] In the bug reporting governance (JS 8) use case, the SE designer can eliminate scenarios where a data-caused bug could occur in the future by sending additional constraints to the DE side, where these constraints are integrated to the schema. [PS5.1] Develop plugins for Confluence and JIRA [PS5.2] Make use of collected process-related data.
    - A4: [Seshat1] We will implement of graphical user interface software to author and edit data and the data will be communicated and captured in the DE life cycle.
- (B) Interlinking/Fusing
    - B2: [PS4.1] Extract data from Confluence and JIRA [PS4.2] Create RDF data from the extracted data. [DBP1] Refers to the fact overlap and conflict evaluation (DS1.3) and in the interlink evaluation (DS3.2). DS1.3 refers to validation by fusing data from different DBpedia language editions and Wikidata in order to identify overlaps and conflicts. DS3.2 refers to tools that validate external links to other datasets.
    - B3: [PS4.1] Extract data from Confluence and JIRA [PS4.2] Create RDF data from the extracted data [PS4.4] Link Development Process Data with Data Model Integrity Information.
- (C) Classify/Enrich: There are few synchronisation points where DE use cases exploit SE tools, possibly because classification in DE is a well-studied task.

68 *Methodology*

- o A5: [Seshat2] The graphical user interface software widgets on the SE side will be continuously updated and maintained as the DE schemas evolve.
- (D) Quality Analysis
    - o D2: [DBP3] Quality analysis for mapping (DS2.1), ontology (DS2.2) and instance data (DS3.1).
    - o [WKD4] When a quality-related schema change is introduced and accepted in the DE Life Cycle, the changes are communicated to the SE Life Cycle, where the software is accepted. There is a protocol for accepting quality changes.
    - o D4: [PS1.1] Constraints for Internal Actions [PS1.2] Rules for Reasoning and Inferencing [PS1.3] Constraints for Specific Schemas [PS2.3] Validate Thesaurus Against Schema.
    - o D5: [DBP4] Schemas refers to reports, generated by the automated mapping validation tool (DS5.1) and erroneous fact report to the Wikimedia community (DS5.2). [PS1.1] Constraints for Internal Actions [PS1.2] Rules for Reasoning and Inferencing [PS1.3] Constraints for Specific Schemas [PS2.3] Validate Thesaurus Against Schema.
- (E) Evolve/Repair
    - o E4: [PS5.3] Integrate Data Constraints Information with PPT Data Migration and Deployment Strategy.
    - o E5: [PS3.1] Formulation of Constraint Violation Repair Strategies [PS3.2] Creation of Repair User Interfaces.
- (F) Search/Browse/Explore
    - o F2: [PS5.1] Develop plugins for Confluence and JIRA [PS5.2] Make use of collected process-related data.
    - o F3: [PS5.1] Develop plugins for Confluence and JIRA [PS5.2] Make use of collected process-related data.
    - o F4: [DBP5] These integration points use the generation of DataID as a core and auto generate tool for browsing and querying based on the DataID file. Browsing is achieved by auto generating a download page for a DBpedia release and querying by providing a Docker image that contains the release stored in a triple store.

- (G) Extract
  - G2: [PS4.1] Extract data from Confluence and JIRA [PS4.2] Create RDF data from the extracted data.
  - G3: [PS4.1] Extract data from Confluence and JIRA [PS4.2] Create RDF data from the extracted data.
  - G4: [DBP6] Extraction of two additional Wikimedia projects: Wikimedia Common (DS1.1) and Wikidata (DS1.2), implementing tools in the SE domain that extract the data.
- (H) Store/Query
  - H2: [PS5.1] Develop plugins for Confluence and JIRA [PS5.2] Make use of collected process-related data [PS5.3] Integrate Data Constraints Information with PPT Data Migration and Deployment Strategy.
  - [WKD5] This integration point appears in the schema change (JS 7) use case. Once the schema change is in place in the DE Life Cycle, new instance by the DE expert to the SE expert. The new data are used to execute test scenarios on how the new schema is affects the existing software, to formulate new requirements for the design and implementation phases.
  - H4: [DBP7] These integration points use the generation of DataID as a core and auto generate tool for browsing and querying based on the DataID file. Browsing is achieved by auto generating a download page for a DBpedia release and querying by providing a Docker image that contains the release stored in a triple store.

A1–F1: The planning phase of the software engineering life cycle does not contain any synchronisation points. Possibly because there are few tools for this stage (in general) artefacts produced at this stage are informal and documentary, and not useful to Data Engineering processes.

3.7 Sample Synchronisation Point Activities

As example, tools from the synchronisation table and details of the changes made are included below, in the Model Catalogue tool and Semantic Booster. The aim of the following sections is to demonstrate the methodology using the example tools. The implication of iteration in the life cycles is also discussed.

3.7.1 Model Catalogue: Analysis and Search/Browse/Explore

The Model Catalogue Tool has been developed for use cases supporting model driven software engineering. The main purpose is to capture, document, and disseminate models including software systems, data standards and data interchange formats, amongst others. The interface of the Model Catalogue is shown in Figure 3.8. The tool helps end users to analyse the available models and understand requirements for capturing new data against existing models. In the standard version, models can be imported

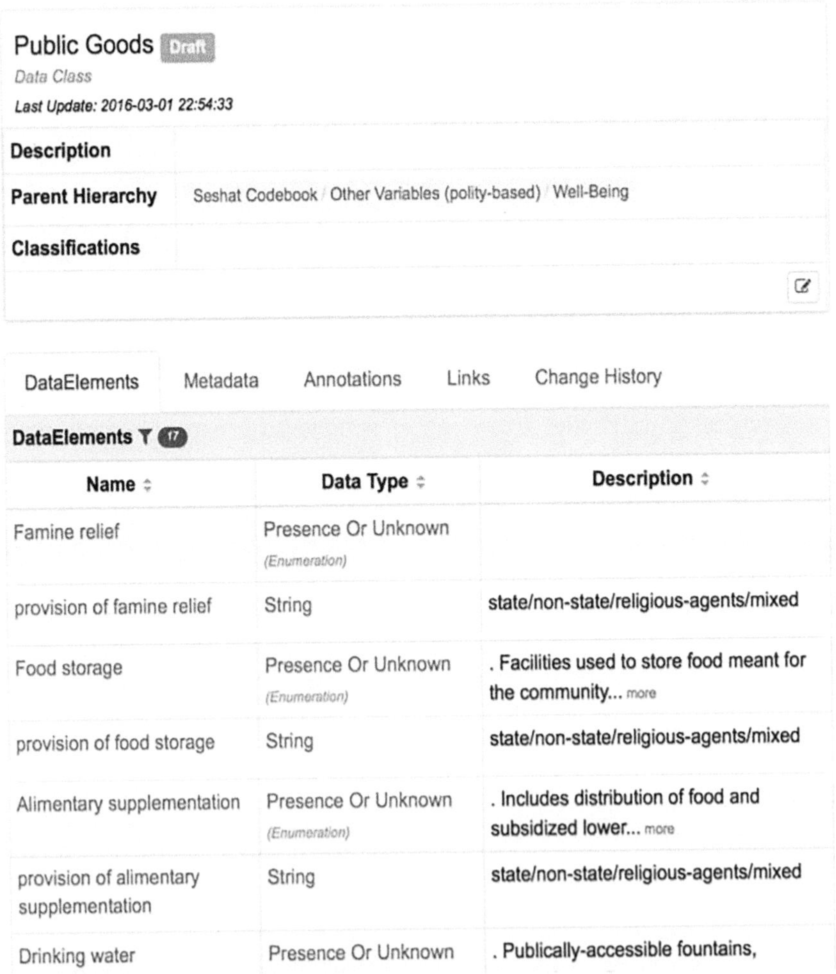

Figure 3.8 Model catalogue interface: browsing the SESHAT code book.

from formalisms such as UML and XSD. Models may be interlinked and reuse elements from related models. Some of the output formats include Booster for software generation and Microsoft Word for documentation of the model.

The catalogue tool has been adapted to support similar data engineering use cases, and thus bridge between data engineering and software engineering domains. The main addition has been the import and export of models in standard data engineering formats, such as RDFS and OWL. For the data engineer, the tool can be used to explore how their data models are used in practice in software. The models can be updated and changed without relying on software engineers to create new versions of software. Models exported using the catalogue will retain interlinks between models in the two domains. This allows more streamlined integration of semantic metadata into working software.

The synchronisation point is bi-directional. The models can capture a software model from a data engineering model or use the model to capture data in the data engineering domain. Multiple iterations of the software engineering and data engineering life cycles will typically result in new versions of the model; the changes will need to be synchronised after each iteration. A feature to compare the changes in models in the model catalogue is planned to support this activity.

3.7.2 Model Catalogue: Design and Classify/Enrich

In model-driven development, the model catalogue tool also supports the creation of new models for capturing emerging designs for data standards, software systems, and so on. The tool supports definition of new data classes and data elements that form the basis of data models. The tool has features such as model versioning, annotation, collaborative editing and communication between developers. The models can be built using existing models in the catalogue or imported from partial models that exist in semi-structured and human-readable formats such as spread sheets, CSV or text documents.

The model catalogue has been adapted for data engineering activity: classify data and enrich data models by linking elements with existing model elements. Model classes can be refined and developed in the catalogue, capturing new and emerging structures in a data model, which leads to more precise understanding of the domain. Data engineers can use the catalogue to link between concepts in separate data engineering standards, and decide where links are semantically appropriate.

72 *Methodology*

Similar to the "Analysis and Search/Browse/Explore" synchronisation point, this sync point is bi-directional. Iterations of the software and data engineering life cycles can result in new versions of the models. The model catalogue compare feature will support synchronisation of independent changes in models across both life cycles.

3.7.3 Semantic Booster: Implementation and Store/Query

Booster is a tool for the model-driven generation of information systems. High-level specifications are developed in Booster notation, which models the system implementation. Booster performs a series of translations and refinements on the model to generate a working system and Application Programming Interface (API) backed by a standard relational database. A user interface to the system is provided as an example of how the API may be used. The tool is used in the software engineering life cycle at the implementation phase.

Semantic Booster is a set of modifications to the Booster framework to support some data engineering life cycle activities. The changes add support for semantic annotation to standard Booster specifications, as shown in Figure 3.9: Example Semantic Booster System with Annotations. The Booster translations have been adapted to present the data as triples, with a SPARQL endpoint. The data in such a Booster system can be accessed and queried using standard data engineering toolsets. In combination with the design activity supported in the model catalogue, data engineering tools can be generated automatically using semantic Booster.

This synchronisation point is unidirectional at the model level, as MDE provides an implementation for the data engineering domain. The created implementation will be used by subsequent stages in the data engineering domain. As the mapping into triple form created by Booster uses a live version of the data, subsequent data engineering life cycle phases will access the version of latest data. Any modifications to the data must performed via the Booster generated API.

3.7.4 Semantic Booster: Maintenance and Search/Browse/Explore

In model-driven software development, maintenance and adaption of existing systems is a challenging task. Any changes to a Booster specification must be reflected in the implemented system, which can require re-generation of

3.7 Sample Synchronisation Point Activities   73

```
examplesystem.boo2
1 @prefix rdfs: http://w3.org/.../rdf-schema#
2 @prefix snomed: http://snomed.info/sct/
3
4 @mappings
5 //a snomed `Information system software'
6 examplesystem   snomed:706594005
7 Patient  snomed:116154003
8 Patient.name  rdfs:label
9 Patient.name  snomed:371484003
10 Patient.nhsno  snomed:395451000000101
11
12 system examplesystem
13
14 class Patient {
15    attributes
16      name : String
17      nhsno : String
18 }
19
20 class Clinic {
21    attributes
22      name : String
23 }
```

Figure 3.9 Example semantic booster system with annotations.

the implementation. The Booster approach ensures that any data entering a system are always validated to conform to the constraints. A large or complex change to the model involves the migration and validation of existing system data. Previous experiments with Booster have shown that for some model edits, existing data can be migrated automatically.

The data in Semantic Booster are presented as triples. Using the Booster mechanism for migration, automated migration of triple data in the Booster system becomes possible. Once data have been migrated, tools from the data engineering world can be used to validate the migration for compliance with the semantic rules of the model.

This synchronisation point is bi-directional. In subsequent iterations of the software and data engineering life cycles, the model catalogue will capture changes to the model. Booster will use the changes to automate migration of data stored in the Booster system; the data will be presented both in the API of Booster and as triples.

3.8 Summary

3.8.1 Related Work

That software and data engineering life cycles should be more closely integrated are not a new observation: Cleve et al.[26] took a more concrete approach and also proposed a number of contemporary challenges in system evolution, based on higher levels of tool support; better tooling for co-evolution of databases and programs; more agile coding techniques; and aligning data orientation through Object-Relational Mappings.

A more general-purpose approach to integrating life cycles elicits a number of broader challenges: software-engineering aims of software quality, agility and development productivity may conflict with data engineering aims of data quality, usability, and user productivity. Such is the importance of this integration work, the NESSI has identified "Collaborative Service Engineering based on the convergence of software and data" and "Integration of Big Data Analytics into Business Processes" as EU research priorities.[27] Further challenges relating more specifically to Big Data applications have been identified by Chen and Zhang:[28] in particular, those relating to data capture and storage, curation and analysis are of relevance here: hardware as well as software limitations can impact the effectiveness of Big Data techniques and highlighted opportunities may be missed.

Auer et al.[29] identified challenges within the domain of life cycles for Linked Data. These include extraction, authoring, natural-language queries, automatic management of resources for linking, and Linked Data visualisation. Typically seen as concerns for data life cycles, they all have a major impact on software development: the authors mentioned component integration, the management of provenance information, abstraction to hide complexity, and artefact generation from vocabularies or semantic representations.

---

[26] A. Cleve, T. Mens, J.-L. Hainaut, Data-intensive system evolution, Computer 43(8), pp. 110–112, 2010.

[27] NESSI, Strategic research and innovation agenda, Tech. rep., NESSI, version 2.0, April, 2013.

[28] C. P. Chen, C.-Y. Zhang, Data-intensive applications, challenges, techniques and technologies: A survey on big data, Information Sciences 275 pp. 314–347, 2014.

[29] S. Auer, J. Lehmann, A.-C. N. Ngomo, A. Zaveri, Introduction to linked data and its lifecycle on the web, in: Reasoning Web. Semantic Technologies for Intelligent Data Access, pp. 1–90, Springer, 2013.

Mattmann et al.[30] used their experience of data-intensive software systems across a range of scientific disciplines to identify seven key challenges:

- data volume: scalability issues that apply not just to the hardware of the system, but may affect the tractability and usability of the data;
- data dissemination: distributed systems bring challenges of interoperability and can lead to complex system architectures;
- data curation: supporting workflows and tools for improving the quality of data, in a way that allows subsequent inspection or analysis;
- use of open source: complex technologies will depend upon reliable, reusable components supporting generic functionality;
- search: making the data collected available in a usable fashion to users, including access to related metadata;
- data processing and analysis: boiling down to workflows, tasks, workflow management systems, and resource management components;
- information modelling: the authors state that "the metadata should be considered as significant as the data".

The authors split these challenges into further subcategories and pointed out the many interdependencies between these problems. Zaveri et al.[31] took a broader view, highlighting inadequate tool support for Linked Data quality engineering processes. Where tool support does exist, these tools are aimed at knowledge engineers rather than domain experts or software engineers.

Anderson[32] agreed with this issue, describing a more wide-ranging lack of support for developers of data-intensive systems. He also identified "the necessity of a multidisciplinary team that provides expertise on a diverse set of skills and topics" as a non-technical issue that can be addressed by projects dealing with large, distributed datasets. A technical equivalent to this issue is to understand notions of iteration with respect to the data modelling – Anderson argued that domain knowledge is required to understand data collection and curation. Subsequently, he also argues for technical knowledge

---

[30] C. A. Mattmann, D. J. Crichton, A. F. Hart, C. Goodale, J. S. Hughes, S. Kelly, L. Cinquini, T. H. Painter, J. Lazio, D. Waliser, et al., Architecting data-intensive software systems, in: Handbook of Data Intensive Computing, pp. 25–57, Springer, 2011.

[31] A. Zaveri, A. Rula, A. Maurino, R. Pietrobon, J. Lehmann, S. Auer, Quality assessment for linked data: A survey, Semantic Web 7 (1) pp. 63–93, 2016.

[32] K. M. Anderson, Embrace the challenges: Software engineering in a big data world, in: Proceedings of the First International Workshop on BIG Data Software Engineering, pp. 19–25, IEEE Press, 2015.

in order to match frameworks with requirements; emphasising the need for a multi-disciplinary team.

Some solutions to these challenges have been identified – most notably in the area of model-driven software engineering, DSLs, and generative programming. These approaches, in combination with Linked Data languages and schemas, enable self-describing data structures with rich semantics included within the data itself. Aspects of program logic previously encapsulated in software are now embedded in data models, meaning that the alignment between data and software engineering becomes even more important. But these approaches can lead to further problems: Qiu et al.[33] identified two issues: firstly the interaction between domain experts and application developers, and secondly that changes to schema code may not always impact application code in a straightforward manner. In this document, we attempt to tackle these two issues explicitly.

3.9 Conclusions

We have described a flexible methodology for integrating software and data engineering life cycles, identified a number of barriers to harmonisation, and made recommendations in order to better implement the combined methodology, and reduce the impediments. The methodology reflects the observed practices and experiences of the ALIGNED consortium – across a range of application domains, development practices, and experiences, both for the development of new solutions and the evolution of existing ones. We outlined the application of the methodology in each of the use cases, describing the particular challenges and requirements faced by each, and how the use of the methodology has improved development practice. We also described a number of tools built by the ALIGNED project partners that have been adapted to fit the integration points in the methodology, showing how they may be repurposed, or similar tools may be adapted for application to data-intensive systems.

The use cases presented here represent a small fraction of the potential application domains: further work is to apply the methodology in a wider range of projects, with a different selection of tools, and with different development teams. Further validation may be obtained from more qualitative or quantitative validation: although it is rare for two system developments to

---

[33]D. Qiu, B. Li, Z. Su, An empirical analysis of the co-evolution of schema and code in database applications, in: Proceedings of the 2013 9th Joint Meeting on Foundations of Software Engineering, pp. 125–135, ACM, 2013.

be directly comparable, experienced developers may be able to evaluate the effectiveness of the methodology against previous practice.

As discussed above, the software engineering life cycle is relatively mature and is broadly similar in all developments, but the data engineering processes are less well-defined, and may be more varied in further real-world applications – perhaps differing by domain or toolsets used. Further investigation is necessary to ensure that the methodology presented here is applicable to different data engineering practices.

4

ALIGNED MetaModel Overview

Rob Brennan[1], Bojan Bozic[1], Odhran Gavin[1] and Monika Solanki[2]

[1]Trinity College Dublin, Ireland
[2]University of Oxford, UK

The foundation of our ALIGNED methodology is an RDF-based semantic metamodel or language to describe software and data life cycles, inter-life cycle events, design intent, and domain models. This common framework for software and data engineering enables the following techniques for managing complexity: (1) Model-driven software engineering of data-intensive systems based on Linked Data; (2) Integrating expert-based data curation workflows into the software and data quality cycles; and (3) Providing unified views and governance of both software and data engineering activities when developing data-intensive systems;

This common metamodel for software and data engineering describes data-intensive systems both at a system specification level and in terms of the engineering activities, actors and artefacts.

Figure 4.1 illustrates the ALIGNED metamodels. At the top layer (the generic metamodel), it documents the common concepts used in data-intensive systems as a set of Linked Data vocabularies. The next ALIGNED layer covers the domain-specific metamodels that constitute a vocabulary and constraints for operating in a specific domain. This layer constrains the types of data-intensive systems that can be built in terms of architecture and tools, best practices for data collection and curation and common data assets (e.g., Linked Data datasets to be consumed by applications in this domain). ALIGNED has developed four domain-specific metamodels based on each of our use cases: enterprise information processing (JURION), e-research in the Social Sciences and Humanities (Seshat), crowd-sourced public datasets (DBpedia), and enterprise software development (PoolParty).

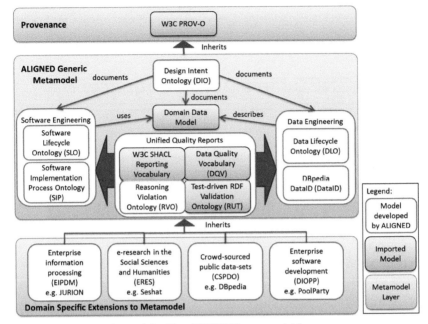

Figure 4.1 The ALIGNED metamodel layers.

Both the generic metamodel and domain-specific model layers are further specified in the following sections of this document.

4.1 Generic Metamodel

As specified in the last section, the ALIGNED metamodel is split into two major layers: the upper or generic layer is described in this section.

4.1.1 Basic Approach

The ALIGNED generic metamodel is structured as a set of complementary vocabularies that can be used to document the development and design of a data-intensive system throughout its life cycle. It extends the W3C PROV Ontology (PROV-O) to define software and data engineering agents, activities and entities. This facilitates the creation of provenance records describing software and data engineering.

The constituent vocabularies defined are as follows:

Software Life cycle Vocabulary (SLO and SIP): This is split into two components: the Software Life cycle Ontology (SLO) and the Software Implementation Process Ontology (SIP). SLO defines a top-level ontology for describing life cycle processes. SIP uses SLO to define the major agents (project roles, classes of software tools, etc.), activities (life cycle stages) and entities (models, code, test cases, etc.) involved in a software engineering project and their relations. SIP is implemented as a set of RDF modules implementing specific processes in the software development and implementation life cycle.

Data Life Cycle Ontology (DLO): Defines the major agents (project roles, classes of software tools, etc.), activities (life cycle stages) and entities (schema, datasets, code, test cases, etc.) involved in a data engineering project and their relations with a special focus on capturing the engineering life cycle.

Design Intent Ontology (DIO): Used to document the design decisions about data-intensive system artefacts such as software components or datasets. The purpose of the DIO ontology is to model the design intent or design rationale while undertaking the design of any artefact. A design intent or design rationale is an explicit documentation of the reasons behind decisions made when designing a system or artefact.

Domain Vocabulary: Describes the domain(s) of a data-intensive system. It is the specific data model or knowledge model used within the data-intensive system. The SLO, DLO, and DIO vocabularies are used to document additional context or constraints for the domain vocabulary to support semantics-driven software engineering, data quality engineering, engineering project governance, and tool integration.

In the subsections below, some basic details about our specification approach are described. Then we provide an overview of the structure and contents of each vocabulary.

4.1.2 Namespaces and URIs

Table 4.1 lists the standard prefixes used for each vocabulary. All have been checked for clashes with prefix.cc.

Table 4.1 Generic metamodel namespace declarations

Generic Metamodel Vocabulary Name	Prefix
Data Life Cycle Ontology	dlo
Design Intent Ontology	dio
Domain Vocabulary	This is defined by the specific data-intensive system rather than by the ALIGNED metamodel.
Software Life Cycle Vocabularies	slo, sip

Each prefix has been registered as a persistent URL (PURL) with purl.org or the W3C community persistent name service. These namespaces will be maintained by TCD servers.

4.1.3 Expressivity of Vocabularies

Since these generic vocabularies are designed to have the widest possible reuse, they only require the use of RDFS semantics. However, full utilisation of the model also requires the use of the W3C PROV ontology and in line with that specification the OWL2 RL profile is used for advanced features of the model.

4.1.4 Reference Style for External Terms

The ALIGNED metamodel vocabularies (DIO, SLO, DLO) must reference terms from each other and from externally defined vocabularies or ontologies. This necessitates an ontology implementation style decision that ranges from full OWL import statements to free-flowing Linked Data with no defined style or structure. For ALIGNED, the consortium has decided to adopt the MIREOT (Minimum information to reference an external ontology term) implementation style guide.[1] This avoids the practical problems with OWL imports and yet provides some structure around the reuse of existing resources.

4.1.5 Links with W3C PROV

The basic strategy for the ALIGNED metamodel is to specialise the W3C PROV ontology to describe software and data engineering activities (processes, tasks), entities (engineering artefacts or concepts) and agents (roles or software tools). Examples include:

- prov:Activity – sub-types defined to describe data or software engineering life cycle stages
- prov:Plan – used to describe engineering workflows
- prov:Entity – to describe software or data engineering artefacts – test case, design, test results, and so on
- prov:SoftwareAgent – to describe software engineering tools
- prov:Role – for software and data engineering roles

This approach means that PROV acts as a common upper ontology for all of our metamodel vocabularies and binds them together into a coherent whole.

---

[1] http://obi-ontology.org/page/MIREOT

It also facilities the creation of provenance records describing software and data engineering. The software and data engineering tools created in ALIGNED generate these PROV records as a way of logging their activities using enterprise Linked Data. This common representation of the domain facilitates tool integration and the creation of unified governance tools for combined software and data engineering.

4.2 ALIGNED Generic Metamodel

4.2.1 Design Intent Ontology (DIO)

The purpose of the DIO ontology is to model the design intent or design rationale while undertaking the design of any artefact. A design intent or design rationale is an explicit documentation of the reasons behind decisions made when designing a system or artefact.

The Design Intent Ontology (DIO)[2] is a generic ontology that provides the conceptualisation needed to capture the knowledge generated during various phases of the overall design life cycle. It provides definitions for design artefacts such as requirements, designs, design issues, solutions, justifications, and evidence and relationships between them to represent the design process and how these things lead to design outcomes. It draws upon the paradigms of IBIS (Interactive Intent-Based Illustration),[3] argumentation, and design rationale. It is linked to W3C PROV by defining the actors in the design process as PROV agents and the design artefacts themselves are PROV entities. It makes few assumptions about the design process used as the definitions of these activities properly belongs in the software life cycle and data life cycle models. Figure 4.2 illustrates the conceptual entities in DIO and their relationships.

4.3 Software Engineering

4.3.1 Software Life Cycle Ontology

The purpose of the SLO is to provide a top-level ontology for describing a process in the life cycle of a software. The ontology conforms to the ISO/IEC 12207 standard for Systems and software engineering – Software life cycle processes. The terminology used in the ontology conforms to ISO/IEC TR 24774:2010(E). All subprocesses will require to import this module.

---

[2] http://purl.org/dio/
[3] http://www.cs.columbia.edu/~doree/IBIS/thesis.html

84 ALIGNED MetaModel Overview

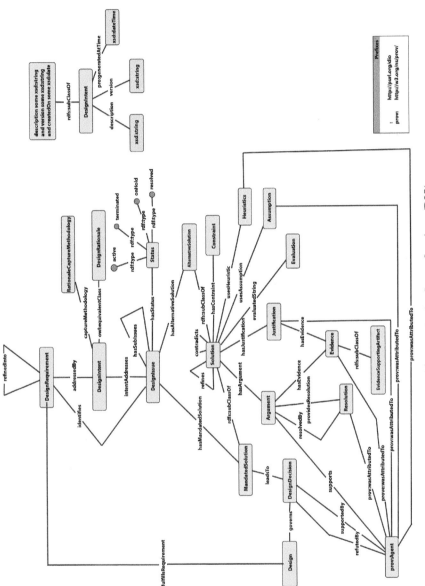

Figure 4.2 The Design Intent Ontology (DIO).

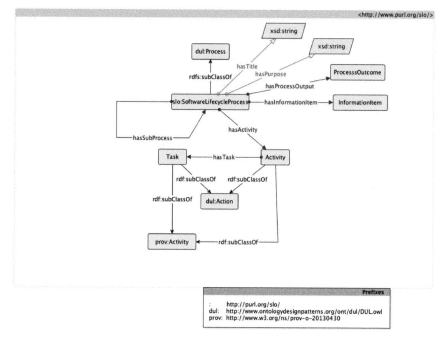

Figure 4.3 The Software Life cycle Ontology.

Figure 4.3 illustrates the conceptual entities in SLO. The core concept is a SoftwareLifecyleProcess, which can be decomposed into sub-processes, tasks and activities. The SIP ontology (see below) builds on this basic framework to describe standard software engineering processes e.g., requirements analysis and architectural design.

4.3.2 Software Implementation Process Ontology (SIP)

The purpose of the SIP is to provide a set of conceptual entities to represent a specified system element implemented as a software product or service.

This ontology imports and builds upon the ALIGNED SLO as the basic description of a process. It also utilises concepts defined in the SEON (Software Evolution ONtologies)[4] and the Software Ontology (SWO).[5]

The basic concepts of the SIP ontology are illustrated in Figure 4.4. It shows the definition of basic software engineering processes and activities

---

[4] http://www.se-on.org/
[5] http://purl.obolibrary.org/obo/swo.owl

86 ALIGNED MetaModel Overview

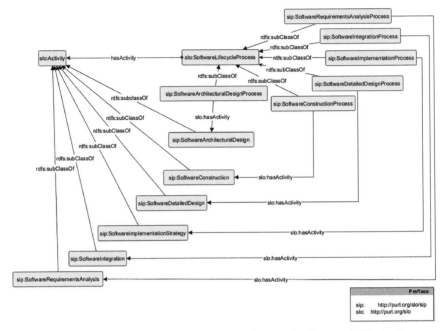

Figure 4.4 Core Concepts of the Software Implementation Process (SIP) Ontology.

such as requirements analysis, design, implementation, integration in terms of SLO activities and processes.

4.4 Data Engineering

4.4.1 Data Life Cycle Ontology

The purpose of the DLO is to provide a set of conceptual entities, agents, activities, and roles to represent the general data engineering process. Furthermore, it is the basis for deriving specific domain ontologies which represent life cycles of concrete data engineering projects such as DBpedia or Seshat.

Figure 4.5 shows the main classes of the data life cycle model. We have used the W3C PROV ontology, in this example represented by the classes Role, Person, Entity, and Activity. We use the Process class which is derived from Activity to implement the Linked Data Stack life cycle stages as subclasses. This allows us to represent LOD activities in our data life cycle metamodel. In addition, we have modelled datasets, data sources, and

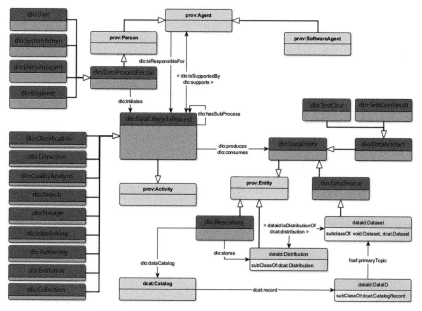

Figure 4.5 Generic data life cycle metamodel (DLO).

data repositories. For datasets, we import the W3C Data Catalog Vocabulary (DCAT)[6] definition of a dataset as it is a broad definition that goes beyond representing only RDF-based datasets.

The full documentation and OWL ontology file of the ALIGNED data life cycle model can be downloaded from http://www.essepuntato.it/lode/owlapi/ https://w3id.org/dlo.

4.5 DBpedia DataID (DataID)

DataID is a multi-layered metadata system, extending both the DCAT and PROV Ontology to provide more specific dataset metadata. Depending on context, type of data and use case, this core ontology can be augmented by multiple existing extensions (e.g., Linked Data, repository descriptions, etc.).

DataID core, as the kernel element of this ecosystem, describes datasets and their different manifestations, as well as relations to agents like persons or organisations, in regard to their rights and responsibilities. Together with

---

[6]http://www.w3.org/TR/vocab-dcat/

88 ALIGNED MetaModel Overview

DLO, DataID core constitutes the data management side of the ALIGNED Suite of Ontologies.[7]

The DBpedia DataID core vocabulary is a metadata system for detailed descriptions of datasets and their different manifestations. Established vocabularies like DCAT, VoID, PROV-O and FOAF are reused for maximum compatibility, in order to establish a uniform and accepted way to describe and deliver dataset metadata for arbitrary datasets and to put existing standards into practice. In addition, DataID can describe the relations of Agents (like persons or organisations) to datasets with regard to their rights and responsibilities.

Due to the growing complexity and different usage purposes, the DataID ontology was modularised into a core ontology and multiple mid-layer ontologies. While the core ontology is mandatory for any of the mid-level ontologies presented, none of those are required for describing data. That being said, in many use cases, some or all the mid-level ontologies will be a useful extension.

The DataID core vocabulary (Figure 4.6) describes datasets (based heavily on the DCAT ontology), as well as their relation to agents like persons or organisations with regard to their rights and responsibilities.

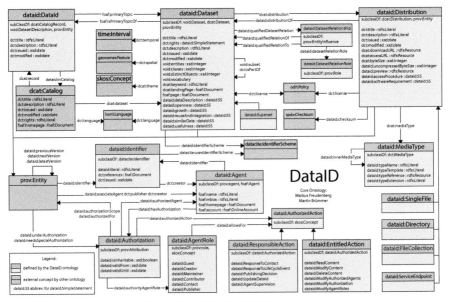

Figure 4.6 The DataID Ontology.

---

[7]http://aligned-project.eu/data-and-models

The full documentation and OWL ontology file of DataID can be downloaded from DBpedia,[8] the DataID landing page at DBpedia,[9] and through the ALIGNED website.

4.6 Unified Quality Reports

4.6.1 Reasoning Violation Ontology (RVO) Overview

The purpose of RVO is to enable a reasoner to describe reasoning errors detected in an input ontology, in order to facilitate the integration of reasoners into semantic Web tool chains.

It is defined as a simple OWL 2 ontology that is amenable to RDFS-based interpretations or use as a Linked Data vocabulary without any dependence on reasoning. A permanent identifier for the ontology has been registered with the W3C permanent identifier community group. The full source of the ontology is published online. This ontology is used to describe RDF and OWL reasoning violation messages in the Dacura Quality Service. These are generated by running an RDF/RDFS/OWL-DL reasoner over an RDF-based ontology model and allowing the Dacura quality service to report any integrity violations detected at schema or instance level. These violations report areas where the input model is logically inconsistent or breaks RDFS/OWL semantics or axioms. Violations may be reported as based on open world or closed world assumptions. The open world is the default OWL semantics and can typically only detect a limited number of problems due to incomplete knowledge. The closed world interpretation assumes that you have provided all relevant aspects of the model and is able to detect a much wider range of violations, e.g., missing or misspelled term definitions. This is often useful during ontology development or in a system that interprets OWL as a constraint language.

RVO will allow machine-readability and interpretation of detailed reasoning error messages. Furthermore, this would enable building tools to verify the OWL DL compliance of an ontology, find out which best practice requirements the ontology meets or violates, track the impact of interpreting the ontology in open and closed world contexts, identify the exact position of violations, and support intelligent visualisation of errors. The structure of the base RVO classes is shown in Figure 4.7.

---

[8]http://dataid.dbpedia.org/ns/core.html
[9]http://dbpedia.org/projects/dbpedia-dataid#Data%20model

90 ALIGNED MetaModel Overview

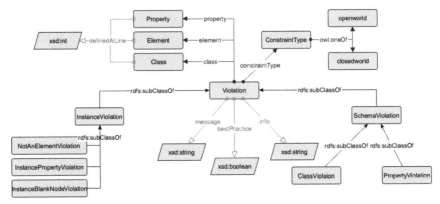

Figure 4.7 Reasoning Violation Ontology (RVO) Base Classes.

RVO class and instance violations are shown in Figure 4.8. Class violations are used for reporting issues regarding the TBox and instance violations ABox in general. Therefore, class violations are reported when e.g., property domains are missing, subsumption errors are detected, or class and property cycles are found. Instance violations show instances which are not elements of valid classes, cardinalities which are incorrect, property constraints that are violated, literals and objects which are confused, and so on.

The full documentation and OWL ontology file for RVO can be downloaded using the LODE documentation service and the persistent URI for the ontology.[10]

Example

This example shows a ClassViolation which is a SchemaViolation and more specifically a ClassCycleViolation. Such specific violation detection results make it possible to provide exact suggestions to ontology developers or repair agents and trigger ontology improvements. Figure 4.9 shows the errors produced by this violation.

Ontology Snippet Producing the Violation:

seshat:Territory seshat:hasValue xsd:DateTime.

---

[10] http://www.essepuntato.it/lode/owlapi/https://w3id.org/rvo

4.6 Unified Quality Reports

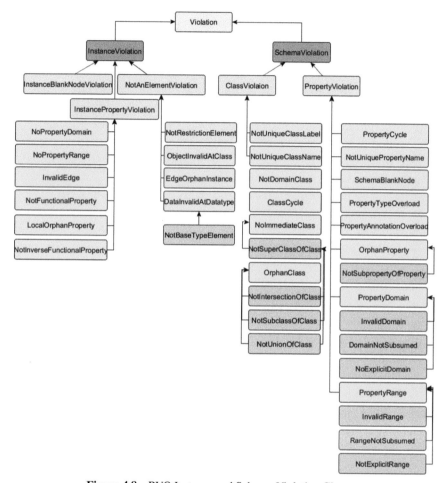

Figure 4.8 RVO Instance and Schema Violation Classes.

Figure 4.9 Resulting RDF Graph after Validation.

4.6.2 W3C SHACL Reporting Vocabulary

The Shapes Constraint Language is a language to validate RDF graphs against a set of constraints. These constraints are formalised as shapes and other constructs expressed in the form of an RDF graph. The language features and approaches occurring in the current specification of SHACL were in part inspired by the SPIN[11] and Shape Expressions (ShEx). The current revision of the specification for SHACL is published by the W3C[12] with complementary material available in a GitHub repository.[13]

SHACL Core defines frequently needed features to formulate common constraints for RDF graphs. SHACL Core Constraints are defined by parameterising Constraint Components that are templates for checks for a specific required property of an RDF nodes (e.g., unique occurrence of a property value associated with a specific property, for instance only one foaf:age value for a given foaf:Person). One or several of such constraints are associated with target RDF nodes to validate against in a SHACL Shape. SHACL shapes are expressed as RDF resources and aggregated in a Shapes Graph. An RDF graph to be checked for conformance against a Shapes Graph (the Data Graph) is provided to a Validation Engine that produces a Validation Report. The Validation Report states whether the Data Graph conforms to the Shapes Graph, listing violations of individual RDF nodes against shapes detected during the validation process in case of non-conformance.

SHACL Example

The following example data graph contains three SHACL instances of the class ex:Person. It is taken from the SHACL documentation.

```
ex:Alice
      a ex:Person ;
      ex:ssn "987-65-432A" .
ex:Bob
      a ex:Person ;
      ex:ssn "123-45-6789" ;
      ex:ssn "124-35-6789" .
ex:Calvin
      a ex:Person ;
      ex:birthDate "1971-07-07"^^xsd:date ;
      ex:worksFor ex:UntypedCompany .
```

The following conditions are shown in the example:

A SHACL instance of ex:Person can have at most one value for the property ex:ssn, and this value is a literal with the datatype xsd:string that matches a specified regular expression.

A SHACL instance of ex:Person can have unlimited values for the property ex:worksFor, and these values are IRIs and SHACL instances of ex:Company.

---

[11] http://spinrdf.org/
[12] https://www.w3.org/TR/shacl/
[13] https://github.com/w3c/data-shapes

A SHACL instance of ex:Person cannot have values for any other property apart from ex:ssn, ex:worksFor and rdf:type.

These conditions can be represented as shapes and constraints in the following shapes graph:

```
ex:PersonShape
      a sh:NodeShape ;
      sh:targetClass ex:Person ;    # Applies to all persons
      sh:property [                  # _:b1
            sh:path ex:ssn ;         # constrains the values of ex:ssn
            sh:maxCount 1 ;
            sh:datatype xsd:string ;
            sh:pattern "^\\d{3}-\\d{2}-\\d{4}$" ;
      ] ;
      sh:property [                  # _:b2
            sh:path ex:worksFor ;
            sh:class ex:Company ;
            sh:nodeKind sh:IRI ;
      ] ;
      sh:closed true ;
      sh:ignoredProperties ( rdf:type ) .
```

The shape declaration above illustrates some of the key terminology used by SHACL. The target for the shape ex:PersonShape is the set of all SHACL instances of the class ex:Person. This is specified using the property sh:targetClass. During the validation, these target nodes become focus nodes for the shape. The shape ex:PersonShape is a node shape, which means that it applies to the focus nodes. It declares constraints on the focus nodes, for example using the parameters sh:closed and sh:ignoredProperties. The node shape also declares two other constraints with the property sh:property, and each of these is backed by a property shape. These property shapes declare additional constraints using parameters such as sh:datatype and sh:maxCount.

Some of the property shapes specify parameters from multiple constraint components in order to restrict multiple aspects of the property values. For example, in the property shape for ex:ssn, parameters from three constraint components are used. The parameters of these constraint components are sh:datatype, sh:pattern and sh:maxCount. For each focus node the property values of ex:ssn will be validated against all three components.

4.6.3 Data Quality Vocabulary

The Data Quality Vocabulary (DQV) is an extension to the DCAT vocabulary which covers data quality, frequency of updates, user correction, persistence, and other properties of the dataset in question. It is designed to improve trust in data. It does not provide a determination of what quality is, but instead

94 ALIGNED MetaModel Overview

seeks to allow data consumers to judge whether the data in a dataset is suitable for their uses, and to publish their opinions and annotations about the dataset and its quality. The vocabulary seeks to do this by making it easier to publish, exchange, and consume metadata at every step of the dataset life cycle. Figure 4.10 shows the DQV ontology.

The quality of a dataset is assessed via certain observed properties. To express these properties, an instance of a dcat:Dataset or dcat:Distribution can be related to five different types of quality information represented by the following classes:

- dqv:QualityAnnotation represents feedback and quality certificates given about the dataset or its distribution.
- dcterms:Standard represents a standard the dataset or its distribution conforms to.
- dqv:QualityPolicy represents a policy or agreement that is chiefly governed by data quality concerns.
- dqv:QualityMeasurement represents a metric value providing quantitative or qualitative information about the dataset or distribution.

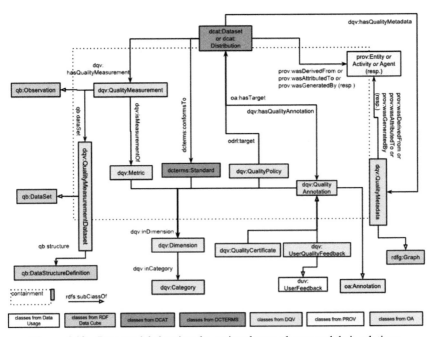

Figure 4.10 Data model showing the main relevant classes and their relations.

- prov:Entity represents an entity involved in the provenance of the dataset or distribution.

DQV defines quality measures as specific instances of Quality Measurements, adapting the daQ quality framework. It relies on quality dimensions and quality metrics. Figure 4.11 shows the interrelation of statements about data quality.

A Quality Dimension (dqv:Dimension) is a quality-related characteristic of a dataset relevant to the consumer (e.g., the availability of a dataset).

A Quality Metric (dqv:Metric) gives a procedure for measuring a data quality dimension, which is abstract, by observing a concrete quality indicator. There are usually multiple metrics per dimension; e.g., availability can be indicated by the accessibility of a SPARQL endpoint, or that of an RDF dump. The value of a metric can be numeric (e.g., for the metric "human-readable labeling of classes, properties and entities", the percentage of entities having an rdfs:label or rdfs:comment) or Boolean (e.g., whether or not a SPARQL endpoint is accessible).

Besides quality measurements, DQV considers certificates, standards, and quality policies, which can also be organised according to dimensions. Quality metadata containers (dqv:QualityMetadata) can group together different quality statements, so that their provenance can be tracked jointly.

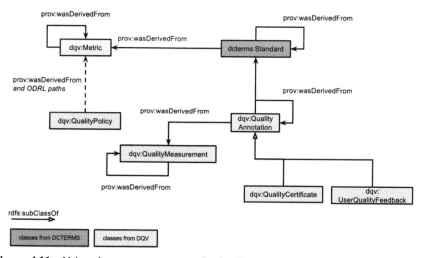

Figure 4.11 Using the property prov:wasDerivedFrom to interrelate quality metrics and other quality statements.

4.6.4 Test-Driven RDF Validation Ontology (RUT)

The RDFUnit ontology describes concepts used in RDFUnit, a test-driven RDF Validation framework that can run automatically generated (based on a schema) and manually generated test cases against an endpoint.[14]

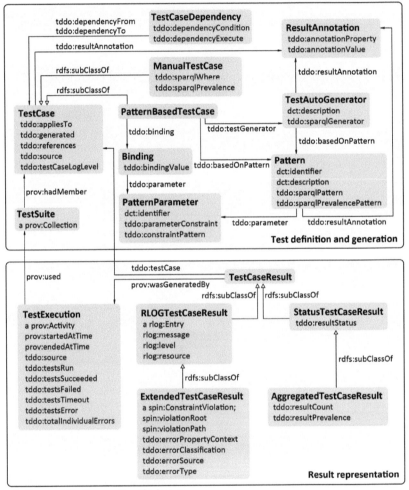

Prefix prov: http://www.w3.org/ns/prov#
Prefix spin: http://spinrdf.org/spin#
Prefix dct: http://purl.org/dc/terms/
Prefix tddo: http://databugger.aksw.org/ns/core#
Prefix rlog: http://persistence.uni-leipzig.org/nlp2rdf/ontologies/rlog#

---

[14]"NLP data cleansing based on Linguistic Ontology constraints" pp. 5–7, http://jens-lehmann.org/files/2014/eswc_rdfunit_nlp.pdf, ESWC, 2014.

Table 4.2 Domain-specific metamodel namespace declarations

Domain-specific Metamodel Vocabulary Name	Prefix
Enterprise information processing	eip
E-research in the Social Sciences and Humanities	sdo
Crowd-sourced public datasets	pds
Enterprise software development	sdev

Domain-Specific Extensions

Namespaces

Table 4.2 lists the standard prefixes used for each vocabulary. All have been checked for clashes with prefix.cc. Each prefix has been registered as a persistent URL (PURL) with purl.org or the W3C community persistent name service. These namespaces will be maintained by TCD servers.

Enterprise Information Processing

The purpose of the Enterprise Information Processing Domain-specific Meta-Model (EIPDM) is to provide a set of concrete entities, agents, activities, and roles to represent the data engineering process. It is based on the general DLO. The initial information gathered to build the domain-specific enterprise information processing metamodel is based on the JURION use case. The JURION use case includes both processes for data and software development and therefore uses the DLO and the SLO.

As the JURION use case includes the both processes of data (D) and software development (S), the model information are marked with their respective process type.

The actors identified in JURION are listed in Table 4.3. The entities identified in JURION are listed in Table 4.4. The activities are listed in Table 4.5.

For functionalities, we have a number of existing models for different kinds of documents. Depending on the document type, there is different mandatory metadata and additional information.

e-research in the Social Sciences and Humanities

The purpose of the ALIGNED E-research in the Social Sciences and Humanities domain-specific metamodel is to provide a set of concrete entities, agents, activities, and roles to represent the specific data engineering process for e-research in the social sciences and humanities. It is based on the Seshat use case within ALIGNED. It specialises the ALIGNED generic DLO and imports the W3C PROV ontology.

Table 4.3 JURION actors

Actor	Description
CMS Expert	Responsible for the technical correctness of process and data
Content Architect	Responsible for the overall process and schemas
Legal Domain Expert	Responsible for ensuring that legal data are correct
Legal Editor	Responsible for editing legal information
Product Owner	Wants the best possible product
Schema Expert	Responsible for executing and documenting schema changes
Software Developer	Review requirements, suggest possible solutions, estimate cost of certain features and bugfixing actions and implement them.
Software Manager	Coordinates all software development teams and projects
Software Testers	Perform manual testing, issue and observe automated test runs
Software Analyst	Studies the application domain and defines requirements based on his experience the software on the one hand, and the domain and customers on the other hand
Customers	Partners and testers

Table 4.4 JURION entities

Entity	Description
Schema Changes	Schema changes are done at regular intervals
Test Cases	Data tests
Text files	In XML, data in Ontowiki, databases – specific with constraints
Controlled Vocabularies	Several controlled vocabularies are maintained in PoolParty
Data Sources	External data sources
Testing Suites	Java unit tests, Jenkins, Performance Tests, Integration Test, Sonarqube
Source Code	Git and SVN repositories
Server Infrastructure	Servers that support the development process
Data/Software Requirements Documents	Mostly unstructured and free-text description of new features

This model adds support for specific external data sources for datasets like wikis, Web pages, and academic paper repositories. It adds new entities to represent candidate data for inclusion in a dataset, reports of historical events and historical interpretations created by domain experts. It extends the set of data life cycle processes to include data curation activities such as data collection and data publishing. Finally, new roles are defined for data consumer, processor and producer tools that help maintain semi-automated data curation pipelines or workflows.

Table 4.5 JURION activities

Activity	Description
Specify and model data	e.g., definition of base URI and schema mapping
Transform data	Transformation process from XML to RDF format
Integrate/Upload data	Integrate new datasets, entities, and so on
Maintain data	Enrich, delete, change, curate
Link data	Mapping with internal or external sources, link sources
Extract data	Generate test data, configure, test, review, e.g., for classification purposes
Use data	e.g., for visualisations, search, and so on
Quality analysis of data	Checking for consistency, integrity, and so on
Plan Software	Requirements planning, application evolution, data requirements for the data development team
Analyse Software	Requirement validation – requirements changes, version tracking, schema/data-based software evolution analysis
Design Software	Design verification, query design, design evolution via mapping evaluation
Implement Software	Code generation and transformation, application verification
Software maintenance	Schema and instance change impact evaluation, bug classification
Publish data	Converting a dataset to a release

Figure 4.12 illustrates the concepts found in the ontology. A full specification of the model is available online at http://www.essepuntato.it/lode/owlapi/ https://w3id.org/sdo.

Seshat Domain Ontologies

The Seshat Domain Ontology Set consists of the following specific ontologies: seshat, xdd, and dacura.

seshat

This ontology describes human societies throughout time. It is used by the Evolution Institute and its partners to describe time-series data collected about all human societies. Figure 4.13 shows the seshat ontology in graphic form.

The most important classes are:

- Polity: A polity is defined as an independent political unit. Kinds of polities range from villages (local communities) through simple and

100 ALIGNED MetaModel Overview

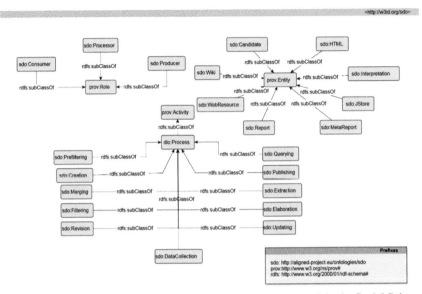

Figure 4.12 The ALIGNED domain-specific ontology for E-research in the Social Sciences and Humanities.

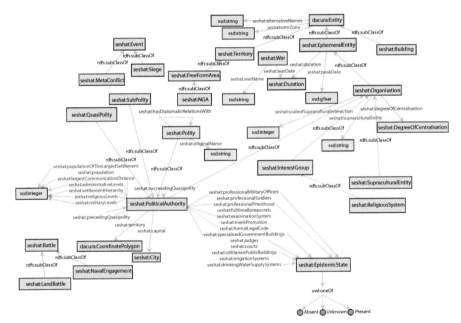

Figure 4.13 The Seshat ontology.

complex chiefdoms to states and empires. A polity can be either centralised or not (e.g., organised as a confederation). What distinguishes a polity from other human groupings and organisations is that it is politically independent of any overarching authority; it possesses sovereignty. Polities are defined spatially by the area enclosed within a boundary on the world map. There may be more than one such areas. Polities are dynamical entities, and thus their geographical extent may change with time. Thus, typically each polity will be defined by a set of multiple boundaries, each for a specified period of time. For prehistoric periods and for geographical areas populated by a multitude of small-scale polities, we use a variant called quasi-polity.
- TemporalEntity: An abstract concept describing anything that must have temporal bounds.
- PointInSpace: This is an abstract class for all points in space.
- Box: Class for boxing datatypes in order to add annotations.

Which have the following properties:

- alternativeName: The name of a seshat Entity. Generally same as the name of the wiki page.
- population: Estimated population of the polity; can change as a result of both adding/losing new territories or by population growth/decline within a region.
- name: The name of a seshat Entity. Generally same as the name of the wiki page.
- peakDate: A property used to define the temporal bounds of a seshatbox:TemporalEntity. For example, corresponds to the Duration for a Polity from the Seshat code book.
- longitude and latitude: In numeric form.
- capitalCityLocation: The latitude and longitude of the capital city.
- type: The xsd datatype of a Box.

xdd

The xdd ontology describes complex datatypes such as polygon, polyline and range types.

```
@prefix xdd: <http://dacura.scss.tcd.ie/ontology/xdd#>.
@prefix rdfs: <http://www.w3.org/2000/01/rdf-schema#> .

xdd:coordinatePolygon    a rdfs:Datatype ;
  rdfs:label "Coordinate Polygon"@en ;
  rdfs:comment "A closed JSON list of coordinates."@en .
xdd:coordinatePolyline    a rdfs:Datatype ;
```

102 ALIGNED MetaModel Overview

```
  rdfs:label "Coordinate Polyline"@en ;
  rdfs:comment "A JSON list of coordinates."@en .
xdd:gYearRange    a rdfs:Datatype ;
  rdfs:label "Year"@en ;
  rdfs:comment "Either a year or a range of years."@en .
xdd:integerRange    a rdfs:Datatype ;
  rdfs:label "Integer"@en ;
  rdfs:comment "Either an integer or a range of integers."@en .
xdd:decimalRange    a rdfs:Datatype ;
  rdfs:label "Decimal"@en ;
  rdfs:comment "A number with an arbitrary number of decimal places, or
a numberrange"@en .
```

dacura

The dacura ontology covers all xsd datatypes, rdf and rdfs literal types used in other ontologies on the platform. The dacura ontology can be seen in Figure 4.14.

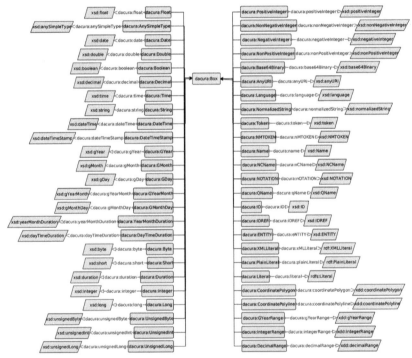

Figure 4.14   The Dacura ontology.

4.6 Unified Quality Reports 103

Figure 4.15 Dacura console usage example.

Usage

The set of seshat domain ontologies is designed to be generic in order to keep the usage as broad as possible. An example of the usage of the seshat domain ontology is the dacura console which is available as a browser plugin and enables the user to harvest social sciences and digital humanities data from websites and store them as RDF triples in a knowledge base.

Figure 4.15 shows the dacura console listing types for a candidate from the seshat domain ontology and available candidates for entities and polities. The user selects an existing collection on the left side of the browser bar and depending on her user role, which can be Data Harvester, Expert Annotator or Architect, she can browse and create new candidates of a certain type and edit existing candidates in addition to automatically harvested candidates from a website.

The usage of the dacura and xdd ontologies is best shown on the backend of the dacura platform as represented in Figures 4.16 and 4.17. The screenshots show the creation of a new candidate of the type Polity. Dacura and xdd ontologies are used to describe the used datatypes for properties of a polity (in this example e.g., original name, polity territory, religious levels, polity population, military levels, etc.). Used datatypes can be strings in open text fields as in "Original name", classes in dropdown boxes as in "Has diplomatic relations with", polygons in google maps as in "Polity territory" and many more datatypes.

Crowd-sourced Public Datasets (CSPDO)

This ontology is used to describe the domain-specific extensions to the ALIGNED data life cycle model ontology (DLO) and SLO for crowd-sourced public datasets based on the DBpedia use case within the ALIGNED

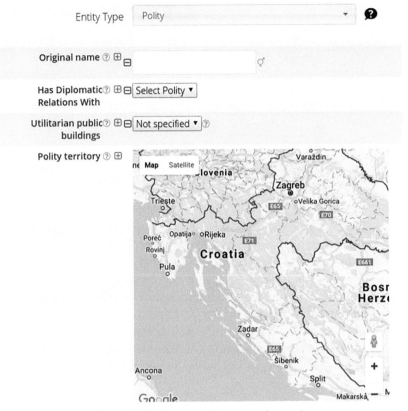

Figure 4.16 New candidate example part 1.

project. Over time, generic features may be migrated to the upper ontology. Figure 4.18 shows the CSPDO ontology.

This ontology is used to describe the domain-specific extensions to the ALIGNED data life cycle model ontology (DLO) and SLO for crowd-sourced public datasets based on the DBpedia use case within the ALIGNED project. Over time, generic features may be migrated to the upper ontology.

This ontology supports extensions needed for DBpedia. Thus, there is a focus on the validation activities. DBpedia is a large-scale extraction project of unstructured and semi-structured data from different Wikipedia language editions to RDF. This extraction is achieved from a modular extraction framework that is customised to handle multilingualism and structural differences

Figure 4.17 New candidate example part 2.

between different Wikipedia language editions. The latest DBpedia release (v. 2016) generated a total of three billion facts from 125 localised versions. As Wikipedia evolves over time, the code should be able to adapt to these changes. However, identifying errors at this data scale becomes very hard and validation workflows must be established that will ensure the quality of the extracted data.

ALIGNED tackled these challenges with data validation and interlink validation tools that communicate their results though the ALIGNED vocabularies.

The actors identified in DBpedia are listed in Table 4.6. The entities identified in DBpedia are listed in Table 4.7. The activities are listed in Table 4.8.

106 ALIGNED MetaModel Overview

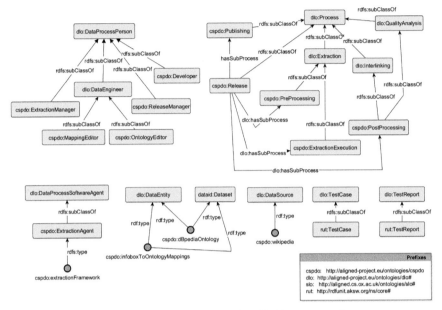

Figure 4.18 The Crowd-sourced Public Datasets ontology.

The DBpedia software and data engineering development process involves the following **actors**:

Table 4.6 DBpedia actors

Actor	Description
Extraction manager	DBpedia team members who run the extraction process for a given DBpedia release
Extraction Agent	Software agents that perform the extraction such as DBpedia live
Mapping editor	Community members who edit the DBpedia mapping wiki
Ontology Editor	DBpedia foundation members that edit the DBpedia ontology
Release manager	DBpedia team members that are responsible for the actions leading to a given release of DBpedia
Developer	DBpedia team members or community who write code for the **extraction framework** or **tools**
User	Users of DBpedia

We can identify the following **entities** (i.e., tools and technologies) that support the DBpedia development workflow:

4.6 Unified Quality Reports

Table 4.7 DBpedia entities

Entity	Description
Wikipedia	Input source for DBpedia
Extraction Framework	The source code used to extract knowledge from **Wikipedia**
Server	The physical/virtual server where an **extraction agent** is running
DBpedia Ontology	Crowdsourced OWL ontology describing DBpedia concepts and properties
Infobox to Ontology mappings	Crowdsourced mappings between the DBpedia ontology and Wikipedia infoboxes
Dataset static dataset (dump) External dataset Live feed	The output that comes after an extraction manager or release manager runs an extraction agent based on the extraction framework on a **Wikipedia** input. The output can be a static dataset, an external dataset (such as links to other datasets) or the DBpedia Live feed
Tools	Scripts or applications that work on DBpedia data
Issue or support question	New feature or support requests and bug reports are filed as tickets in the **extraction framework** Github issue tracker or reported in the DBpedia-related mailing lists

Actors and Entities are connected by the following **activities**:

Table 4.8 DBpedia activities

Activity	Description
Coding	Involves resolution of **issues/error reports** (i.e., bug fixing, feature development), but also refactoring. Done by **developers**, working on the **extraction framework**.
Release Pre-processing step Extraction Post-processing step Publishing	Releasing a DBpedia **dataset** is a complex procedure that involves a lot of pre-processing steps, the actual extraction, additional postprocessing steps and finally the dataset publishing step. This activity is performed by the **release manager** using the **extraction framework** and DBpedia **tools**
Maintain dbpedia.org	The act of maintaining the information website of DBpedia
Support (mailing lists or bugs)	Acting on a user support or new feature request or tackling a bug report

DBpedia Ontology (DBO)

The structure of the DBpedia knowledge base is maintained by the DBpedia user community. Most importantly, the community creates mappings from Wikipedia information representation structures to the DBpedia ontology. This ontology unifies different template structures, both within single Wikipedia language editions and across currently 27 different languages. The complete DBpedia ontology can be browsed online at http://mappings.dbpedia.org/server/ontology/classes/.

DBO is used to describe the data that are extracted with the DBpedia information extraction framework.

Usage

Model Mapper tool

The prototype Model Mapper tool (D3.4) uses CSPDO to record interlink validation processing on the DBpedia release candidate. This enables its activities to be shown in the Unified Governance tool (D5.2), and for other data engineering tools to co-ordinate with it in a toolchain. For example, as shown here, for the exchange of which mappings failed the validation test. This allows another tool to take corrective action on these mappings or to present them to a user.

The RDF shows the description of an interlink validation run which identifies the specific tool used for validation, the three datasets consumed (the linkset, DBpedia and Geonames) and the validation report produced (ex:interlink_validation_report_1). The datasets are identified as DataID datasources and thus could have a large amount of metadata recorded about them. The actor who initiated the interlink validation is recorded and classified as a SysAdmin. The interlink validation report itself identifies two invalid mappings in the mapping set analysed, in the first case both ends of the mapping are incorrect (probably missing from the mapped datasets) and the second mapping error identifies only one mal-formed resource.

```
ex:interlink_val_1 a cspdo:InterlinkValidation ;
    dlo:isSupportedBy ex:interlink_validator ;
    dlo:consumes ex:dbpedia_geonames_interlinks_2015 ;
    dlo:consumes ex:dbpedia_dataset_2015 ;
    dlo:consumes ex:geonames_dataset_20151010 ;
    dlo:produces ex:interlink_validation_report_1 .
ex:dbpedia_dataset_2015 a dlo:DataSource .
ex:geonames_dataset_20151010 a dlo:DataSource .
ex:person_1 a dlo:SystemAdmin ;
    dlo:initiates ex:interlink_val_1 .
```

```
ex:model_mapper a dlo:DataSoftwareAgent ;
    dlo:supports ex:interlink_val_1 .
ex:interlink_validation_report_1 a cspdo:InterlinkValidationReport ;
    prov:wasGeneratedBy ex:interlink_val_1 ;
    ex:invalidMapping1 [ ex:mapId ex:mapping_1 ;
    ex:invalidResource <resource_1> ;
    ex:invalidResource <resource_2> ] ;
    ex:invalidMapping2 [ ex:mapId ex:mapping_2 ;
    ex:invalidResource <resource_3> ] ;
    prov:generatedAtTime ''20151010''^^xsd:date .
```

DBpedia release description

Since 2015, DBpedia releases are described with the DataID ontology. This created the opportunity for application on top of the machine readable dataset metadata. These DataID descriptions are used to automatically generate the DBpedia release download page as well as automate the creation of a triple store loaded with the release data using the Docker container technology.[15]

DBpedia workflow description (planned)

As a future work, we plan to integrate DataID, DLO and PROV to describe DBpedia extraction workflows and keep track of origin and pre-processing steps of each dataset.

4.6.5 Enterprise Software Development (DIOPP)

The aim of the ontology is to integrate the datasets generated through requirements specification and the issues raised during their implementation. This ontology covers the mappings defined between the PoolParty conceptualisation and the DIO ontology. The mappings are further supported by the figures illustrated here. An example illustrating the mapping can be found here.

In the following, we describe PoolParty's requirements for the ALIGNED domain-specific metamodel for enterprise software development.

The actors identified in PoolParty are listed in Table 4.9. The entities identified in PoolParty are listed in Table 4.10. The activities are listed in Table 4.11.

---

[15] https://github.com/dbpedia/Dockerized-DBpedia

For the software life cycle and design intent, the development process involves the following **actors**:

Table 4.9 PoolParty actors

Actor	Description
Project Manager	Responsible for resource planning
Requirements Editor	Specifies requirements for a specific feature in a way that it fits to the application's design (functional and UI)
Product Owner	Knows the market and customers, identifies new features, (informally) specifies requirements, continuous and final inspection of new features
Consultant	Knows the customers and their needs, provide support for existing and training for new customers. May act as Project Managers, Requirements Editors, and Testers
Developer	Review requirements, suggest possible solutions, estimate cost of certain features and bugfixing actions and implement them.
Tester	Perform manual testing, issue and observe automated test runs
Customer	Partners, Integrators

We can identify the following **entities** (i.e., tools and technologies) that support the PoolParty development workflow:

Table 4.10 PoolParty entities

Entity	Description
Issue Ticket	New feature requests and bug reports are filed as tickets in Atlassian Jira. They have assigned, e.g., a creator (a **consultant** in most cases), an actor responsible for resolution (a **developer** in most cases), a **cost** estimation (in days), and **version** information (e.g., which version it occurred) and other metadata like description, dates, comments. Can be organised in Epics, Stories and Issues. Each of these may cover a Requirements Document (see below).
Requirements Documents	Are written using Atlassian Confluence Wiki. Mostly unstructured and freetext description of new features. Are proofread by **product owner** and **developers**.
Source Code	Git and SVN repositories
Server Infrastructure	Servers that support the development process, e.g., **testing** PoolParty or performing demos, scheduled builds for continuous integration, hosting developer chat/continuous integration notifications
Testing Suite	Java unit tests, Selenium Web Browser automation tests, API tests, operated by **testers**
Communication Resource	Skype, GotoMeeting, Chat clients, email

Actors and Entities are connected by the following **activities**:

Table 4.11 PoolParty activities

Activity	Description
Resource planning	Meetings where project managers and product owners decide (based on the issue ticket cost estimations) what features and bug requests will be scheduled for a **sprint** with what **priority**
Sprint	Certain period of time during which a specified set of **issue tickets** should be resolved
Coding	Involves resolution of **issue tickets** (i.e., bug fixing, feature development) but also refactoring. Done by **developers**, creating **source code**.
Staging	Preparing a release version of the software, i.e., creating installation packages and installing them at customer **server infrastructure**
Requirements writing	The activity of creating requirement documents and issue tickets
Communication to customers	Informal **communication** between **Consultant** and **Customer** for initiating requirements writing

4.6.6 Unified Governance Domain Ontologies

The motivation for our work was the current setup at SWC, where Atlassian Confluence wiki-like team collaboration software is used to support requirements engineering, feature specification and discussion, providing documentation of research projects and publishing of technical information. Atlassian JIRA is a ticket system used for issue and change tracking, organising ideas from team members as well as collecting from customers. These loosely coupled tools form the basis for a requirements engineering system.

Following the agile methodology of software development, the data are recorded in Confluence under headings such as "Requirements", "Goal","User Story", "Epic" and "Stakeholders". Additional fields such as "Precondition", "Detailed description", "Acceptance criteria & Test scenario" are included to provide further context to the requirements. A single field, "Comment" captures the opinions/discussion carried out by human agents. The JIRA interface is used without any major modification.

SWC collects the requirements for each version of PPT in the PoolParty development space. Requirements are then linked to pages containing epics and user stories. Most of these pages are structured based on standard templates defined by SWC. The outputs from these template-based pages are largely document-centric and require extensive human intervention to synthesise and synchronise them with PoolParty development tasks.

By using DIO, DIOPP and bespoke mappings to annotate and provide metadata to the content extracted from Confluence and JIRA, SWC is able to create merged repositories of requirements, customer feedback, bug reports and project documentation thereby consolidating PoolParty experiences, customer ideas and market needs in order to integrate them into products. This is a key factor for successful development of SWC products and for raising customer satisfaction and enterprise agility. Questions asked by customers will flow faster into the requirement engineering system. The process will help to generate concise reports on distributed business objects and entities relevant for the development processes, and to coordinate the data management and development workflows required to deliver new versions of the evolving PoolParty product. The serendipitous mining of design intents from requirements and issues will therefore have a significant impact on the full life cycle of PoolParty products from requirements through to development and maintenance.

4.6.7 Semantic Booster and Model Catalogue Domain Ontology

4.6.7.1 Model catalogue

The Model Catalogue can be used to document models and metamodels – adding descriptions and descriptive metadata to concepts and relationships. Search and comparison tools allow modellers and data engineers to understand concepts in the model and better understand the underlying data. In the ALIGNED project, we have been building a repository of the metamodels and domain-specific models for external users.

The catalogue can also be used to provide, and reason about, links between concepts in different domains. For example, showing how a software model reuses and extends concepts from the data life cycle (DLO) will help the users of data understand how data can be linked and compared.

The Model Catalogue may be used as a development platform for metamodels or domain-specific models – a collaborative editing platform enables the easy development of new versions of models, permitting discourse and iteration, controlling versioning and user access. The catalogue is tightly integrated with pipelines for MDE: enabling export of software components – alternate representations, sources or configurations for data entry such as XForms, data transfer such as XSD and XML, or data storage, such as relational database schemas or Booster specifications.

The catalogue can also use the ALIGNED ontologies to capture metadata about the models themselves – for example using PROV to capture provenance information about a dataset, or DIO to capture design intent behind software modelling decisions. The metamodel for the catalogue itself – that constrains the way that models are represented –is being extended in the next phase of the ALIGNED project to incorporate more concepts from the generic ALIGNED models.

4.6.7.2 Booster

As a MDE tool, Booster uses domain-specific models to build systems. Booster can aid the development of tools that build interoperable datasets by extending the ALIGNED metamodels. For example, an abstract model of DLV in Booster may be extended and specialised in a domain-specific model. This will ensure that any data captured and stored in the Booster-generated system will be semantically interoperable with data collected in other systems based on the DLV ontology. By understanding how these domain-specific models extend or instantiate parts of the ALIGNED metamodels, the tool can be configured to specialise the software. For example, data captured and stored in Booster might be automatically linked to public external datasets corresponding to compatible ontologies.

Booster has its own metamodel: instances of which are Booster specifications. The Booster metamodel may be linked to concepts in the ALIGNED ontologies – in particular "design decisions", and parts of a "software life cycle". Currently the textual notation for Booster does not easily support the linking of these concepts, but the design and development of a more advanced metamodel for Booster is underway, allowing explicit links to external ontologies, with support for maintaining and using these links within the generated software components.

4.6.8 PROV[16]

The provenance of digital objects represents their origins. PROV is a specification to express provenance records, which contain descriptions of the entities and activities involved in producing and delivering or otherwise influencing a given object. Provenance can be used for many purposes, such as understanding how data were collected so it can be meaningfully used, determining ownership and rights over an object, making judgements

---

[16]This section contains material derived from "PROV-Overview An Overview of the PROV Family of Documents", https://www.w3.org/TR/prov-overview/ © 2013 W3C.

about information to determine whether to trust it, verifying that the process and steps used to obtain a result complies with given requirements, and reproducing how something was generated.

As a specification for provenance, PROV accommodates all those different uses of provenance. Different people may have different perspectives on provenance, and as a result, different types of information might be captured in provenance records.

- One perspective might focus on agent-centred provenance, that is, what people or organisations were involved in generating or manipulating the information in question. For example, in the provenance of a picture in a news article we might capture the photographer who took it, the person that edited it, and the newspaper that published it.
- A second perspective might focus on object-centred provenance, by tracing the origins of portions of a document to other documents. An example is having a Web page that was assembled from content from a news article, quotes of interviews with experts, and a chart that plots data from a government agency.
- A third perspective one might take is on process-centred provenance, capturing the actions and steps taken to generate the information in question. For example, a chart may have been generated by invoking a service to retrieve data from a database, then extracting certain statistics from the data using some statistics package, and finally processing these results with a graphing tool.

The goal of PROV is to enable the wide publication and interchange of provenance on the Web and other information systems. PROV enables one to represent and interchange provenance information using widely available formats such as RDF and XML. In addition, it provides definitions for accessing provenance information, validating it, and mapping to Dublin Core.

The design of PROV stems from the recommendations of the Provenance Incubator Group which performed an extensive information gathering process including use case cataloging, requirements elicitation and a literature survey. From this process, the following eight recommendations were made:

1. the core concepts of identifying an object, attributing the object to person or entity, and representing processing steps;
2. accessing provenance-related information expressed in other standards;
3. accessing provenance;
4. the provenance of provenance;
5. reproducibility;

4.6 Unified Quality Reports

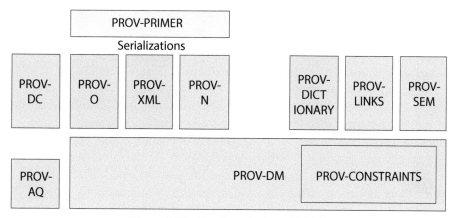

Figure 4.19 The Organisation of PROV.

6. versioning;
7. representing procedures;
8. representing derivation.

Figure 4.19 shows the organisation of PROV and how the documents (roughly) depend on each other. At its core is a conceptual data model (PROV-DM), which defines a common vocabulary used to describe provenance. This is instantiated by various serialisations. These serialisations are used by implementations to interchange provenance. To help developers and users express valid provenance, a set of constraints (PROV-Constraints) are defined, which can be used to implement provenance validators. This is complimented by a formal semantics (PROV-SEM). Finally, to further support the interchange of provenance, additional specifications are provided for protocols to locate and access provenance (PROV-AQ), connect bundles of provenance descriptions (PROV-Links), represent dictionary style collections (PROV-Dictionary) and define how to interoperate with the widely used Dublin Core vocabulary (PROV-DC).

4.6.9 SKOS[17]

The SKOS is a data-sharing standard, bridging several different fields of knowledge, technology and practice. In the library and information sciences,

---

[17]This section contains material derived from "SKOS Simple Knowledge Organization System Reference", https://www.w3.org/TR/skos-reference/ © 2009 W3C.

a long and distinguished heritage is devoted to developing tools for organising large collections of objects such as books or museum artefacts. These tools are known generally as "knowledge organization systems" (KOS) or sometimes as "controlled structured vocabularies". Several similar yet distinct traditions have emerged over time, each supported by a community of practice and set of agreed standards. Different families of knowledge organisation systems, including thesauri, classification schemes, subject heading systems, and taxonomies are widely recognised and applied in both modern and traditional information systems. In practice, it can be hard to draw an absolute distinction between thesauri and classification schemes or taxonomies, although some properties can be used to broadly characterise these different families. The important point for SKOS is that, in addition to their unique features, each of these families shares much in common and can often be used in similar ways. However, there is currently no widely deployed standard for representing these knowledge organisation systems as data and exchanging them between computer systems.

The W3C's Semantic Web Activity has stimulated a new field of integrative research and technology development, at the boundaries between database systems, formal logic and the World Wide Web. This work has led to the development of foundational standards for the Semantic Web. The RDF provides a common data abstraction and syntax for the Web. The RDF Vocabulary Description language (RDFS) and the OWL together provide a common data modelling (schema) language for data in the Web. The SPARQL Query Language and Protocol provide a standard means for interacting with data in the Web.

These technologies are being applied across diverse applications because many applications require a common framework for publishing, sharing, exchanging and integrating ("joining up") data from different sources. The ability to link data from different sources is motivating many projects, as different communities seek to exploit the hidden value in data previously spread across isolated sources.

The SKOS therefore aims to provide a bridge between different communities of practice within the library and information sciences involved in the design and application of knowledge organisation systems. In addition, SKOS aims to provide a bridge between these communities and the Semantic Web, by transferring existing models of knowledge organisation to the Semantic Web technology context, and by providing a low-cost migration path for porting existing knowledge organisation systems to RDF.

The SKOS is a common data model for knowledge organisation systems such as thesauri, classification schemes, subject heading systems and taxonomies. Using SKOS, a knowledge organisation system can be expressed as machine-readable data. It can then be exchanged between computer applications and published in a machine-readable format in the Web. The SKOS data model is formally defined as an OWL Full ontology. SKOS data are expressed as RDF triples and may be encoded using any concrete RDF syntax (such as RDF/XML or Turtle). The SKOS data model views a knowledge organisation system as a concept scheme comprising a set of concepts. These SKOS concept schemes and SKOS concepts are identified by URIs, enabling anyone to refer to them unambiguously from any context, and making them a part of the World Wide Web. SKOS concepts can be labelled with any number of strings, in any given natural language. One of these labels in any given language can be indicated as the preferred label for that language, and the others as alternative labels.

4.6.10 OWL[18]

The OWL is a language for defining and instantiating Web ontologies. Ontology is a term borrowed from philosophy that refers to the science of describing the kinds of entities in the world and how they are related. An OWL ontology may include descriptions of classes, properties, and their instances. Given such an ontology, the OWL formal semantics specifies how to derive its logical consequences, i.e., facts not literally present in the ontology, but entailed by the semantics. These entailments may be based on a single document or multiple distributed documents that have been combined using defined OWL mechanisms.

One question that comes up when describing yet another XML/Web standard is "What does this buy me that XML and XML Schema don't?" There are two answers to this question.

- An ontology differs from an XML schema in that it is a knowledge representation, not a message format. Most industry-based Web standards consist of a combination of message formats and protocol specifications. These formats have been given an operational semantics, such as, "Upon receipt of this PurchaseOrder message, transfer Amount dollars from AccountFrom to AccountTo and ship Product". But the specification is

---

[18]This section contains material derived from "OWL Web Ontology Language Overview", https://www.w3.org/TR/owl-features/ © 2004 W3C.

not designed to support reasoning outside the transaction context. For example, we will not in general have a mechanism to conclude that because the Product is a type of Chardonnay it must also be a white wine.
- One advantage of OWL ontologies will be the availability of tools that can reason about them. Tools will provide generic support that is not specific to the particular subject domain, which would be the case if one were to build a system to reason about a specific industry-standard XML schema. Building a sound and useful reasoning system is not a simple effort. Constructing an ontology is much more tractable. It is our expectation that many groups will embark on ontology construction. They will benefit from third-party tools based on the formal properties of the OWL language, tools that will deliver an assortment of capabilities that most organisations would be hard pressed to duplicate.

The OWL language provides three increasingly expressive sublanguages designed for use by specific communities of implementers and users.

OWL Lite supports those users primarily needing a classification hierarchy and simple constraint features. For example, while OWL Lite supports cardinality constraints, it only permits cardinality values of 0 or 1. It should be simpler to provide tool support for OWL Lite than its more expressive relatives, and provide a quick migration path for thesauri and other taxonomies.

OWL DL supports those users who want the maximum expressiveness without losing computational completeness (all entailments are guaranteed to be computed) and decidability (all computations will finish in finite time) of reasoning systems. OWL DL includes all OWL language constructs with restrictions such as type separation (a class cannot also be an individual or property, a property cannot also be an individual or class). OWL DL is so named due to its correspondence with description logics, a field of research that has studied a particular decidable fragment of FOL. OWL DL was designed to support the existing Description Logic business segment and has desirable computational properties for reasoning systems.

OWL Full is meant for users who want maximum expressiveness and the syntactic freedom of RDF with no computational guarantees. For example, in OWL Full, a class can be treated simultaneously as a collection of individuals and as an individual in its own right. Another significant difference from OWL DL is that a owl:DatatypeProperty can be marked as an owl:InverseFunctionalProperty. OWL Full allows an ontology to augment the meaning of the pre-defined (RDF or OWL) vocabulary. It is unlikely that any reasoning software will be able to support every feature of OWL Full.

4.6.11 RDFS[19]

The RDF is a framework for expressing information about resources. Resources can be anything, including documents, people, physical objects, and abstract concepts.

RDF is intended for situations in which information on the Web needs to be processed by applications, rather than being only displayed to people. RDF provides a common framework for expressing this information so it can be exchanged between applications without loss of meaning. Since it is a common framework, application designers can leverage the availability of common RDF parsers and processing tools. The ability to exchange information between different applications means that the information may be made available to applications other than those for which it was originally created.

In particular, RDF can be used to publish and interlink data on the Web. For example, retrieving http://www.example.org/bob#me could provide data about Bob, including the fact that he knows Alice, as identified by her IRI (International Resource Identifier). Retrieving Alice's IRI could then provide more data about her, including links to other datasets for her friends, interests, and so on. A person or an automated process can then follow such links and aggregate data about these various things. Such uses of RDF are often qualified as Linked Data.

Triples

RDF allows us to make statements about resources. The format of these statements is simple. A statement always has the following structure:

<subject> <predicate> <object>

An RDF statement expresses a relationship between two resources. The subject and the object represent the two resources being related; the predicate represents the nature of their relationship. The relationship is phrased in a directional way (from subject to object) and is called in RDF a property. Because RDF statements consist of three elements, they are called triples.

Here are examples of RDF triples (informally expressed in pseudocode):

Example 1: Sample triples (informal)

<Bob> <is a> <person>.
<Bob> <is a friend of> <Alice>.

---

[19]This section contains material derived from "RDF Schema 1.1", https://www.w3.org/TR/rdf-schema/ © 2004–2014 W3C.

120   *ALIGNED MetaModel Overview*

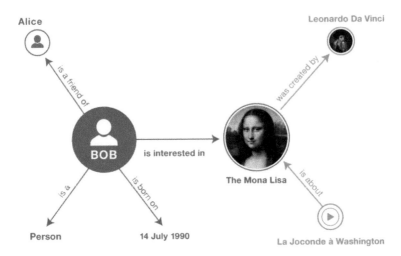

\<Bob\> \<is born on\> \<the 4th of July 1990\>.
\<Bob\> \<is interested in\> \<the Mona Lisa\>.
\<the Mona Lisa\> \<was created by\> \<Leonardo da Vinci\>.
\<the video "La Joconde Washington"\> \<is about\> \<the Mona Lisa\>

The same resource is often referenced in multiple triples. In the example above, Bob is the subject of four triples, and the Mona Lisa is the subject of one and the object of two triples. This ability to have the same resource be in the subject position of one triple and the object position of another makes it possible to find connections between triples, which is an important part of RDF's power.

We can visualise triples as a connected graph. Graphs consist of nodes and arcs. The subjects and objects of the triples make up the nodes in the graph; the predicates form the arcs. Figure 4.20 shows the graph resulting from the sample triples.

Once you have a graph like this you can use SPARQL to query for e.g., people interested in paintings by Leonardo da Vinci.

The RDF Data Model is described in this section in the form of an "abstract syntax", i.e., a data model that is independent of a particular concrete syntax (the syntax used to represent triples stored in text files). Different concrete syntaxes may produce exactly the same graph from the perspective of the abstract syntax. The semantics of RDF graphs are defined in terms of this abstract syntax.

4.6.12 RDF[20]

The RDF is a language for representing information about resources in the World Wide Web. It is particularly intended for representing metadata about Web resources, such as the title, author, and modification date of a web page, copyright and licensing information about a Web document, or the availability schedule for some shared resource. However, by generalising the concept of a "Web resource", RDF can also be used to represent information about things that can be identified on the Web, even when they cannot be directly retrieved on the Web. Examples include information about items available from online shopping facilities (e.g., information about specifications, prices, and availability), or the description of a Web user's preferences for information delivery.

RDF is intended for situations in which this information needs to be processed by applications, rather than being only displayed to people. RDF provides a common framework for expressing this information so it can be exchanged between applications without loss of meaning. Since it is a common framework, application designers can leverage the availability of common RDF parsers and processing tools. The ability to exchange information between different applications means that the information may be made available to applications other than those for which it was originally created.

RDF is based on the idea of identifying things using Web identifiers (called Uniform Resource Identifiers, or URIs), and describing resources in terms of simple properties and property values. This enables RDF to represent simple statements about resources as a graph of nodes and arcs representing the resources, and their properties and values. To make this discussion somewhat more concrete as soon as possible, the group of statements "there is a Person identified by http://www.w3.org/People/EM/contact#me, whose name is Eric Miller, whose email address is em@w3.org, and whose title is Dr." could be represented as the RDF graph below:

This illustrates that RDF uses URIs to identify:

- individuals, e.g., Eric Miller, identified by http://www.w3.org/People/EM/contact#me

---

[20]This section contains material derived from "RDF Primer", https://www.w3.org/TR/rdf-primer/ © 2004 W3C.

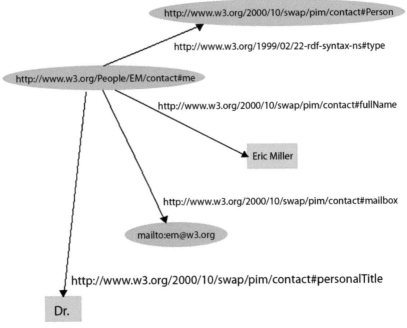

Figure 4.20 An RDF Graph Describing Eric Miller.

- kinds of things, e.g., Person, identified by http://www.w3.org/2000/10/swap/pim/contact#Person
- properties of those things, e.g., mailbox, identified by http://www.w3.org/2000/10/swap/pim/contact#mailbox
- values of those properties, e.g., mailto:em@w3.org as the value of the mailbox property (RDF also uses character strings such as "Eric Miller", and values from other datatypes such as integers and dates, as the values of properties)

RDF also provides an XML-based syntax (called RDF/XML) for recording and exchanging these graphs. Example 2 is a small chunk of RDF in RDF/XML corresponding to the graph in Figure 4.20:

Example 2: RDF/XML Describing Eric Miller

```
<?xml version="1.0"?>
<rdf:RDF xmlns:rdf="http://www.w3.org/1999/02/22-rdf-syntax-ns#"
     xmlns:contact="http://www.w3.org/2000/10/swap/pim/contact#">

<contact:Person rdf:about="http://www.w3.org/People/EM/contact\#me">
<contact:fullName>Eric Miller</contact:fullName>
```

4.6 Unified Quality Reports 123

```
<contact:mailbox rdf:resource="mailto:em@w3.org"/>
<contact:personalTitle>Dr.</contact:personalTitle>
</contact:Person>

</rdf:RDF>
```

Note that this RDF/XML also contains URIs, as well as properties like mailbox and fullName (in an abbreviated form), and their respective values em@w3.org, and Eric Miller.

Like HTML, this RDF/XML is machine processable and, using URIs, can link pieces of information across the Web. However, unlike conventional hypertext, RDF URIs can refer to any identifiable thing, including things that may not be directly retrievable on the Web (such as the person Eric Miller). The result is that in addition to describing such things as Web pages, RDF can also describe cars, businesses, people, news events, and so on. In addition, RDF properties themselves have URIs, to precisely identify the relationships that exist between the linked items.

5

Tools

Kevin Feeney[1], Christian Dirschl[2], Katja Eck[2], Dimitris Kontokostas[3], Gavin Mendel-Gleason[1], Helmut Nagy[4], Christian Mader[4] and Andreas Koller[4]

[1]Trinity College Dublin, Ireland
[2]Wolters Kluwer Germany, Germany
[3]University of Leipzig, Germany
[4]Semantic Web Company, Austria

5.1 Model Catalogue

5.1.1 Introduction

Careful management of metadata is essential for the effective reuse of data and the correctness of any software designed for processing the data. Metadata may capture best practice in a domain and as such the reuse of metadata can proliferate best practice. Traditionally, most metadata is usually captured implicitly, and embedded in the software or system that use the data. Motivation for developing a tool for managing and curating metadata includes data and software interoperability, documenting metadata as models and building a platform for automatically generating software systems from models.

Without easily available metadata, determining the compatibility of datasets and software systems becomes challenging. The problem is compounded where there are multiple datasets and software systems, as the metadata for each needs to be documented manually, ex post facto, in order to determine compatibility. There are several challenges to collecting, managing and documenting metadata. Metadata may have only been documented conceptually, and there may be ambiguities in the metadata that require further clarification.

The Model Catalogue is an online tool that supports the capture and documentation of metadata as generic and reusable models. The tool facilitates collaboration between metadata creators and potential users. The system defines a core language for describing metadata, which enables sharing, documentation and reuse of metadata. The tool uses standards-based concepts for registration, versioning and a standard four level architecture with an API for interoperability with external tools.

Metadata is data that describes data by capturing essential relationships, classifications and atomic data elements. Metadata is important for data reuse and underpins the software that stores, processes and analyses any dataset. This information captures the meaning and guides interpretation of the data. However, in typical usage, metadata is captured implicitly and embedded within the software system that uses the data. This is problematic because metadata is essential to reuse the data outside the original context.

Metadata is also essential to interoperability. The compatibility of two datasets may only be determined by examination of the datasets' metadata. This can be an arduous task where the metadata is embedded with the software and the difficulty can be compounded where there are several datasets involved. Software interoperability, where independent systems can share messages and data, relies on compatible metadata for the involved systems.

When metadata is separated from the use of the data, two important efficiencies are made possible. Firstly, the metadata can be explored independently from the data, so the compatibility of datasets and interoperability of software can be determined without analysis or re-collection of the data. Metadata can also be captured in a generic form, as metadata models and software systems for managing data can be generated from sufficiently detailed metadata models. Other benefits to a metadata-oriented approach include reasoning and discussion about the underlying model with experts, reuse of metadata and creating a map of data across the systems in an organisation.

Metadata can also be used to encode the established best practice in a domain. The definitions of how data should be structured, what data should be captured and the intended use for the data can be captured in metadata. For example, metadata can specify the resolution and level of granularity for data capture. When metadata is embedded directly with the use of the data, reuse of the best practice can be a challenge. Where metadata is encoded as generic models and the models are documented independently of any system, the essential information about a domain becomes more readily available for reuse. Software developers can use these metadata models to encode the best practice of a domain. Reusing metadata models helps proliferate established best practice.

Generic models of metadata can encode the best practice of a domain. This can be taken further by allowing potential data users to reuse subsets of a generic data model. This means that only the relevant data elements and classifications from a model can be reused and repurposed in a new context. However, reusing, changing or merging models for a new context is challenging unless elements have exactly the same meaning in both contexts. Subtle differences between models' elements can make the reuse of models and data problematic, especially when changes in versions are not documented or tracked. Each element of metadata must be documented in order to determine the compatibility of that data element for reuse in a new context.

5.1.2 Model Catalogue

The Model Catalogue is a toolkit for creating, sharing, and updating data models. The system uses a layered architecture, described below, which allows for a number of possible Graphical User Interfaces. The data models are descriptions or specifications of data artefacts, objects, or implementations.

A data model may describe or specify:

- a dataset or database holding data of interest
- a request for data from a collection of databases or datasets
- a standard for developers to work to
- a form used for data entry
- a message carrying data from one system to another
- a report offered or required
- a workflow or pathway in which data are collected and used.

A data model will be simpler than the artefact it describes. It need not consider every aspect of the artefact or implementation, only the data items of interest and the relationships between them.

A data model will be more comparable. It is easier to compare data models, written in a single modelling language, than to compare artefacts implemented using a range of different technologies.

A data model will be more re-usable. It is safer to produce a new artefact by copying parts of a model than by copying parts of the existing implementation. There are additional advantages – cost, consistency – if the new artefact can be generated automatically.

5.1.2.1 Architecture

The Model Catalogue has been built in a traditional layered architecture, facilitating access through both manual and programmatic means. This structure is shown in Figure 5.1. At the base layer is a relational database: in the current implementation, we use PostgreSQL as a stable, tried-and-tested open source solution. To ensure consistency of the underlying data, we insist

128 *Tools*

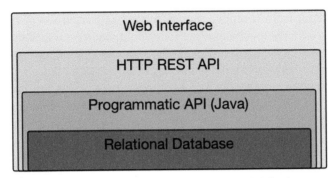

Figure 5.1 The layered architecture of the Model Catalogue.

that all data access and manipulation is through a programmatic API: this is currently implemented in Java and is used by the higher levels of the stack, but can also be used by external tools built with Java.

At the next level up is a Web-based REST API. This can be used to programmatically access remote deployments of the catalogue, and is language-independent: it can be used by any sophisticated toolset to interact with a publicly available catalogue. The final layer is the human-readable Web interface. This provides an attractive view of the data to facilitate a range of use cases, accessible on modern Web browsers using standard interface patterns for security and interaction. The Model Catalogue provides a generic API so that any Java-based tool, including Eclipse, can be integrated programmatically.

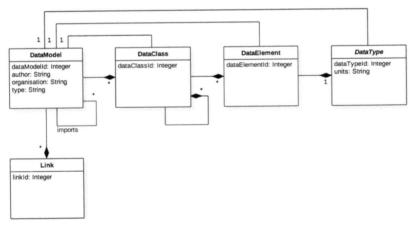

Figure 5.2 Core concepts – data model components – within the Model Catalogue.

5.1 Model Catalogue

At the core level, the catalogue contents are structured in a simple hierarchy, a subset of which is shown in Figure 5.2. At the top level is a Data Model, which may be versioned and published. A model contains a number of Data Classes, which provide categorisation or structuring. At the lowest level are Data Elements, which describe individual data points. Each Data Element has a Data Type, which may be either: a Primitive, such as String or Integer; an Enumerated Type, where allowed values may be defined in the context of this model or taken from a larger terminology; or a Reference Type, denoting a pointer to data from another Data Class. Figure 5.3 shows this hierarchy.

Components of data models may be linked: a link between two elements can represent that two data items are equivalent, that one is derived from another, or that one is different to another. For example, one element in a dataset might be linked to a definition within a data-standard to assert that the guidelines have been followed in the collection of that data item. Another item might be linked to that same data standard definition with a "different to" annotation to assert that although these data points might look the same, there is a subtle difference that may be explained in the item's description.

The Model Catalogue Web interface currently supports the fundamental use case requirements: browsing, searching and editing/updating models. These are described in the next sections.

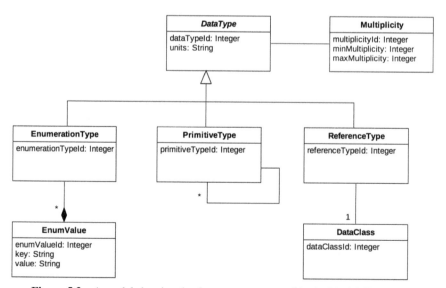

Figure 5.3 A model showing the datatypes represented in the Model Catalogue.

5.1.2.2 Searching and browsing the catalogue

There may be many models in the catalogue with the same name. For this reason, the contents of the catalogue can be searched or browsed using system metadata:

- model name
- editors – the catalogue users with write access
- status – draft or finalised
- catalogue version
- creation date
- last edit date
- imports from – the list of models which use the datatypes defined within this model

or by user-supplied metadata:

- owner(s) – responsibility
- author(s) – credit
- organisation(s) – authority
- external version name/label
- external release name/label

Other searching and browsing requirements will be addressed using annotations and classifications. Users with write access to a model can add annotations against templates provided for that model type. Users with read access can classify models and model components to which they have read access.

The two-panel view of the catalogue, as shown in Figure 5.4, provides a familiar interface, with the structured model contents in the left-hand pane, and the currently viewed data model component in the larger right-hand pane. A view of a data model displays metadata about that model and a list of all child data classes; similarly, the view of a data class shows metadata about that class, along with all contained data elements. The view of a data element, as illustrated in Figure 5.5, shows detailed information about the datatype, including any enumerated values, along with the description, and metadata about its place within the model, its current publication status, and when it was last updated.

The Web interface currently offers a basic keyword search across different component types within the catalogue. This helps potential users of collected datasets find data items that may be useful to their work. Figure 5.6 shows this keyword search in action for the Seshat Code Book model.

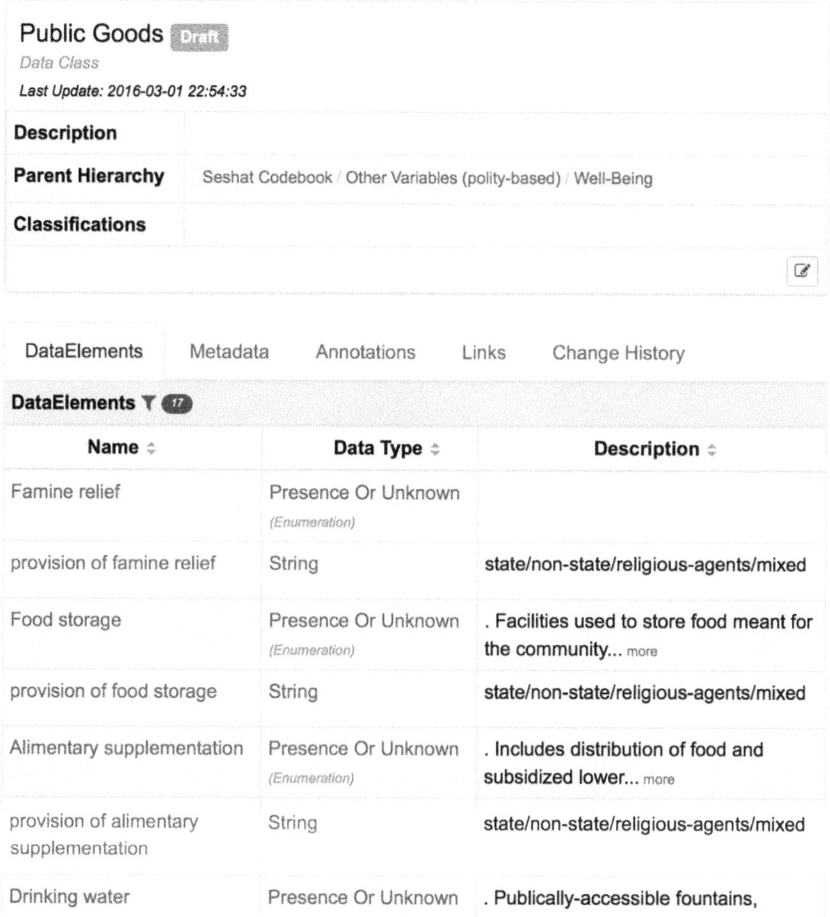

5.1 Model Catalogue

Public Goods *Draft*
Data Class
Last Update: 2016-03-01 22:54:33

Description	
Parent Hierarchy	Seshat Codebook / Other Variables (polity-based) / Well-Being
Classifications	

DataElements | Metadata | Annotations | Links | Change History

DataElements (17)

Name	Data Type	Description
Famine relief	Presence Or Unknown (Enumeration)	
provision of famine relief	String	state/non-state/religious-agents/mixed
Food storage	Presence Or Unknown (Enumeration)	. Facilities used to store food meant for the community... more
provision of food storage	String	state/non-state/religious-agents/mixed
Alimentary supplementation	Presence Or Unknown (Enumeration)	. Includes distribution of food and subsidized lower... more
provision of alimentary supplementation	String	state/non-state/religious-agents/mixed
Drinking water	Presence Or Unknown	. Publically-accessible fountains,

Figure 5.4 Model Catalogue interface: browsing the Seshat code book.

5.1.2.3 Editing the catalogue contents

Editing a draft data model is a simple process whereby most fields may be edited in place. Figure 5.7 shows the editing of basic data element details: name and description. All modifications are recorded. A change log is created and preserved for each model, showing the time, date, and user responsible for each change.

Although people may currently use the catalogue to collaboratively edit a model, simultaneous updates can cause confusion. In the next version of the system, at the point of opening a component for editing, a user will be alerted

Figure 5.5 Model Catalogue interface: data element view.

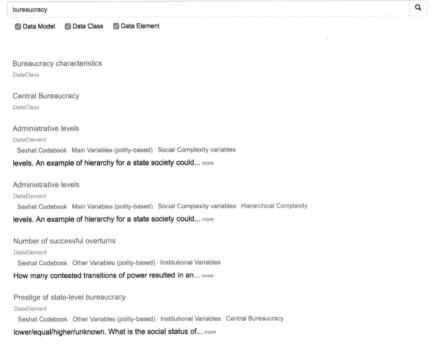

Figure 5.6 Model Catalogue interface: keyword search for a data item.

if that same component is currently open for editing by other user(s). The identity of the other user(s) will be displayed, together with the time at which they started their edit. At the point of saving an edit upon a component, a user will be alerted if the component has been updated, by other user(s), since the

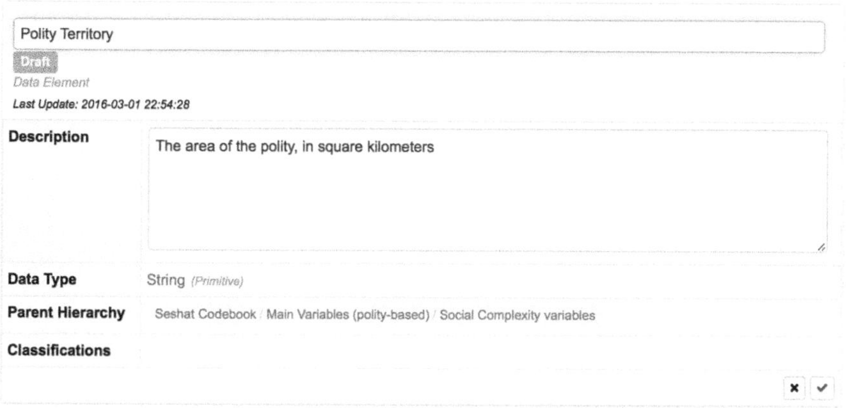

Figure 5.7 Model Catalogue interface: editing a data item.

edit began. In either case, the current can choose to proceed with the edit or the save.

Once finalised, those with "write access" may further annotate a model: while annotations are potentially useful and valuable, they are not taken to contribute to the finalised interpretation of the model or model component.

The contents of a model, and any associated links, cannot be changed after finalisation. However, the lists of users with write or read access can be updated by any of those with write access.

Finalisation cannot be undone: instead, a new version must be created. This can be done by any of those with write access to the finalised model: the result is a copy of the model with a link to the existing, finalised model indicating that it is indeed a "new version".

There is no need to create a new version of a model that has not been finalised: a user with write access can simply update the contents of the model in its current "draft" state; an edit log is maintained automatically.

If two users with write access to a finalised model both create new versions, then the development will branch. A branched development may be merged by a user with write access to finalised models in both branches, creating a model that is a new version of both.

A user who does not have write access to a finalised model may create a copy of the model (or a component) that is not a new version but is instead "based upon" that item. There is no requirement that the new item should have the same intended interpretation.

134   *Tools*

When a new version of a finalised model is created, the result is a draft model with a complete copy of the finalised model contents and metadata, including all links and annotations.

Where a model or a component is the target of a classifier or label, whether the classification concerned is updated to include the new version depends upon the properties of the classification (and not the model). The options are:

- add new version to classification alongside existing version
- add new version, remove existing version
- ignore new version

In the last case, the new version can then be added manually, if required.

5.1.2.4 Administration

The Model Catalogue administrator(s) can register new users as editors or readers. They can manage models for which there are no longer any other users with write access.

An editor can create new models and add or delete annotations for existing models to which they have write access. They can create classifications referring to any items to which they have read access. They can explore models to which they have read access.

A reader can add or delete annotations for existing models to which they have write access. They can explore models to which they have read access.

Where there are several users with write access to the same model, the possibility of conflict arises. Any conflict may be addressed through interaction with the administrator(s), who are able to modify any aspect of the catalogue contents.

Any editor can create a user group. Any member of a user group can add or remove members. User groups can be included in model access control lists.

5.1.2.5 Eclipse integration and model-driven development

Core parts of the catalogue functionality are integrated with the Eclipse Modeling Framework (EMF). EMF is fundamental to the majority of model-driven development tools within Eclipse and is also used as the basis for DSLs and transformations. This allows existing model-driven tools within Eclipse to take advantage of the catalogue in order to reuse components of models, increasing the speed of development, and allowing data linking and interoperability between tools built within the framework.

5.1 Model Catalogue   135

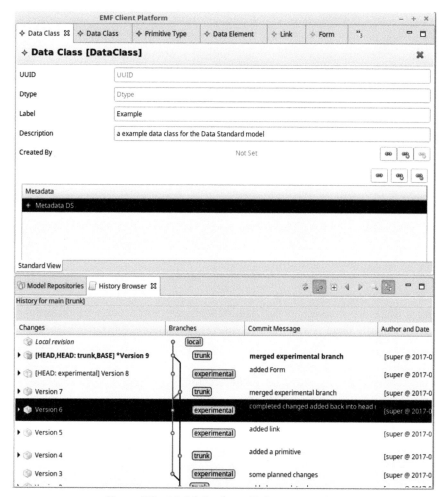

Figure 5.8   Model Catalogue Eclipse Integration.

Furthermore, the EMF integration allows new Model Catalogue components to be built in a MDE fashion – the screenshot in Figure 5.8 shows an automatically generated interface for interacting with the catalogue data, including automatic change management to track multi-user updates. Another auto-generated component stores all versions of every model to disk.

The Model Catalogue also has a plugin architecture, providing extension points through which new functionality can be built and dynamically

136 *Tools*

integrated. For example, two key extension points are those of Importer and Exporter: developers can write their own importers and exporters using the Model Catalogue Java API to automatically document models in their own language, or to use the catalogue as an interface for compiling new models from existing ones. A developer may choose to write a data model importer that documents the usage of a no-SQL database, and an exporter that generates queries to retrieve that data and insert it into an SQL database.

Further plugins are being developed for bespoke types of data model, and custom interfaces that can be used to display and edit particular types of model in a more familiar fashion. For example, a graphical editor for UML diagrams, or a builder tool for designing data entry forms. This plugin architecture also allows custom configurations of the catalogue to be deployed – using just those plugins necessary for the context: providing a better user experience and requiring minimal system resources.

5.1.2.6 Semantic reasoning

Semantic links are created in the catalogue to associate parts of different models – typically Data Elements – to assert that they are similar in meaning or use. This allows descriptions of meaning to be reused: by asserting: "this element is the same as that one" a modeller may take advantage of definitions in other models, reducing the effort in documentation. These links also give us the formal notion of "semantic interoperability": that data from two sources may be combined for a particular purpose.

To reason about this semantic interoperability property, using off-the-shelf reasoning tools, it is useful to view the data in terms of triples. In order to do this, the D2RQ tool[1] allows users to expose internal relational data as RDF triples (see screenshot in Figure 5.9). Although the mapping requires some further customisation for easier use, the mapping is sufficient to reason about key properties of the semantic links: circularities in the transitive "same-as" link, and contradictions in definitions using the "same-as" and "not-same-as" links.

A side effect of making this representation available is that the catalogue contents can be linked to other open datasets. For example, catalogue metadata may be linked to other published artefacts with Dublin Core, provenance information may be attached with PROV, linking to existing tagging and

---

[1] http://d2rq.org

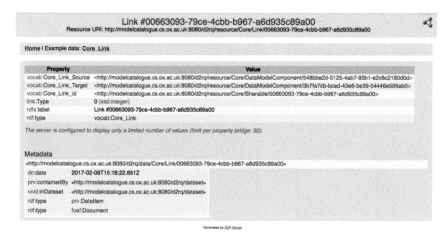

Figure 5.9 Screenshot showing RDF representation of catalogue contents.

folksonomies through the Modular and Unified Tagging Ontology (MUTO), and design intent may be linked to model components using DIO.

To assist with this linking, namespaces can be added to metadata elements within the catalogue. These can be used to indicate fields for linking in the RDF representation, or can be used by plugins to configure generated artefacts, such as adding constraints to systems generated with Semantic Booster, or shaping XSD outputs to match existing specifications.

5.1.2.7 Automation and search

Previous implementations of metadata registries have shown that it can be difficult to encourage users to carefully document the whole data model to a level that is sufficient for potential users of the data. In particular, when dealing with models at scale, even simple tasks like finding a data element in another model to link to, or comparing multiple versions, can be complicated and time-consuming. In order to improve usability, a number of features assist modellers in using the tool effectively and efficiently.

At the heart of this effort is greater power in searching across many hundreds or thousands of data elements, in order to find related items, create semantic links between items, and import or reuse whole model components. Lucene and Solr[2] help with indexing, and allow faster and more flexible searching using keywords, related terms and intelligent suggestions. The

---

[2] http://lucene.apache.org

speed improvements offered by these tools make the Model Catalogue as a whole scalable for domains with large numbers of complex data models.

Finding similar elements to link to can reduce the time it takes to document a data element. To allow users to find similar items, an autosuggest feature will find potential matches across all models, or a particular model, based on datatypes, element names, and text matching in the description.

Using the semantic reasoning described above can also assist users in creating semantic links, using the transitive properties to help them find related data elements not already explicitly linked. Such reasoning can also help find relations between larger model components: for example, linking two data classes where the component data elements of each class are already linked.

Comparing different models is also something that users need extra assistance with – especially comparing multiple incremental versions of the same model. To aid users in this activity, there is a Web interface, supported by back-end API methods. Viewing two models side by side, with differences highlighted, provides a user-friendly experience that will be familiar to those with experience using traditional "diff" style tools.

5.1.3 Semantic Booster

5.1.3.1 Introduction

The data belonging to an organisation is often its most valuable asset: traditionally payroll information and customer details, but more frequently entire business models are based on the gathering and dissemination of information. The software responsible for maintaining the integrity of this data, and its consistent interpretation, will be critical to the ongoing function of the business. Moreover, organisations need to evolve and adapt, and it will be essential for the software and data to follow changes to business rules, and for the semantics of the data to remain clear and unambiguous.

Building software that is both robust and adaptable brings with it many challenges. The typical development process for robust systems for use in safety-critical applications will be slow and labour-intensive. Agile development processes are key to maintaining and evolving software, but are not effective when dealing with large complex datasets, or where the guarantee of software correctness is reliant on more than simple testing.

Automation and abstraction provide some solutions to these difficult problems. By automating part – or indeed all – of the code generation process, the influence of human error can be reduced, and the subsequent speed-up can decrease the time necessary to adapt or evolve generated, working code. By using suitable abstractions to model the software's intended function, correctness may be more carefully clarified, and the scope of updates or evolutions may be more immediately realised.

The MDE approach attempts to combine both automation and abstraction. Models may be domain-specific: comprehensible to non-technical domain-experts, with automated processes generating software components to match. In practice, however, such MDSE tools are either too specific, where models are used for not much more than customising or configuring a particular software artefact, such as in the generation of embedded systems; or too general purpose, where a wide variety of specifications may be expressed, but without the formality required for robust implementation, and in most cases where code generation must be supplemented by custom hand-written code.

The Booster tool has been written in an attempt to find the sweet spot between these two extremes. Models describe information systems: software components focussed on the correct management of business-critical data. The modelling language takes an object-oriented approach to modelling business concepts, but is supplemented with a formal, mathematical notation for describing relationships between entities, integrity constraints, business rules, and constraints upon interaction with data. The compilation process is complete: working implementations are generated with no manual intervention or addition required.

In order to make Booster more widely applicable – in particular to the domain of data-intensive systems – some key enhancements are necessary. Booster models are mostly without semantics: the meaning of entities or attributes is not recorded. This means that data collected and maintained within a Booster system may not be immediately re-usable within a different context. Although Booster is able to maintain and migrate data in the face of changing specifications, the meaning or context of these data may be lost. Booster has been integrated with the Model Catalogue, allowing metadata to be linked to each data item stored. As well as increasing the value and utility of the data, it allows domain experts to more carefully specify the functionality of the system, as well as permitting new notions of correctness.

5.1.3.2 Semantic Booster

The Booster tool takes as input a formal specification, written in the Booster language, and generates a complete working implementation. The Booster

140 Tools

language takes inspiration from the UML, incorporating familiar object-oriented notions of classes, attributes and associations. This language is supplemented with a formal constraint language, inspired by the mathematics notations of Z and B, used in formal methodologies. These constraints can be used to define integrity constraints and business rules, in the form of class invariants, and pre- and post-conditions for methods.

The Booster language is supported by a custom editor written for the Eclipse IDE. This provides a number of features that aid developers, such as syntax highlighting, auto-suggestion, document outline, and code validation "as-you-type". Figure 5.10 shows a Booster specification being edited within the IDE.

Once a specification is completed, an automatic generator can be executed to generate an implementation. The generator consists of a number of stages, implemented as a pipeline (see Figure 5.11). In the first stage, the model is elaborated – this flattens the class hierarchy and in-lines any references to other parts of the model, essentially making explicit any default assumptions. The second stage of the pipeline is to apply a number of heuristics, to generate simple code from each constraint in the model. These heuristics have been

```
1  system defaultBooster
2
3  set UserRole { Administrator, User }
4
5  /* The Default User class.
6  */
7  class User {
8    attributes
9      username : String
10     lastName (ID) : String
11     firstName (ID) : String
12     emailAddress : String
13     passwd : Password
14     enabled : Boolean
15     created : DateTime
16     role : UserRole
17     auditTrail : SET ( AuditModOp . auditForUser ) [*]
18   methods
19     create {(currentUser.role = Administrator or card(User) = 0) & u! : User' ;
20          u!.firstName' = firstName? &
21          u!.lastName' = lastName? &
22          u!.emailAddress' = emailAddress? &
23          u!.username' = username? &
24          u!.passwd' = passwd? &
25          u!.role' = role? &
26          u!.created' = CurrentDateTime &
27          u!.enabled = true }
28     enableAccount { currentUser.role = Administrator & enabled = false & enabled' = true}
29     disableAccount { currentUser.role = Administrator & enabled = true & enabled' = false}
30     changeUserDetails { (currentUser.role = Administrator or currentUser = this) &
31                         lastName' = lastName? &
32                         firstName' = firstName? &
33                         emailAddress' = emailAddress? }
34     changePassword { (currentUser.role = Administrator or currentUser = this) & passwd' = passwd? }
35     changeUsername { (currentUser.role = Administrator or currentUser = this) & username' = username
36     newAuditModOn (NA) {
```

Figure 5.10 A Booster specification edited with the Eclipse IDE.

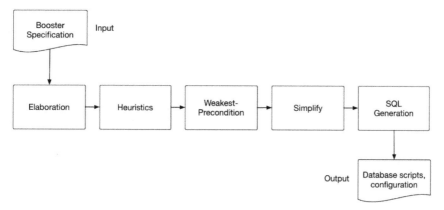

Figure 5.11 The Booster generation pipeline.

defined based on experience of developing information systems, and the code itself is written in an abstract, mathematical notation suitable for subsequent analysis.

The third stage of the pipeline is to generate additional code based on all constraints across the entire model. This process is similar to a 'weakest-precondition' calculation in formal methods and ensures the correctness of the final system: all business rules and integrity constraints are guaranteed to be considered and upheld in the final system.

The fourth stage of the pipeline is simplification: the previous steps generate large amounts of code, and much of it may be simplified to produce more efficient programs. The final stage is to generate a database implementation – the original implementation generates MySQL. This implementation includes database tables to store the core information, stored procedures to implement all data update methods, and additional system metadata to provide an object-relational mapping suitable for external users to interact with the system.

The completed database implementation can be used in conjunction with a bespoke API and user interface to provide a complete working system. This structure is shown in Figure 5.12; the Web-based Booster interface is shown in a screenshot in Figure 5.13.

The Booster approach embodies an approach in which the integrity of the data is all important. The rigorous calculations and code generation in the development pipeline ensure that for any form of update to the data, all business rules and constraints are considered, guaranteeing that no data integrity constraints will be invalidated as a result of any subsequent change.

142   Tools

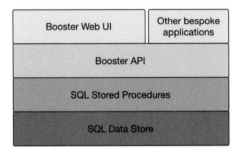

Figure 5.12   The architecture of a Booster information system.

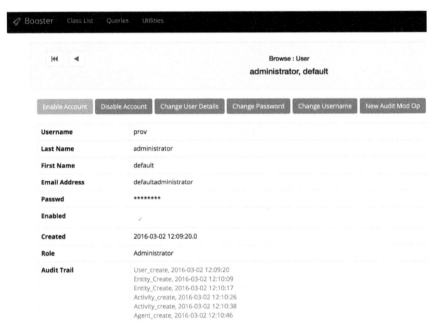

Figure 5.13   The Booster Web-based user interface.

Access to the data is through a carefully managed API, which ensures that data are only manipulated in the manner specified in the original model.

In order to apply the Booster toolset to data-intensive systems, a number of enhancements have been made. By integrating Booster with the Model Catalogue, the stored semantic metadata can be used to enrich and inform the development of Booster specifications, and to ensure the consistency and reusability of the data held within Booster systems.

5.1 Model Catalogue 143

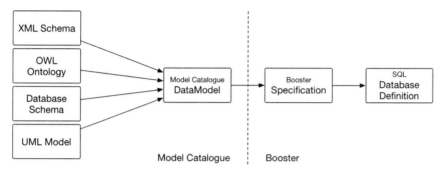

Figure 5.14 Generating Booster systems from Model Catalogue models.

Booster specification generation from Model Catalogue models

The first enhancement has been to build functionality to generate Booster specifications automatically from models described in the Model Catalogue. By using model components from the catalogue in our specification, we can ensure that the generated software can conform with existing data standards, or UML specifications, or can match data formats described using XML schema or OWL ontologies. This automation also allows domain-experts to begin generating software components without the need for development effort. Figure 5.14 shows the Semantic Booster pipeline, where platform-specific representations can be loaded into the Model Catalogue, and Booster systems generated via the Booster compiler, with no manual intermediate steps.

The structure of a model in the Model Catalogue is in many ways similar to the structure of a Booster specification. However, a number of transformations are required to take the tree-structured model and turn it into the flatter specification required for Booster. As well as this structural transformation, some more practical changes to Booster were required – in particular allowing all specification constructs to take a human-readable name, in addition to the standard identifiers required by the constraint language. By hiding system identifiers below the API level, the resulting information system is easier to use for subject-matter experts, and ambiguity may be reduced.

At the outermost level, a DataModel in the Model Catalogue is translated into a System in Booster. Every EnumerationType within the DataModel is converted into a Booster Set, using the human-readable names, and generating system identifiers if necessary. All DataClass components from the model, at any level in the hierarchy are converted into Booster Class declarations. Where one DataClass is contained within another in the catalogue DataModel, a bi-directional optional-to-one association between the two classes is created in Booster, corresponding to the notion of ownership, or composition in UML.

Every DataElement in the catalogue is translated into an Attribute in the Booster model, with multiplicities maintained. Those elements with a PrimitiveType datatype in the catalogue are mapped to the appropriate Booster primitive type. Similarly, EnumerationType elements in the catalogue get mapped to equivalent Set valued attributes in Booster. Finally, ReferenceType valued attributes are converted to bi-directional associations to the relevant class in Booster.

As part of this transformation, a basic collection of update methods is generated. For each Booster class, methods are created for creating, updating and destroying objects of that class. In addition, for every bi-directional association created by the transformation, methods are created for adding and removing links. Where these associations correspond to composition or aggregation, appropriate constraints are added to maintain the ownership properties. Figure 5.15 shows the generated specification for part of the PROV-DM Core Structures model.

The resulting Booster specification is suitable for generating a functional system capable of entering and storing data corresponding to the original model. In many cases, this may be sufficient, in particular with the addition of a bespoke user interface to enact particular workflows on top of the generated methods. However, the specification may also be used as the basis for a more elaborate Booster system, by using the Booster functionality for importing and overriding through inheritance. In this way, constraints and business rules may be added, along with additional methods and attributes, to provide a richer implementation, but by the nature of inheritance in Booster, still compliant with the original data model.

Model Catalogue information through the Booster interfaces

Once a Booster system has been used to capture and store data, it is important that this information can be reused. In many cases, this may require an understanding of the context of collection. This is especially helpful where data are to be combined from multiple systems, and it is important that only data items with similar definitions are combined. In order to allow such contexts to be available alongside the data, Semantic Booster needs to include functionality to allow integration with the data stored within the Model Catalogue.

It is vital that the Model Catalogue remains as the single source for metadata, rather than copies of the data being moved into the Booster-generated

```
 5⊖ class Entity {
 6⊖    attributes
 7        name (ID) : String
 8        wasDerivedFrom : [ Entity . derivations ]
 9⊖       derivations : SET( Entity . wasDerivedFrom )
10        wasGeneratedBy : [ Activity . derivesEntities ]
11⊖       wasAttributedTo : SET( Agent . attributedEntities )
12⊖       uses : SET( Activity . used )
13
14⊖    methods
15        Create { e! : Entity' & e!.name = name? }
16        Update { this.name = name? }
17        _____: Entity' }
18        Target Constructor: /13
          Press 'F2' for focus om { wasDerivedFrom' = entity? }
19        UnsetWasDerivedFrom { wasDerivedFrom' = null }
20        AddDerivations { entity? : derivations' }
21        RemoveDerivations { entity? : derivations' }
22        SetGeneratedBy { wasGeneratedBy' = entity? }
23        UnsetGeneratedBy { wasGeneratedBy' = null }
24        AddWasAttributedTo { agent? : wasAttributedTo' }
25        RemoveWasAttributedTo { agent? : wasAttributedTo' }
26        AddUses { activity? : uses' }
27        RemoveUsers { true }
28
29 }
```

Figure 5.15 Excerpt from the Booster system generated from Prov-DM Core.

system. Although most parts of the metadata are frozen on publication (and subsequent implementation), other metadata components such as comments and links may be added after the system has been deployed. The approach taken has been to create links between components in the Booster specification such that the metadata can be seamlessly retrieved.

Metadata concerning the Model Catalogue itself is placed in a table alongside the Booster data: the Web URL to locate the catalogue, the version number and name of the catalogue, along with any lists of any user credentials required to access private models. Each Booster specification component corresponding to an element from the Model Catalogue is stored with a GUID, a link to the relevant catalogue metadata, and the credentials required for access. This includes the system itself, every class and attribute, datatypes, and enumeration values.

This model catalogue data is stored in the database alongside the data, but accessed through a bespoke set of SQL stored procedures. These are exposed through the API layer so that external applications can access them.

146   Tools

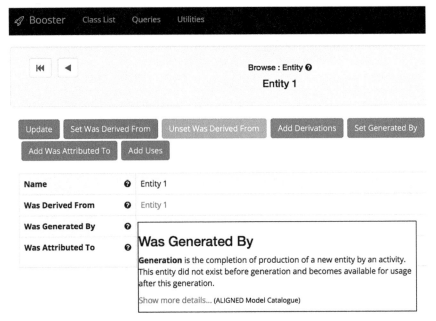

Figure 5.16   Model Catalogue information in the Booster interface.

The data are also propagated through the Booster Web interface, so that users of the data can access the appropriate metadata. A new REST API call has been added to the Model Catalogue, displaying a snippet of HTML with a link for more information. This is shown in Figure 5.16.

Semantic Booster: Booster data as triples

A key requirement of Semantic Booster is that system data must be accessible as RDF Triples. Data systems generated by the original Booster tool have employed a standard relational schema for data persistence, and standard stored procedures for operations on data. In Semantic Booster, a new target had been developed: "Generate Triple Map" is to be used alongside the standard relational generation (Figure 5.17).

This additional transformation generates a mapping specified in the W3 standard R2RML language.[3] This mapping reflects the Booster specification, and includes standard simplifications for Booster constructs of associations and inheritance, creating a lightweight wrapper for the Booster database schema. The R2RML standard has made this extension tractable:

---

[3] https://www.w3.org/TR/r2rml

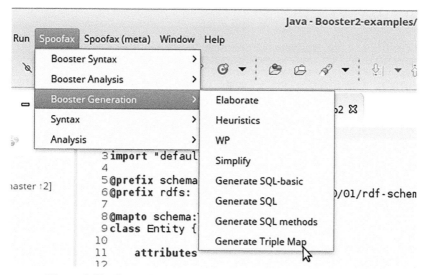

Figure 5.17 Semantic Booster – generation menu in the Eclipse IDE.

the functionality has been enabled using an additional transformation, rather than a ground-up rewrite of the pre-existing transformations.

The R2RML schema can be used to present a 'live' view of the data or to extract the whole dataset as a data 'dump'. Furthermore, the data are opened to a range of existing 'semantic Web' tools which can deal with the data in this triple format.

In the live view, any changes made by Booster operations are automatically available in the triples. No complex update or synchronisation is required as the triple-view of the data is derived directly from the relational data, and existing Booster functionality is not affected.

User-Specified semantic mappings

Typically, RDF triples have associated-type information in a semantic schema, typically via an RDF schema or an OWL ontology. The schema gives users a grounding by which to explore, classify, and query the objects, data and relationships in a dataset. Typical queries over a semantic dataset might find all elements with a particular semantic type, relationship type or a pattern of types, using the SPARQL query language. The schema and types of a data system are defined on a per-system basis by data engineers.

For a system such as Semantic Booster to be useful to data engineers, the types and schemas of the system must be customisable. In an R2RML mapping, each element must have a type, and a default set of types is created by

148 Tools

```
  |● Entities.boo2    ● IPG.boo2    ● Booster2.sdf3    [] Entities.mapping.ttl ⊠
 1 @prefix rr: <http://www.w3.org/ns/r2rml#>.
 2 @prefix rdf: <http://www.w3.org/1999/02/22-rdf-syntax-ns#>.
 3 @prefix rdfs: <http://www.w3.org/2000/01/rdf-schema#>.
 4 @prefix xsd: <http://www.w3.org/2001/XMLSchema#>.
 5 @prefix map: <#>.
 6 @prefix Entities: <booster2/Entities#>.
 7 @prefix schema: <http://schema.org/>.
 8 @prefix rdfs: <http://www.w3.org/2000/01/rdf-schema#>.
 9
10 map:Entity
11
12    rr:logicalTable [ rr:tableName "`Entities`.`Entity_all`"; ];
13
14    rr:subjectMap [rr:template "booster2/Entities/{`EntityId`}";
15                   rr:class Entities:Entity;
16                   rr:class schema:Thing;
17        ];
18
19    rr:predicateObjectMap [
20           rr:predicate Entities:name;
21           rr:predicate schema:name;
22           rr:predicate schema:legalName;
23           rr:objectMap [ rr:column "`name`"; ];
24        ];
25
26    rr:predicateObjectMap [
```

Figure 5.18 Semantic Booster – generated R2RML file.

the "Generate Triple Map" target. A sample generated R2RML file is shown in Figure 5.18. In Semantic Booster, the types of each element can also be specified via annotations in the Booster specification. An annotation syntax has been added to the standard Booster notation, as shown in Figure 5.19, lines 6, 10–11 and 14–15. Annotations on each class, attribute and relationship of a Booster specification can be specified by the engineer. The generated R2RML mappings, and in turn the data, will hold the type information from the annotation for those data elements. As with semantic data, URIs are used for types and a prefix mechanism is provided so that URIs may be shortened in a specification, shown in Figure 5.19 lines 3–4. This annotation mechanism is itself extensible, so that enhancements to the R2RML specification may be further customised to enhance the relational-triple mapping

Integration with the Model Catalogue
In order that the Booster tool can be applied in the context of large-scale data engineering, it is important that data managed by a Booster system can be adequately understood, re-purposed, and combined with other data sources.

```
● Entities.boo2 ⊠   ● IPG.boo2      ● Booster2.sdf3
 1 system Entities
 2
 3 @prefix schema: http://schema.org/
 4 @prefix rdfs: http://www.w3.org/2000/01/rdf-schema#
 5
 6 @mapto schema:Thing
 7 class Entity {
 8
 9     attributes
10         @mapto schema:name
11         @mapto schema:legalName
12         name (ID) : String
13     |
14         @mapto rdfs:comment
15         @mapto schema:note
16         notes : String
17     methods
18         create{
19             entity! : Entity' &
20             entity!.name' = name? &
21             entity!.notes' = notes?
22         }
23 }
```

Figure 5.19 Booster specification with semantic annotations.

To facilitate this, Booster is integrated with the Model Catalogue. A typical software development process using the combined toolset might start with the import of a data standard, or the metadata for an existing dataset, into the Model Catalogue. These descriptions can be reused and extended in the definition of a new data model, and exported as a new Booster specification. Figure 5.20 shows Booster with Model Catalogue semantic annotations. The Booster specification contains hooks back to the original definitions, such that the generated system can store links back to the definitions, and provide them to users at the data-entry interface. This can improve the quality of data entry, and ensure consistency across multiple systems using equivalent definitions. Finally, an existing Booster specification may be re-imported into the catalogue and annotated for further reuse.

The round-tripping provided by this new functionality allows the Model Catalogue to be used as an enhanced IDE for model-driven development. The

150   Tools

Figure 5.20 Booster user interface showing semantic annotations from the Model Catalogue.

extended reuse of data components promotes greater reuse of data, and can lead to improvements to the quality of data, and the adherence to standard definitions.

Eclipse Booster IDE

During the second phase of the ALIGNED project, the Booster tool has been upgraded to use the latest versions of the Spoofax language engineering workbench – from version 1.x to the newly released v2.0. Spoofax provides rich editing support for Booster specifications inside Eclipse and includes syntax highlighting, specification outline views and type checking. The upgrade brings amongst other changes: improved compilation and transformation times, simpler project layouts and support for running transformations without Eclipse, via the Sunshine tool. Running Booster transformations outside of Eclipse is a key component of the Semantic Booster Development and Deployment Kit (sBDK) discussed in Section 4. The enhanced underlying environment has allowed the development of more powerful editors and syntax checking, and allows the Booster tool itself to be more easily updated to meet new requirements or support further functionality.

Semantic Booster development and deployment kit: sBDK

In order to support developers and end users of Semantic Booster systems, a number of features have been created and combined to form a Semantic Booster Development and Deployment Kit (sBDK).

Semantic Booster Docker Container

Creating a useable system using Semantic Booster and Eclipse can be operationally challenging, requiring many non-trivial steps to setup a workable system. The complexity has only increased since the additional developments for the ALIGNED project partners. Several interdependent software packages must be installed and configured correctly for a Semantic Booster system to work as intended.

To simplify and streamline the creation and maintenance of data-intensive systems with Semantic Booster, a Docker container has been developed. Docker[4] is a framework for building applications as containers, making them portable and easily deployed without complicated configuration or system set-up. Docker is used to automate deployment of all software needed for a Semantic Booster system whilst isolating the software from the host operating system. The container takes a single parameter: a Booster specification that the user has previously created: either using the tool support provided in the Semantic Booster Eclipse editor, or exported as an artefact from the Model Catalogue. The Docker container will execute a number of scripts to configure and create tools to access and manage the system. A Web-based Booster editor is provided, along with the Booster Native Data exploration tool and a number of semantic data exploration tools.

Additional parameters can be provided to the Docker initialisation to perform additional system configuration, including the pre-population of the generated Semantic Booster system and the ability to persist, backup and restore the database between different runtimes of the system: features vital to the smooth operation of the system within the ALIGNED use case. Parameters have been added to the container to regenerate only the methods of a Semantic Booster system. The Semantic Booster Docker container is available on GitHub and is in use by the Wolters Kluwer IPG use case.

Semantic Booster web editor

A new feature for Semantic Booster is the Web-based editor for Semantic Booster specifications. The editor is less sophisticated than the existing Eclipse-based tooling and therefore is expected to only be used for iteration to an existing specification or system. The Semantic Booster Web editor is shown in Figure 5.21. The editor also provides automatic generation and redeployment functionality, such that a system previously created by the Docker

---

[4]http://docker.com

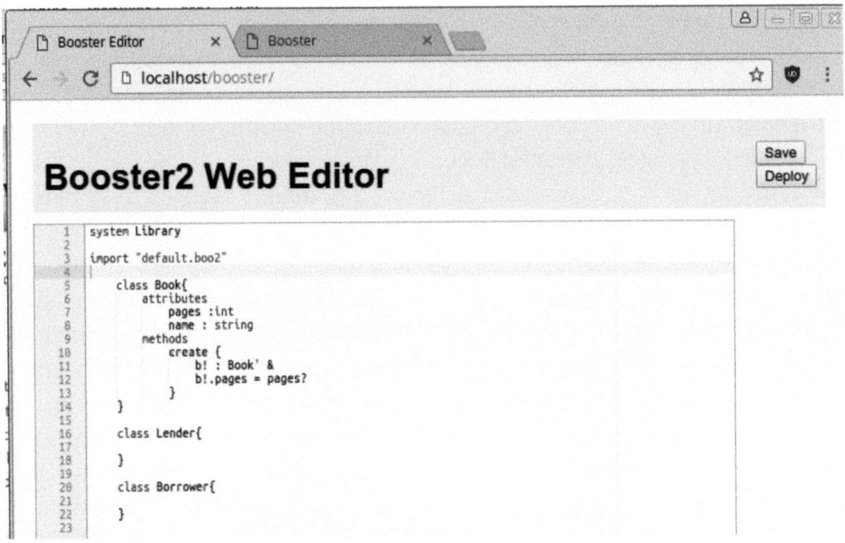

152 *Tools*

```
Booster2 Web Editor                               Save
                                                  Deploy
 1  system Library
 2
 3  import "default.boo2"
 4
 5     class Book{
 6         attributes
 7             pages :int
 8             name : string
 9         methods
10             create {
11                 b! : Book' &
12                 b!.pages = pages?
13             }
14     }
15
16     class Lender{
17
18     }
19
20     class Borrower{
21
22     }
23
```

Figure 5.21 Semantic Booster Web-based editor.

container can be recreated. This uses the Sunshine mechanism of the Spoofax language, to execute Booster transformations outside the Eclipse IDE.

Default Booster web explorer
As well as the Web editor, the Docker container also deploys the default Booster Web Data Explorer (Figure 5.22) from where the data in a Booster system can be explorer and operations invoked. The data explorer can be configured to link to a model catalogue for providing metadata annotations, and is configured to expose information available through the semantic mappings.

Semantic data exploration: D2RQ, Snorql
The D2RQ semantic Web data exploration tool and the Snorql SPARQL query tool (illustrated in Figure 5.23) are pre-configured in the Docker container. These tools use the R2RML mapping created by Semantic Booster to present the data in a Booster system as triples. This allows for interoperability between the software engineering and data engineering worlds. D2RQ provides a data browser – an RDF equivalent to the default Booster data explorer. Snorql provides a standard SPARQL endpoint to the RDF triples: as a W3C standard for querying semantic Web data, this allows a variety of

5.1 *Model Catalogue* 153

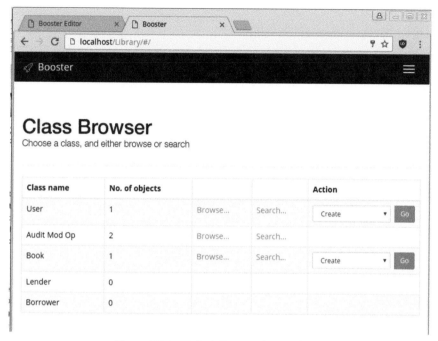

Figure 5.22 Default Booster data explorer.

standard semantic Web tools to interact with the data stored within a Booster system.

The mapping generated is specific to the Booster specification and considers the semantic mapping annotations defined for classes and attributes. In addition, some simplifications are made to better represent Booster inheritance and association.

Semantic data visualisation tool: D3Sparql

A second semantic Web-based data exploration tool has been included in the Semantic Booster Docker container: D3Sparql. Semantic Web data, as triples forms a conceptual graph of data. D3Sparql visualises the results of a SPARQL query as an interactive graph, as shown in Figure 5.8. End users can make use of the tool to interact with the RDF data: domain experts may make use of pre-defined SPARQL queries to provide dashboard-style views or high-level summary metadata; technical users may make use of the built-in Snorql tool to develop complex queries for searching or investigation. Figure 5.24 shows D3Sparql in action.

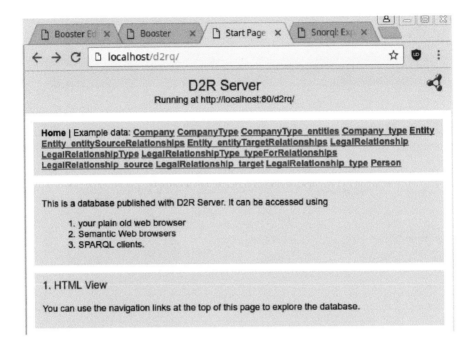

Figure 5.23 D2RQ and SNORQL for exploration of Semantic Booster data.

5.2 RDFUnit

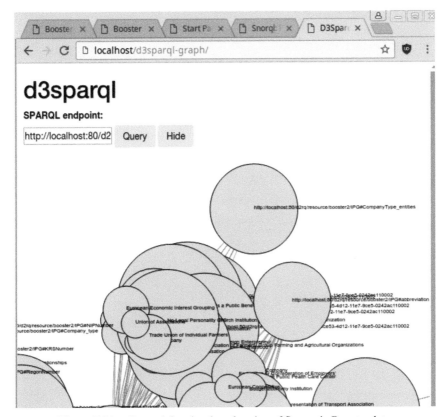

Figure 5.24 d3sparql for visual exploration of Semantic Booster data.

Semantic data validation tool: RDFUnit

The RDFUnit tool has been integrated into the Semantic Booster Docker Container, as shown in Figure 5.25. RDFUnit is used for the validation of data and allows end users to validate RDF triple data against a suite of specifications of data quality constraints, written in SPARQL. The tool checks that the constraints hold post-hoc, in contrast to the correct by construction approach of Semantic Booster. RDFUnit is configured to use the D2RQ endpoint, allowing constraints to be written against up-to-date Booster data.

5.2 RDFUnit

RDFUnit is an RDF validation framework inspired by test-driven software development. The test case definition language of RDFUnit is SPARQL,

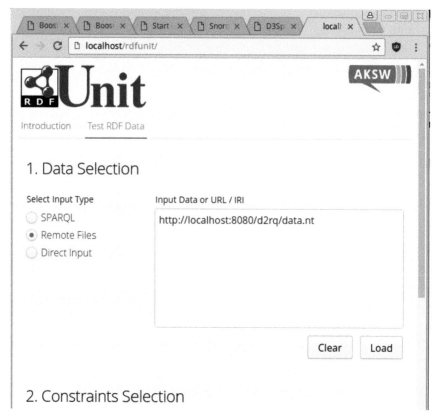

Figure 5.25 RDFUnit Web interface.

which is convenient to directly query for identifying violations. For rapid test case instantiation, a pattern-based SPARQL-Template engine is supported where the user can easily bind variables into patterns. RDFUnit has a Test Auto Generator (TAG) component. TAG searches for schema information and automatically instantiates new test cases. Schema information can be in the form of RDFS or OWL axioms that RDFUnit translates into SPARQL under Closed World Assumption (CWA) and Unique Name Assumption (UNA). Other schema languages such as SHACL[5], IBM Resource Shapes[6] or Description Set Profiles[7] are also supported. For a full overview of

---

[5]https://www.w3.org/TR/shacl/
[6]https://www.w3.org/Submission/shapes/
[7]http://dublincore.org/documents/dc-dsp/

5.2 RDFUnit

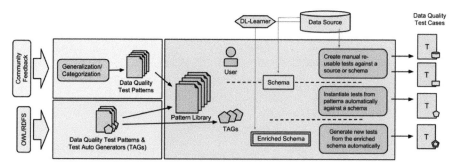

Figure 5.26 RDFUnit architecture.

RDFUnit's data testing and verification capabilities see Kontokostas et al. Figure 5.26 shows the RDFUnit architecture.[8]

5.2.1 RDFUnit Integration

The following subsections describe three ways RDFUnit-based data testing and verification can be integrated into software engineering workflows. The JUnit runner with annotations provides a very easy and well integrated option but does not give room for flexibility beyond testing an input dataset to a fixed schema. The JUnit XML Report gives room for greater flexibility by utilising the complete RDFUnit command line options. Finally, the custom Maven-based integration gives software engineers a way to fine-tune the way they want to automate their data testing and verification options.

JUnit Runner integration with Java annotations

JUnit allows other testing frameworks to extend JUnit with custom Runners[9] tailored for specific testing. A custom JUnit Runner was implemented, RdfUnitJunitRunner[10], which can be used to define JUnit tests for validating RDF datasets against a schema, by adding Java annotations to a JUnit test.

---

[8]D. Kontokostas, P. Westphal, S. Auer, S. Hellmann, J. Lehmann, R. Cornelissen, and A. Zaveri. Test-driven Evaluation of Linked Data Quality, Proc. 23rd International Conference on World Wide Web, pp. 747–758, DOI 10.1145/2566486.2568002, 2014.
[9]https://github.com/junit-team/junit4/wiki/test-runners
[10]https://github.com/AKSW/RDFUnit/tree/master/rdfunit-junit

An example RDFUnit/JUnit test is the following:

```
@RunWith(RdfUnitJunitRunner.class)
@Schema(uri = "schema.ttl")
public static class TestRunner {
  @TestInput
  public RDFReader getInputData() {
                       return new RdfModelReader(
                       RdfReaderFactory.createResourceReader(
                             "/inputmodels/data.ttl" ).read()); }
}
```

Where data.ttl is an RDF data file (using the @TestInput annotation) tested by a JUnit test against schema.ttl (using the @Schema annotation).

For every automatically generated RDFUnit test, a separate JUnit test is generated that validates the input dataset for a specific violation. The reporting of validation errors is integrated with JUnit reports, thus providing the means to display them through IDEs like IntelliJ or with build tools like Maven.

5.2.1.1 JUnit XML report-based integration

JUnit uses a specific XML schema[11] to communicate the test results to IDEs or build tools. For cases when defining an RDFUnit/JUnit test is not an option (i.e., the files are not accessible from the build system with Java code), the RDFUnit results can be converted to the JUnit XML Schema. In these cases, developers can run RDFUnit as a command line tool or through custom code, expecting validation results in the JUnit XML Schema. Build systems, such as Bamboo, can then be configured to look at specific locations for such XML files and report the RDFUnit validation results with the existing unit test error reporting tools.[12] Figure 5.27 shows the RDFUnit report in the IntelliJ IDE. Figure 5.28 shows the report in the Bamboo build system.

5.2.1.2 Custom apache maven-based integration

When the input data or schema graph are not simple input files, but generated through custom procedures, the aforementioned methods are not easy to apply. For those cases, RDFUnit can be used as a Java library, fine-tuned for custom input or more sophisticated Jenkin reports. This was the case with the JURION demo, where RDFUnit was used to validate if the output of specific XSLT scripts adhered to the JURION Schema. All results were archived to

---

[11] https://svn.jenkins-ci.org/trunk/hudson/dtkit/dtkit-format/dtkit-junit-model/src/main/resources/com/thalesgroup/dtkit/junit/model/xsd/junit-4.xsd

[12] https://github.com/AKSW/RDFUnit/wiki/Using-RDFunit-with-Bamboo

5.2 RDFUnit 159

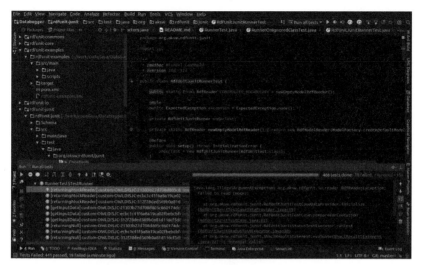

Figure 5.27 RDFUnit report from the IntelliJ IDE.

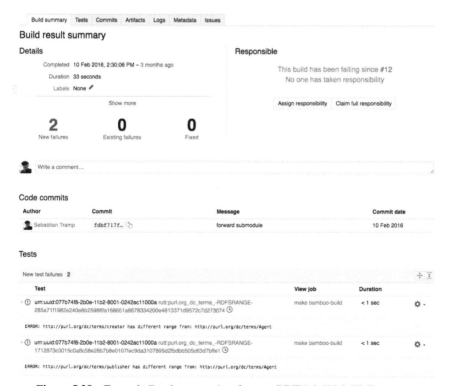

Figure 5.28 Example Bamboo overview from an RDFUnit JUnit XML report.

160 Tools

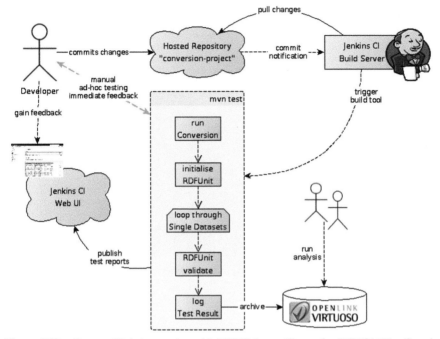

Figure 5.29 Custom JUnit integration with RDFUnit as a library for JURION Use Case in ALIGNED.

a triple store for post-processing analysis (see Image 3). A research paper was published that describes this use case in detail.[13] Figure 5.29 shows the integration of RDFUnit in JURION.

5.2.1.3 The shapes constraint language (SHACL)

The Shapes Constraint Language is a language to validate RDF graphs against a set of constraints. These constraints are formalised as shapes and other constructs expressed in the form of an RDF graph. The language features and approaches occurring in the current specification of SHACL were in part inspired by the SPIN[14] and Shape Expressions (ShEx).[15] The current

---

[13] Dimitris Kontokostas, Christian Mader, Michael Leuthold, Christian Dirschl, Katja Eck, Jens Lehmann and Sebastian Hellmann. Semantically Enhanced Quality Assurance in the JURION Business Use Case. ESWC 2016, Crete. Available at: http://link.springer.com/chapter/10.1007/978-3-319-34129-3_40

[14] http://spinrdf.org/

[15] E. Prud'hommeaux, J. E. Labra Gayo and H. Solbrig. Shape expressions: an RDF validation and transformation language, 10th International Conference on Semantic Systems, pp. 32–40, 2014.

revision of the specification for SHACL is published by the W3C[16] with complementary material available in a GitHub repository.[17]

SHACL Core defines frequently needed features to formulate common constraints for RDF graphs. SHACL Core Constraints are defined by parameterising Constraint Components that are templates for checks for a specific required property of an RDF nodes (e.g., unique occurrence of a property value associated with a specific property, for instance only one foaf:age value for a given foaf:Person). One or several of such constraints are associated with target RDF nodes to validate against in a SHACL Shape. SHACL Shapes are expressed as RDF resources and aggregated in a Shapes Graph. An RDF graph to be checked for conformance against a Shapes Graph (the Data Graph) is provided to a Validation Engine that produces a Validation Report. The Validation Report states whether the Data Graph conforms to the Shapes Graph, listing violations of individual RDF nodes against shapes detected during the validation process in case of non-conformance.

5.2.1.4 Comparison of SHACL to schema definition using RDFUnit test patterns

The original declarative approach to create data tests with RDFUnit (i.e., formulating data constraints without composing SPARQL queries or SPARQL query fragments directly) involves selecting and parameterising an RDFUnit Test Pattern. These test patterns bear several conceptual and functional similarities to both SHACL Shapes and SHACL Constraint Components. Test patterns also define parameters to be set to transform a test pattern into a concrete test case. In contrast to SHACL, the parameters of an RDFUnit test pattern do not only specify the expected constraints for applicable RDF nodes but also often influence the sets of RDF nodes the test is applied to. The clearer separation of these concerns in SHACL increases modularity and thus allows more flexible reuse of parts of shape definitions. As an additional advantage, SHACL Shapes can be defined recursively, i.e., more complex shapes can be composed by combining simpler shapes. For example, a shape S1 can define that all its values nodes for a property P must conform to a shape S2. Also, multiple individual restrictions can be combined to a conjunction (target nodes must conform to S1 and S2), different acceptable shape alternatives can be expressed by disjunction (target nodes must conform to S1 or S2) and shape constraints can be inverted/negated (target nodes must not

---

[16] https://www.w3.org/TR/shacl/
[17] https://github.com/w3c/data-shapes

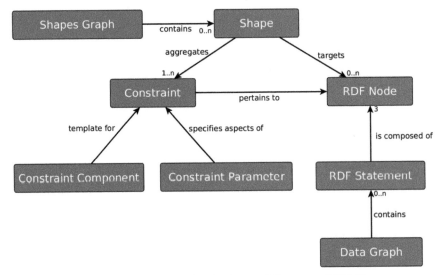

Figure 5.30 Overview for Fundamental Concepts of SHACL.

conform to S1). SHACL provides for a range of ways to define the focus nodes for a shape, i.e., class membership, explicit nodes, subjects and objects of a predicate. In the current set of RDFUnit's test patterns, targeted notes are predominantly only scoped by class membership or property appearance. In general, however, scoping in RDFUnit is more flexible as it is defined directly in SPARQL and there are no limitations. For example, there is no way to define in SHACL the constraint that all entities must have a label. Figure 5.30 shows an overview of the fundamental concepts of SHACL.

5.2.1.5 Comparison of SHACL to auto-generated RDFUnit tests from RDFS/OWL axioms

In addition to a direct instantiation of test patterns in manual test suites, RDFUnit already offers capabilities to create data test suites automatically from RDFS and OWL axioms pertaining to the vocabulary used in the RDF data to be tested (Test Auto Generators). This approach enables utilisation of modelling efforts of RDF vocabulary providers that specified aspects of intended semantics of their vocabularies in RDFS or OWL. As both of these modelling languages, when interpreted in line with the corresponding W3C specifications, are more tailored towards inference as opposed to constraint formulation, basic principles of the semantics and assumptions of the language have to be modified for data quality testing scenarios (especially the application of the Closed World Assumption and a weakened Unique Name

Assumption). However, no unanimous and detailed specification for such alternative semantics has been formulated and standardised to date.

In contrast, the semantics of each language element for SHACL Core is clearly defined in the corresponding W3C Proposed Recommendation and was designed specifically for the purpose of prescriptive constraint formulation. Furthermore, SHACL semantics are solely based on the notion of RDF Graphs, a conceptually much simpler and more approachable model for new adopters of Semantic Web technologies compared to OWL based on Description Logics.

5.2.1.6 Progress on the SHACL specification and standardisation process

ALIGNED was actively involved in co-authorship and revision of the W3C Working Group for SHACL. Since then, several stages of the W3C Recommendation Track Process[18] were passed, by opening the draft for reviewers' comments, discussing or addressing these, and putting forward an implementation report[19] about several prototypical implementations of SHACL. Hence, SHACL is now documented as a W3C Proposed Recommendation.

5.2.1.7 SHACL support in RDFUnit

RDFUnit currently contains implementations for all non-complex core constraint components (i.e., excluding logical constraint components and qualified cardinality restrictions). All variants of target declarations are implemented as well. This provides a substantial, albeit incomplete, subset of SHACL Core that already allows formulating graph constraints for many use cases. RDFUnit also supports SHACL SPARQL, i.e., defining custom constraints and constraint components by SPARQL fragments. This approach provides the whole flexibility and expressive power of that query language.

To evaluate both correctness of the validation logic of the implemented SHACL subset, a runner for the SHACL test suite[20] has been implemented. The tests can be run via a custom JUnit Runner within an IDE for quick feedback cycles about improvements or regressions on conformance during the continued work for a complete coverage of SHACL feature in RDFUnit. Additionally, an RDF document reporting on the test outcomes for the SHACL test suite using the EARL[21] vocabulary can be generated, in a format

---

[18] https://www.w3.org/2004/02/Process-20040205/tr.html
[19] http://w3c.github.io/data-shapes/data-shapes-test-suite/
[20] https://github.com/w3c/data-shapes/tree/gh-pages/data-shapes-test-suite/tests
[21] https://www.w3.org/TR/EARL10-Schema/

```
@prefix doap: <http://usefulinc.com/ns/doap#> .
@prefix earl: <http://www.w3.org/ns/earl#> .

<http://aksw.org/Projects/RDFUnit>

  a doap:Project ;
  a earl:Software ;
  a earl:TestSubject ;
  doap:developer <aksw.org/DimitrisKontokostas> ;
  doap:name "RDFUnit" ;
.
[
  a earl:Assertion ;
  earl:assertedBy <aksw.org/MarkusAckermann> ;
  earl:result [
    a earl:TestResult ;
    earl:mode earl:automatic ;
    earl:outcome earl:passed ;
  ] ;
  earl:subject <http://aksw.org/Projects/RDFUnit> ;
  earl:test <urn:x-shacl-test:/core/complex/personexample> ;
].
```

Figure 5.31 Excerpt of an EARL test report for the SHACL test suite.

suitable to automatically generate a compliance overview for the implementation report for SHACL. Figure 5.31 shows a sample from an EARL test report.

5.3 Expert Curation Tools and Workflows

Linked Data and semantic technologies enable the creation of rich, integrated knowledge models which describe particular domains using standardised languages such as RDF, RDFS and OWL. These technologies have obvious attractions for dataset curators as not only do they provide a range of ways in which complex relationships between entities can be specified and embedded in the data, but they also render the data amenable to sophisticated analytic techniques such as inference and other automated reasoning approaches and allow the construction of sophisticated queries which automatically combine data from different sources into a unified knowledge model. However, datasets are commonly curated by domain experts who have an intimate knowledge of the real-world domain being modelled but rarely have expertise or training in knowledge engineering or semantic technologies. Furthermore, all indications suggest that semantic modelling is a skill that is more difficult to acquire than computer programming, requiring considerable investment of time and effort. If we want to make this technology accessible to domain experts we need to develop paradigms, processes, workflows,

APIs, and software tools which bridge the gap between their knowledge of the domain and the complexities of the underlying semantic models that they are manipulating.

This section describes the process and workflow models, developed within the ALIGNED project that are designed to bridge the gap between dataset curators, Linked Data and semantic Web technologies. The goal of this work is to define semi-automated methods and tools that involve human expert curators in the loop, while minimising their workload and the requirement that they understand the underlying semantic technologies. It builds upon and extends ALIGNED's system integrity enforcement framework by generating data curation workflows and dedicated user interfaces where domain experts can efficiently verify and approve data as part of a publication pipeline which incorporates both automated and human-based quality controls. It uses and extends ALIGNED's meta-modelling work, utilising the metamodel's schema validation ontology while providing fine-grained workflow models for core curation activities (e.g., adding instance data, updating schema, etc).

5.3.1 Requirements

This section defines the workflow requirements that a linked-data curation system must support in order to support domain experts in curating high-quality datasets. Common features identified in ALIGNED's use cases have been translated into the low-level system requirements necessary to provide a data curation system which can support user-level requirements.

5.3.1.1 Graduated application of semantics

Much of the Linked Data available for harvesting is loosely structured, often schema-free and based on reuse of terms from common vocabularies. However, in order to provide dataset quality enforcement, it is necessary to produce a rich, highly structured, and precisely defined schema and ensure that all instance data comply with the schema. Therefore, a Linked Data curation platform should provide support for the management of loosely structured linked-data documents and their gradual transformation into highly structured, schema-conformant high-quality knowledge graphs.

5.3.1.2 Graph – object mapping

Graphs are the knowledge representation form that underlies all Linked Data and semantic technologies. However, when it comes to human management of data, object models are ubiquitous – for example: entity relationship models, UML data models, database records, structured JSON documents. This amounts to a fundamental paradigm difference: in the object model a dataset is conceived as a collection of objects/entities, each of which has a collection of properties, which may be complex and may include links to other objects in the dataset; in the graph model, the dataset is conceived of as a collection of nodes with labelled, directed edges connecting them. In order to support dataset management by non-knowledge engineers, it is necessary to provide a curation interface which allows them to treat the dataset as if it was a collection of objects and takes care of mapping these objects to the underlying graph representation. This object interface should support, at a minimum the following functions:

5.3.1.3 Object/document level state management and versioning

In order to provide a functional object-based data curation interface, the system should provide basic state management and versioning support on a per-object level. That is to say that the system should provide the dataset curator with the ability to change the state of a data-object (e.g., by deleting it, or publishing it) and have the system accurately map this to a modification of the triples making up the object's representation in the underlying knowledge graph. It should also be possible to view and link to previous versions of particular data-objects.

5.3.1.4 Object-based workflow interfaces

In order to allow curators to manage updates to the graphs that they manage, object-based user interfaces must be provided which should display graph-updates as updates to specific data objects. Similarly control interfaces must be provided which allow curators to change the state of specific data objects and automatically translate that into graph updates.

5.3.1.5 Integrated, automated, constraint validation

A core focus of ALIGNED's research has been the development of constraint validation and error detection services and tools to support automated data quality analysis and enforcement. To make these services accessible to domain expert curators, they must be integrated into workflows which correctly trigger the appropriate validation processes in response to curator-driven actions which cause updates to the underlying graph. These processes should be, to as great an extent as possible, invisible to the curator.

5.3 Expert Curation Tools and Workflows

5.3.1.6 Result interpretation

The major exception to the above is that, in certain cases, the results of constraint validation will indicate a situation which requires user-intervention (e.g., an error in the dataset schema) and must be reported back to the user. In such cases the system should, to as great an extent as is possible, map the error from the underlying graph model into the object model used by the curation platform.

5.3.1.7 Deferred updates

From a workflow point of view, automated and human tasks have very different characteristics: fast, synchronous, reliable (from an execution point of view) and typically semantically simple, versus slow, asynchronous, unreliable and often semantically complex. The most basic feature that is necessary to support these characteristics of human processes is deferred updates. That is to say that the curation system should allow updates to curated dataset to be deferred – stored and executed at a temporal distance which may be considerable. This is necessary to support the most basic content approval pipeline – where updates to the dataset must be approved by the curator before they are actually carried out. Deferred updates complicate curation processes considerably, because they can be invalidated by updates that happen between their definition and their acceptance. However, they are necessary in order to provide simple, efficient interfaces for curators, allowing them to, for example, simply click approve, to enact a complex, multi-faceted graph update that has been requested by another process in the system (whether human or automated).

5.3.2 Workflow/Process Models

This section describes the system dynamic models that have been developed in the ALIGNED project in order to produce a data curation system which meets the above requirements and is capable of providing a practically useful curation service for domain expert dataset curators. These models define how human curation actions and activities are integrated with automated processes to provide quality control of the dataset. Figure 5.32 shows the symbols used in the following diagrams.

5.3.2.1 Process model 1 – linked data object creation

JSON objects submitted to the API are stored as objects in a Linked Data document store and require approval by human curators and validation by the

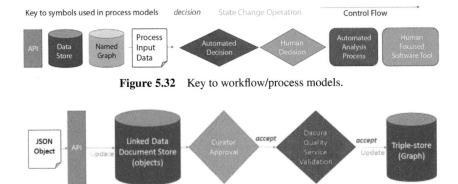

Figure 5.32 Key to workflow/process models.

Figure 5.33 Process Model 1 – Object Creation.

automated Dacura Quality Service before being published to the triple-store. This process model is shown in Figure 5.33.

5.3.2.2 Process model 2 object – linked data object updates

Process model 2 shows how updates to Linked Data objects which are accepted by the curator but fail validation are either saved to the deferred update queue (if the object is published) or executed on the Linked Data object store (if not published). If the update passes validation, it is saved to both the Linked Data object store and the triple-store. This allows objects to be iteratively updated without having to pass DQS validation (DQS results are returned in an informational capacity).

Updates to JSON documents are subject to curator approval. Those updates which receive a 'pending' status from the approval process are saved to the deferred update queue. Those updates which receive an 'accept' status are processed by the automated DQS validation service. If the update receives a 'reject' status from this process, and the document being updated is in a 'published' state, then the update is saved to the deferred update status. If the update receives an 'accept' status, or the document being updated is not in a published state, the update is executed on the document in the Linked Data store. If the update receives an 'accept' status and the document is in a published state, the update is executed on the triple-store version of the document. This process model is shown in Figure 5.34.

5.3.2.3 Process model 3 – updates to deferred updates

Deferred updates which are approved by the human curator (state changed from 'pending' to 'accept') are first analysed by an automated consistency

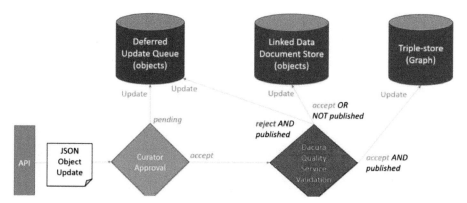

Figure 5.34 Process Model 2 – Object Update.

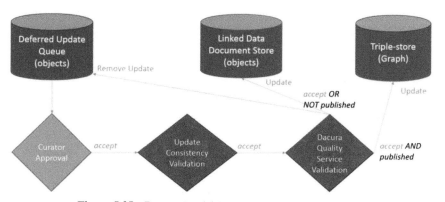

Figure 5.35 Process Model 3 – Updates to deferred update.

checking process, which ensures that no intervening updates have rendered the deferred update invalid. If the update is validated by this process, it is validated by the DQS process. If it correctly validates or if the updated object is not in 'published' state, the update is removed from the queue and the object is updated in the document store. If the update is validated and the updated object is in the "published" state, the graph representation of the object is updated in the triple store. This process model is shown in Figure 5.35.

5.3.2.4 Process model 4 – schema updates

Schema updates received by the API are analysed by the automated ontology dependency analysis process, which identifies the list of ontologies needed to validate schema updates and the list of ontologies needed to validate

170 Tools

Figure 5.36 Process model 4 – Schema Updates.

instance updates. The validity of the resulting graphs is checked using the automated Dacura Quality Service process and if successful, and the ontologies are published to the respective graphs. This process model is shown in Figure 5.36.

5.3.2.5 Process model 5 – validating schema updates

Schema updates are validated by the DQS process in two stages. First, the updates to the schema are validated by the DQS, with the schema graph serving as the instance graph and the schema schema graph serving as the schema. If this update is validated, the update to the schema is validated against the instance graph, with the schema graph serving as the schema. If either validation process fails (status: reject), the updates to the schema schema graph are rolled back. If both successfully validate, the schema graph is updated. This process model is shown in Figure 5.37.

5.3.2.6 Process model 6 – named graph creation

Instance data objects can be configured to map to a graph representation that spans multiple named graphs. Each named graph needs a schema against which the instance data will be validated. Temporary graphs are first constructed to validate the submitted ontology with the DQS, if it validates successfully, three named graphs are created – and the relevant ontologies are published to the schema schema graph and the schema graph. Updates to the graph's schema follow Process Model 4 – schema updates. This process model is shown in Figure 5.38.

5.3 Expert Curation Tools and Workflows 171

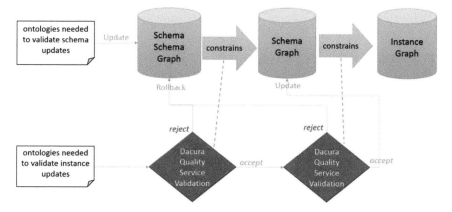

Figure 5.37 Process model 5 – Validating schema updates.

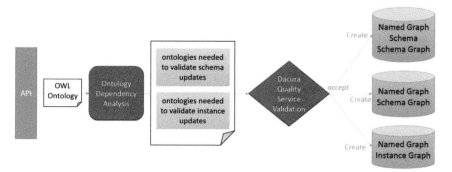

Figure 5.38 Process Model 6 – Named Graph Creation.

5.3.2.7 Process model 7 – instance data updates and named graphs

This process extends and specialises Process Model 1 and Process Model 2, by providing fine-grained detail of the "update" triple-store operation. Updates to instance data objects may map to a graph representation that is distributed across multiple named graphs. Instance data updates to each named graph are validated sequentially by the automated DQS service (using the schema graph that has been configured for that named graph – see Process Model 6 – Named Graph Creation). If the update is validated across all graphs, and the updated object is in published state, the update is published

172 Tools

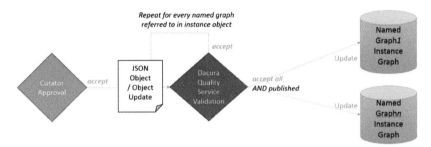

Figure 5.39 Process model 7 – instance data updates in named graphs.

to all relevant named graph instance graphs. This process model is shown in Figure 5.39.

5.4 Dacura Approval Queue Manager

The Dacura Approval Queue Manager is a Web-based GUI tool which allows dataset curators to interact with the object creation, object updating and deferred updating processes. It allows curators to view the approval queue of new objects and updates to objects and to approve or reject object creation requests and object update requests in bulk. The Dacura Approval Queue Manager can be seen in Figure 5.40.

5.5 Dacura Linked Data Object Viewer

The Dacura Linked Data Object Viewer is a Web-based GUI tool which allows dataset curators to view the Linked Data objects that they are

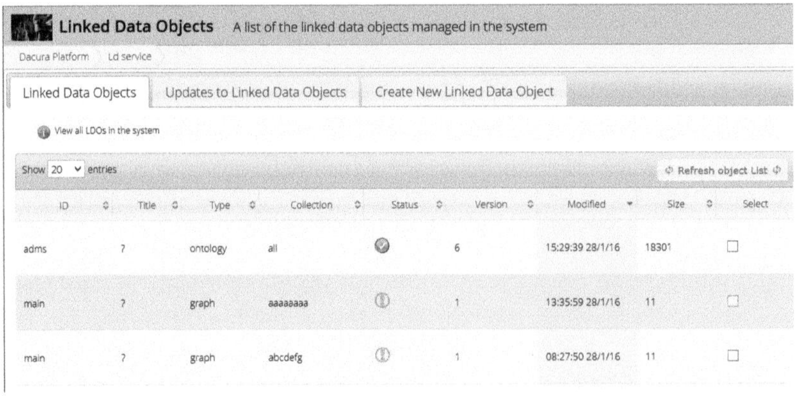

Figure 5.40 Screenshot of Dacura Linked Data Approval Queue Manager Tool.

5.5 Dacura Linked Data Object Viewer

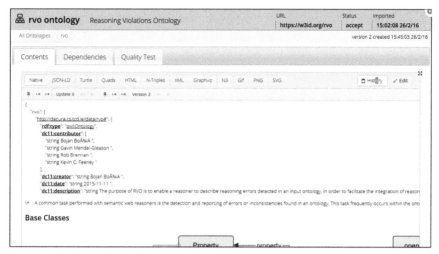

Figure 5.41 Screenshot of Dacura Linked Data Object Viewer Tool showing version browsing toolbar.

managing, browse their history, and manage their metadata and contents on an individual object basis, while maintaining a correct mapping to the object's underlying graph representation. The Dacura Linked Data Object Viewer can be seen in Figure 5.41.

5.5.1 CSP Design of Seshat Workflow Use Case

We formally specified our workflow in CSPM, a dialect of CSP (Communicating Sequential Processes) with the assistance of FDR4 The CSP Refinement Checker[22]. CSPM gives a very rich language for the specification of processes and communication, but we limit ourselves in the model to a very restricted subset with a view to later creating simple user-interfaces for the specification of alternative workflow approaches. In addition, we hope to use the specification to explore properties of the workflow model and potentially create refinements with versioning in a later iteration.

In natural language, the document curation use case can be described thus:

A user may load a candidate object into the system. It is then in a 'pending' state. From the pending state, a candidate may be checked by the DQS (Dacura Quality Service) server. The DQS server will either pass or fail the candidate. If the candidate passes DQS's inspection, it is sent to an 'ok' state. From the 'ok' state, it is possible to review or edit. If the user chooses to

---
[22] https://www.cs.ox.ac.uk/projects/fdr/

review, they may either accept as is and it is placed into an 'accepted' state. From an 'accepted' state, the candidate may be published. If the candidate reviewer likes, they may edit the document, sending it back to a 'pending' state. If the DQS system fails to pass a candidate, it is sent to a 'fail' state from which the user must edit the candidate before it can go back to 'pending'. Additionally, from an 'ok' state which is edited, the candidate is passed back to a 'pending' state.

5.5.2 Specification

We show in Table 5.1 the specification in CSPm of the above natural language description. We show the model with only one document for presentation purposes as little changes by increasing the number of available DOCIDS.

Table 5.1 CSPm specification of workflow

```
DOCIDS = {0 .. 2}
channel load, check, fail, edit,
    pass, store, accept, decline, review, publish, unpublish
: DOCIDS
WFS(i) = load.i -> PENDING(i)

PENDING(i) = check.i -> DQS(i)

DQS(i) = fail.i -> FAILED(i)
   |~|pass.i -> OK(i)

FAILED(i) = edit.i -> PENDING(i)

OK(i) = edit.i -> PENDING(i)
   [] review.i -> REVIEW(i)

REVIEW(i) = accept.i -> ACCEPTED(i)
   [] edit.i -> PENDING(i)

ACCEPTED(i) = publish.i -> PUBLISH(i)
   [] edit.i -> PENDING(i)

PUBLISH(i) = unpublish.i -> ACCEPTED(i)

CHOOSEDOC = |~|i : DOCIDS @ WFS(i)
assert CHOOSEDOC :[deadlock free [F]]
```

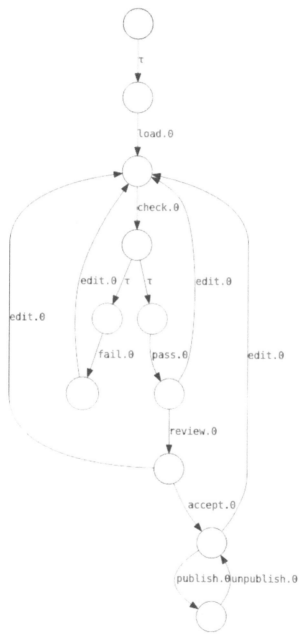

Figure 5.42 Automatically generated workflow diagram from CSPm specification.

176 Tools

The specification can be read as having a number of states which are parameterised by the document id which they refer to and having one or a number of actions which can be taken from those states. In the case of DQS(i), there are two possible actions which are chosen by an internal choice. In all other instances, the choice is an external (user) choice.

The example of Figure 5.42 shows the automatically generated workflow diagram from the CSPm specification. It is clear from inspection that the document is always available for transitions to some new state and we can see clearly how the workflow takes place.

5.6 Dacura Quality Service

The Dacura Quality Service (DQS) is a service for managing a triple-store and ensuring its ongoing consistency. The triple-store is an RDF graph which is stored using the ClioPatria server. ClioPatria provides a durable representation of the graph which can be accessed and updated transactionally. These features constitute the 'A', 'I' and 'D', of ACID (Atomic, Consistent, Isolated, Durable) which are generally considered fundamental design principles for enterprise databases.

DQS extends this feature set with the 'C', Consistency. Consistency of the graph is described using OWL. This ontology is interpreted as constraints over the graph. Failure to meet the constraints specified in OWL leads to a counter-example of satisfaction of the constraints, or a witness of failure. These witnesses are then reported over the API to the client which provides the client with manual or automatic remedial actions.

The DQS software is provided as a plugin to ClioPatria and interaction with DQS takes place over an HTTP (HyperText Transfer Protocol) API (Application Programming Interface). The API exchanges information about triples and witnesses of failure in the widely used JSON object format.

The DQS service is used by Dacura to ensure that data quality of curated data is consistent on an ongoing basis. Since the data must be amenable to constant update by data practitioners, and the data must be available for analytics in a consistent and coherent format, it is imperative that basic data consistency constraints be maintained. DQS provides a straightforward framework for assisting Dacura in maintaining these constraints.

5.6 Dacura Quality Service

Dacura is the main consumer of the Dacura Quality Service. However, it is completely modular and therefore could be used in other projects which would like to manage consistency constraints using OWL ontologies.

5.6.1 Technical Overview of Dacura Quality Service

The DQS Service is implemented as a plugin for ClioPatria, which is written in the prolog programming language. Prolog provides a seamless interface to the RDF triple-store as a predicate which can then be combined for the purposes of reasoning. A number of reasoning tasks are carried out by a list of predicates, which can be accessed by calling a number of pre-defined HTTP endpoints, carrying appropriate JSON POST variables which describe the relevant graphs, updates to those graphs, and various reasoning activities which should be undertaken.

Currently, the primary consumer of the DQS service is the Dacura platform which provides a user interface to the service, allowing the user to select the relevant ontologies, and instance data to be checked, and the various constraints which should be checked. The interface for schema checking is shown in Figure 5.43.

Figure 5.43 Dacura platform Quality Test Interface that calls the DQS.

5.6.2 Dacura Quality Service API

The API is structured as a series of HTTP endpoints which are accessed through POST requests. The POST requests have a number of variables communicated in JSON and with some standard translations for RDF URIs and literals. We first describe this general format of RDF encoding in quads, and then the specific format of some shared POST variables.

5.6.2.1 Resource and interchange format

Inserts and deletes in the DQS system are managed through supplying quads which specify the RDF triples, and their associated graph. These are encoded in JSON which is a widely accepted format.

5.6.2.2 URI

An RDF URI resource is described as a JSON string. For instance, the following string represents the "label" property:
"http://www.w3.org/1999/02/22-rdf-syntax-ns#label"

5.6.2.3 Literals

Literals are composite objects which cannot be represented directly as a string. The format for a literal is formatted as one of the two:
{"data":"2015-06-08T12:30:00","type": "http://www.w3.org/2001/XMLSchema#dateTime"}
or
{"data": "This is a string", "lang":"en"}

5.6.2.4 Literal types

xdd:coordinatePolygon

The coordinate polygon type is represented in as a list of doubles. An informal grammar is as follows:
xdd:coordinatePolygon := [float1, float2, ... floatn]

xdd:coordinatePolyline

The coordinate polygon type is represented in as a list of doubles identically to a coordinate polygon but with a semantics of a non-closed region. An informal grammar is as follows:
xdd:coordinatePolyline := [float1, float2, ... floatn]

xdd:gYearRange
The xdd:gYearRange is a (possibly degenerate) range of years, with the first year smaller than or equal to the second.
xdd:gYearRange := [gYear] | [gYear1, gYear2]

xdd:integerRange
The xdd:integerRange is a (possibly degenerate) range of integers, with the first integer smaller than or equal to the second.
xdd:integerRange := [integer] | [integer1, integer2]

xdd:decimalRange
The xdd:decimalRange is a (possibly degenerate) range of decimal numbers of arbitrary precision, with the first number smaller than or equal to the second.
xdd:integerRange := [decimal] | [decimal1, decimal2]

5.6.2.5 Quads
Quads are described as lists of strings or JSON representations of resources.

```
[
"resource1",
"resource2",
"resource3",
"graph"
]

[
"resource4",
"resource5",
{
"data": "Hello world",
"lang": "en-utf8"
},
"graph"
]

[
"resource6",
"resource7",
{
"data": "2015-06-08T12:30:00",
"type": http://www.w3.org/2001/XMLSchema\#dateTime
},
"graph"
]
```

5.6.2.6 POST variables

There are a number of post variables whose format is shared amongst the various endpoints. Many endpoints require a "pragma" JSON object to be posted in the post variables, which specifies the instance graph, "instance", the schema graph, "schema" and associated tests. It also takes a "commit" flag, which will store the changes if the tests are successful.

pragma: {"tests":"all","schema":"schemaGraphName","instance":"instanceGraphName", "commit": "true"}

In order to perform updates, we specify all quads (as described above) which are to be deleted, and then inserted. Deletes happen prior to inserts. Modification of either schema, instance or both, is possible merely by specifying the appropriate schema and instance graphs.

update: {"insert": QUADS, "delete": QUADS}

Example:

update: {"insert":[["resource1", "resource2","resource3", "instance"],["resource6", "resource7", {"data":"2015-06-08T12:30:00", "type": "http://www.w3.org/2001/XMLSchema#dateTime"}, "instance"]]}

5.6.2.7 Tests

A number of the API endpoints require that tests be passed to define which constraints are considered when consistency is required of the triple store. The tests are divided into two categories. One for schema constraints, all of which are suffixed with "SC", and one for instance constraints which are suffixed with "IC".

Users can specify a JSON list of constraints for the "test" field of a pragma, or send the string "all" which will run every available test. Specifying tests which are not available has no effect. We give the exhaustive list of tests below.

5.6.2.8 Required schema tests

These tests for class cycles in the subsumption hierarchy for classes and properties respectively. They are required for any further tests to take place as non-cyclicity is assumed in the other predicates.

"classCycleSC", "propertyCycleSC"

5.6.2.9 Schema tests

These three tests check to see if there is a class for a given URI, which does not need to be inferred, or that a given property has a defined range and domain which is not inferred.

"noImmediateClassSC", "noImmediateDomainSC", "noImmediateRangeSC"

These three tests check uniqueness of definitions. In particular, the first is useful to avoid overlapping labels which can lead to confusion in interfaces which utilise the labels for display.

"notUniqueClassLabelSC", "notUniqueClassSC", "notUniquePropertySC"

Does the schema contain blank nodes?

"schemaBlankNodeSC"

Annotations can be used to black out various properties such that they are not reasoned over, but this test will issue an error if this is being done.

"annotationOverloadSC"

A class (property respectively) is used without definition (inferred or otherwise)

"orphanClassSC", "orphanPropertySC"

Check for invalid ranges or domains.

"invalidRangeSC"., "invalidDomainSC"

Check to see if domain and range subsumption leads to inconsistency.

"domainNotSubsumedSC", "rangeNotSubsumedSC"

Check to see if properties are used as both datatype and object properties simultaneously in violation of OWL.

"propertyTypeOverloadSC"

Instance Tests

Check to see if property has no defined domain (range respectively).

"noPropertyDomainIC", "noPropertyRangeIC"

Check to see if blanknodes are being used?

"instanceBlankNodeIC"

Check to see if edges are valid under the given schema rules. Related classes, properties and restrictions as well as a number of assertions are all checked against the edges of the instance graph for conformance.

"invalidEdgeIC"

Check to see if instances have no defined class.

"edgeOrphanInstanceIC"

Check functional (inverse functional) property assertions for correctness relative the instance graph.

"notFunctionalPropertyIC", "notInverseFunctionalPropertyIC"

Check that properties are defined.

"localOrphanPropertyIC"

5.6.2.10 Errors

The DQS API returns errors which are specified in a JSON format and which are described in the Reasoning Violations Ontology (RVO). RVO has been developed in the ALIGNED project and is fully described in Chapter 3 and is available online[23]. Further details have been published at the third Workshop on Linked Data Quality. [24]

5.6.2.11 Endpoints

/dacura/schema
POST variables: pragma, update
Requires: pragma.schema, pragma.tests, pragma.commit

Endpoint for schema updates.

/dacura/instance
POST variables: pragma, update
Requires: pragma.instance, pragma.schema, pragma.tests, pragma.commit

Endpoint for simultaneous schema and instance updates.

---

[23] http://aligned-project.eu/data/rvo_documentation.html

[24] "Describing Reasoning Results with RVO, the Reasoning Violations Ontology", Bojan Bozic, Rob Brennan, Kevin Feeney and Gavin Mendel-Gleason, 3rd Workshop on Linked Data Quality, co-located with ESWC 2016, Crete, 30 May 2016.

/dacura/schema_validate
POST variables: pragma
Requires: pragma.schema, pragma.tests
Endpoint for testing validity of an already existing schema

/dacura/validate
POST variables: pragma
Requires: pragma.instance, pragma.schema, pragma.tests
Endpoint for testing validity of already existing instance/schema pair.

/dacura/test
POST variables: N/A
Requires: N/A
Runs the internal testing suite.

/dacura/entity
POST variables: entity, schema, instance
Requires: entity, schema, instance
Returns all entities in the given instance graph for the given schema.

/dacura/entity_frame
POST variables: class, schema, instance
Requires: class, schema, instance
Returns the frame associated with a given entity instance, filled with its respective values. The 'class' post variable is the URI of a valid class in the schema provided by the post variable 'schema'.

/dacura/class_frame
POST variables: class, schema
Requires: class, schema
Returns the frame associated with a given entity instance, filled with its respective values. The 'lass' is the URI of a valid class in the given schema.

/dacura/class
POST variables: schema
Requires: schema
Endpoint for obtaining information on all defined classes in a given schema.

/dacura/dacura_entity_property_frame
POST variables: schema, instance, property, entity
Requires: schema, instance, property, entity

184 Tools

Endpoint returns a filled frame for a given entity and property when supplied with the entity URI, the schema and instance graphs and the necessary property URI.

/dacura/subsumes
POST variables: schema,class
Requires: schema, class

Endpoint returns a list of all classes which are subsumed by the supplied class.

DQS is now relatively stable and most changes will involve bug-fixes. The most recent source code is released open-source as a plugin, available at https://github.com/GavinMendelGleason/dacura. The Dacura system will continue to maintain and update the plugin as it is required for important data curation functionality in Dacura.

5.7 Linked Data Model Mapping

5.7.1 Interlink Validation Tool

The Interlink Validation Tool is designed to be used in a scenario where a specific source dataset is being maintained. This source dataset contains interlinks to external target datasets. As the source dataset and the target datasets evolve over time, the maintainers of the source dataset need to ensure that none of the existing interlinks have become invalid due to the evolution of the datasets.

The Interlink Validation Tool was initially validated in the ALIGNED DBpedia use case. It was identified that DBpedia does not include interlink validation during its release process (activities involved when a new version of DBpedia is to be released). This can result in invalid interlinks being published in the DBpedia release, reducing overall dataset quality. The Interlink Validation Tool provides a lightweight approach to reduce the number of invalid interlinks that could get published in a dataset. While the tool does not repair interlinks, it does highlight, which interlinks have become invalid and which resource (the source dataset resource or target dataset resource) has caused it to become invalid. This information can then be used by other tools in a software and data engineering toolchain, to help in the interlink repair process. The tool was deployed in the DBpedia environment for the v.2015-10 release and discovered 53,418 invalid interlinks[25].

---
[25] https://sourceforge.net/p/dbpedia/mailman/message/34980754/

5.7 Linked Data Model Mapping

As an input, the tool takes a set of interlinks between the source dataset and a target dataset. The resources in the interlinks are compared to their respective datasets to discover which interlinks are still valid and which are invalid. The tool outputs a set of valid and invalid interlinks along with two log files. One log file is a human readable log indicating, which set of interlinks have been checked and which interlinks were discovered as invalid. The other log file records similar information but is encoded in RDF and uses the ALIGNED Metamodel, especially the DLO and the DBpedia use case specific ontology (crowd-sourced public datasets) to describe the activities, entities and agents in the log. This means that the RDF logs (produced by the Interlink Validation Tool) can be consumed by the ALIGNED Unified Governance Tools in an ALIGNED tool chain.

The tool has already been deployed live in the DBpedia environment for the v.2015-10 release and discovered 53,418 invalid interlinks[26].

5.7.1.1 Interlink validation

The tool validates interlinks between two or more datasets through the use of standard SPARQL[27] query templates. RDF interlinks are typically expressed as a single triple linking resources in one (source) dataset with resources in another (target) dataset. Interlinks can be validated in two ways:

Source resources of the interlinks are only checked against their respective dataset. While this is useful when it is not possible to access the target dataset, it does only validate the source resources meaning that target resources could still be invalid.

Both source and target resources are checked against their respective datasets.

Figure 5.44 shows the process of interlink validation. Since SPARQL queries are used to validate interlinks, a SPARQL query endpoint and a local triple-store are required. It is assumed that the interlinks to be validated are stored in a named graph in the local triple-store and the source dataset is stored in a separate named graph. The query templates work by accessing the source resources (subject of the triple) and target resources (object of the triple). Validation is done in the following way:

When only the source resources of the interlinks are to be validated, they are compared to the source dataset to see, if those resources exist. If the source

---

[26] https://sourceforge.net/p/dbpedia/mailman/message/34980754/
[27] The query language for RDF, https://www.w3.org/TR/rdf-sparql-query/

186 Tools

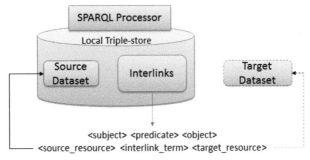

Figure 5.44 Interlink Validation Process.

resource of an interlink does not exist in the source dataset, then that interlink is classified as invalid.

When both the source and target resources of the interlinks are to be validated, then these resources are compared to their respective source and target datasets to see, if they exist. If either the source or target resource of an interlink does not exist in their respective dataset, then that interlink is classified as invalid. External target datasets can be accessed through loading them temporarily into the local triple-store or through remote access via a federated SPARQL query.

While this approach for validating interlinks is lightweight (relying on standard SPARQL queries), it does have a drawback. This approach cannot detect interlinks that have become invalid due to a resource merge or resource split event,[28] that can occur to resources in evolving datasets. A resource merge is done, when two or more resources from a dataset merge into a single resource and a resource split is done, when a single resource splits into two or more resources. A situation can arise where a resource merge or split takes place and the original resource identifier does not change. Therefore, a resource may have changed semantically, but is syntactically still the same. It is this particular situation, where a resource changes semantically but not syntactically, where this approach for detecting invalid interlinks will fail. In practice, given the dynamic nature of data on the Web, supporting distributed maintenance of data, without detecting resource merge or split events is still very valuable.

---

[28]Dos Reis, J. C., Pruski, C., Da Silveira, M. and Reynaud-Delaître, C. "Analyzing and supporting the mapping maintenance problem in biomedical knowledge organization systems." In Proc. of the Workshop on Semantic Interoperability in Medical Informatics collocated with the 9th Extended Semantic Web Conference, pp. 25–36, 2012.

5.7 Linked Data Model Mapping

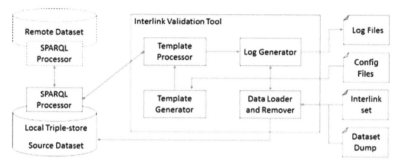

Figure 5.45 Operation of the Interlink Validation Tool. The arrows indicate the flow of information/data among the different components.

5.7.1.2 Technical overview

In this subsection, the operation of the tool is described. Figure 5.45 displays the operation of the tool and its different components.

The interlink validation tool reads in two configuration files. The first configuration file contains parameters about accessing the source dataset in the local-triple-store and some details about the source dataset itself (that will be used in the log files). The second configuration file is where a user specifies the details about each set of interlinks (between the source dataset and multiple external target datasets) that are to be validated. The parameters in this file are described in detail in the next section but they allow a user to set: the name of the external target dataset, the location of the interlink set to be validated, flags specifying validation behaviour and scope, and federated SPARQL query details.

When all the parameters are set, the tool can be run. The tool first gets the location of the interlink set from the second configuration file and loads it into a named graph in the local triple-store.

Next, based on the third parameter in the second configuration file, a SPARQL query template is generated – for example, if the external dataset is to be accessed via a federated query, then a federated query call will be included in the query template, with the external SPARQL endpoint URI provided by parameter 4.

If the third parameter specifies that an external dataset is to be accessed from a dataset dump file, then this dump file is retrieved and loaded into a named graph in the local triple-store.

Next, the template processor sends the query template to the SPARQL endpoint of the local triple-store for execution. The source resources are always checked against the source dataset. The execution results are then

188   Tools

returned to the template processor, which sends the results to the log generator.

Then based on the parameters specified in both configuration files, and the execution results sent from the template processor, two log files are produced. One log file is a human readable log, describing which set of interlinks have been checked and which interlinks were discovered as invalid. The other log file records similar information, encoded in RDF and uses the ALIGNED Design life cycle Ontology to describe the logs.

Finally, the tool removes all temporary created data loaded into the local triple-store. The tool will repeat this process for all interlink sets specified in the second configuration file.

User guide

This subsection provides a guide on how to use the Interlink Validation Tool.

The tool is a Java program designed to be run in a UNIX environment via the command line. The current prototype of the tool is designed to be used with a Virtuoso[29] triple-store only. The tool consists of three files and three directories:

```
The 'interlink_validator.java' file
The 'iv_config.txt' file
The 'external_datasets.txt' file
The 'valid' directory
The 'invalid' directory
The 'temp' directory
```

5.7.1.3 Configuration via iv_config.txt

The iv_config.txt file contains seven parameters that need to be set. These parameters are:

Parameter1 (p1=): The file path to Virtuoso's isql utility. This is necessary to be able to load data into the local triple-store and execute SPARQL queries.

Parameter2 (p2=): Virtuoso's dba password. Similar to the above point, this is needed in order to access Virtuoso's triple-store.

Parameter3 (p3=): The graph name where the local dataset is stored in the triple-store. This specifies the location of the local dataset where the source dataset resources in the interlinks to be validated will be compared against.

Parameter4 (p4=): The file path where the (human readable) log file will be generated. If this parameter is not the log file, then it will be generated in the same directory as well.

---

[29]http://virtuoso.openlinksw.com/

Parameter5 (p5=): The file path where the RDF log file will be generated. If this parameter is not the log file, then it will be generated in the same directory as well.

Parameter6 (p6=): A URI to provide reference to the source dataset. This is used to refer the source dataset and is used in the RDF log generated by the tool.

Parameter7 (p7=): A URI to provide reference to the source dataset can be accessed. The reference can be a Web page containing dump files or a SPARQL endpoint. This is also used in the RDF log generated by the tool.

Each parameter is to be provided on a separate line in this file.

5.7.1.4 Configuration via external_datasets.txt

The external_datasets.txt file contains parameters to be set. Up to six parameters can be set for each set of interlinks that are to be validated:

Parameter1 (p1=): Provide a name for the external dataset that will appear in the log files.

Parameter2 (p2=): Provide a URI or file path to the file containing the interlinks that are to be validated.

Parameter3 (p3=): State whether the external dataset will be accessed through:

A federated query {F}.

A federated query with a named graph {FG}

A named graph {G] in the local triple-store.

A dump file {D}

None {N}, which means that only the source dataset resources in the interlinks will be validated.

Parameter4 (p4=): Depending on the setting done in p3, the following options are available:

If "F" was stated for parameter 3, provide the external dataset SPARQL endpoint URI.

If "FG" was stated for parameter 3, provide the external dataset SPARQL endpoint URI along with the named graph URI (see Parameter5).

If "G" was stated for parameter 3, provide the graph name where the external dataset is stored, in the local triple-store.

If "D" was stated for parameter 3, provide a URI or a file path to the dump file of the external dataset.

If "N" was stated for parameter 3, parameter 4 can be left blank.

Parameter5 (p5=): Depending on the setting done in p3, the following options are available:

If "FG" was stated for parameter 3, then provide the named graph URI, where the external dataset is stored.

Parameter6 (p6=): [Optional Parameter] Provide a URI to reference the external target dataset. This will be used in the RDF log generated by the tool.

One set of parameters must be provided per line in the file and each parameter must be separated by a " " (blank space).

5.7.1.5 Execute the interlink validator tool

When the two configuration files (iv_config.txt and external_datasets.txt) have been configured, the tool can be executed. To execute the tool, use the following command:

java InterlinkValidator

After a successful execution, two log files will be generated in the specified location. In addition, files containing the valid interlinks and the invalid interlinks will be generated in the respective directories.

5.7.2 Dacura Linked Model Mapper

The Dacura Linked Data Model Mapping service has been developed to help users to create rich ontological models from semi-structured HTML input and then to automate the harvesting of instance data that conform to the model, again sourced from semi-structured HTML input. This process involves a series of structural and semantic mappings to be applied on both sides – in generating the model and generating the instance data input mapping.

The service is designed to be used in a scenario where a data model is implicitly defined in a HTML page with markup used to identify labels of

5.7 Linked Data Model Mapping

properties (e.g., <h3> or <strong> tags). This is a common scenario where a wiki or other CMS is used to collate a structured dataset. Unfortunately, from a machine's point of view, the data are semi-structured at best – the structure is designed primarily to be human-interpretable and we cannot even assume that the HTML is well-formed, never mind that there will be consistency in tags used or their attributes. Nevertheless, in almost all real cases, it will be possible to identify some pattern used in the HTML that can be mapped to a feature of the model.

The service was developed to support the ALIGNED Seshat use. Seshat researchers have collected a large quantity of data using a wiki. The data as a list of variables, organised into sections delineated by a variety of HTML tags (Figure 5.46) with variables identified by a label between special characters and variable values having a special syntax, which captures uncertainty and disagreement and temporal scoping, followed by free html containing citations and commentary on the value.

The wiki worked well as a tool for collecting a large volume of data by a distributed team of researchers – over 150,000 facts were collected on the wiki. However, the process of extracting and cleaning data from the wiki for analysis became overwhelming over time. Thus, the goal of the Linked Data model mapper service is to automate the process of importing both the model and the instance data from the wiki to generate a structured, semantic format that is ready for analysis.

The tool allows users to map from a semi-structured wiki data-model to a rich structured semantic model. However, it cannot create structure from nothing – thus if the user wishes to use a highly structured data model with complex containment relationships, this should be defined by the user before importing the model by creating the necessary classes and properties to bind the object's basic containment structure together.

When the service is used to add a new property to the data model, the system generates a location pattern which is associated with the property. This pattern is then used to locate and import instance data elsewhere on the wiki. The service uses ALIGNED metamodel ontologies RVO, PROV, and the Seshat domain ontology. The tool has been deployed live in the Seshat use-case and was used to create Seshat's first public release of data in April 2017.[30]

---

[30] http://dacura.scss.tcd.ie/seshat/

Main Variables (Polity)

General variables

- **RA** ◆ ▼ The name of the research assistant or associate who coded the data. If more than one RA made a substantial contribution, list all.

- **Expert** ◆ ▼ The name of the historical or archaeological expert who supervised coding and checked/improved/approved the result. If more than one expert was involved, list all.

- **UTM zone** ◆ ▼ List only one, usually where the capital city is located

- **Original name** ◆ ▼ Generally same as the name of this page

- **Alternative names** ◆ ▼ Used in the historical literature; also supply the most common name used by the natives

- **Peak Date** ◆ ▼ The period when the polity was at its peak, whether militarily, in terms of the size of territory controlled, or the degree of cultural development. This variable has a subjective element, but typically historians agree when the peak was.

Temporal bounds: The next three coding positions define the temporal bounds of the polity. These codes take into account that such temporal bounds may be fuzzy and allow us to capture this fuzziness. For example, some polities such as the Medieval German Empire or China under the Zhou Dynasty began as reasonably coherent states, but with time gradually lost cohesion, the degree to which the center exercised control over regional subpolities. Because this process was gradual, there was no sharp temporal boundary. The Degree of centralization' variable allows us to capture these transitions (by coding time periods when the polity transitions, for example, from a 'confederated state' to 'loose' and finally to 'nominal' degree of centralization. Similarly, polities may have a fuzzy starting date, if they originate as subpolities under a disintegrating overarching polity. These transitions are captured by the variable 'Supra-polity relations'.

- **Duration** ◆ ▼ The starting and ending dates covered by this coding sheet. Briefly explain the significance of each date. For example, the starting date could be the establishment of a long-ruling dynasty, while the ending date may be the year when the polity was conquered by an aggressive neighbor. In cases when starting and/or ending dates are fuzzy, as explained above, use the earliest possible starting date and the latest possible ending date. This approach will result in a temporal overlap, so that some NGAs for some periods will be coded as belonging to two polities simultaneously (e.g., to a disintegrating overarching polity and to the rising regional subpolity). Such overlap is acceptable, and will be dealt with at the analysis stage.

- **Supra-polity relations** ◆ ▼ unknown/ none/ alliance/ nominal allegiance/ personal union/ vassalage/
  - 'alliance' = belongs to a long-term military-political alliance of independent polities ('long-term' refers to more or less permanent relationship between polities extending over multiple years)
  - 'nominal allegiance' = same as 'nominal' under the next variable (Degree of centralization) but now reflecting the position of the focal polity within the overarching political authority
  - 'personal union' = the focal polity is united with another, or others, as a result of a dynastic marriage
  - 'vassalage' = corresponding to 'loose' category in the Degree of centralization

- **Degree of centralization** ◆ ▼ unknown/ nominal/ loose/ confederated state /unitary state
  - 'nominal' = regional rulers pay only nominal allegiance to the overall ruler and maintain independence on all important aspects of governing, including taxation and warfare. (example: Japan during the Sengoku period)
  - 'loose' = the central government exercises a certain degree of control, especially over military matters and international relations. Otherwise the regional rulers are left alone (example: European 'feudalism' after the collapse of the Carolingian empire)
  - 'confederated state' = regions enjoy a large degree of autonomy in internal (regional) government. In particular, the regional governors are either hereditary rulers, or are elected by regional elites or by the population of the region; and regional governments can levy and dispose of regional taxes. Use this category for the more centralized 'feudal states'.
  - 'unitary state' = regional governors are appointed and removed by the central authorities, taxes are imposed by, and transmitted to the center

- **Capital** ◆ ▼ The city where the ruler spends most of its time. If there were more than one capital supply all names and enclose in curly braces. For example, {Susa, Pasargadae, Persepolis, Ecbatana, Babylon}. Note that the capital may be different from the largest city (see below).

- **Language** ◆ ▼ List the language(s) used polity-wide for administration, religion, and military affairs. Also list the language spoken by the majority of the population, if different from the above.

Map

Figure 5.46 Example of seshat code book page.

5.7.3 Model Mapper Service

This section provides an overview of the service: first, we outline how the modelling tool creates mappings between a model and HTML patterns. Then we outline how the harvesting tool uses these patterns to automate the harvesting of semi-structured data.

5.7.3.1 Modelling tool – creating mappings

Dacura's Modelling tool, shown in Figure 5.47, enables users to create the structure of the dataset from existing sources. In this case, we used the Seshat code book page as the basis of our model. Dacura associates the imported properties with the pattern of the HTML that they were imported from. It uses this pattern later to automatically find and import data from the rest of the wiki into the structured model, as shown in Figure 5.48.

5.7.3.2 Importing semi-structured data with data harvesting tool

The Dacura data harvesting tool can be run on any Web page. When it loads, it attempts to fit the data on the page to the shape of the model using the patterns associated with the model that were created upon import. It uses Dacura's quality control API to test different possibilities in order to identify the best fit, as shown in Figure 5.49. It can even automatically correct mistakes.

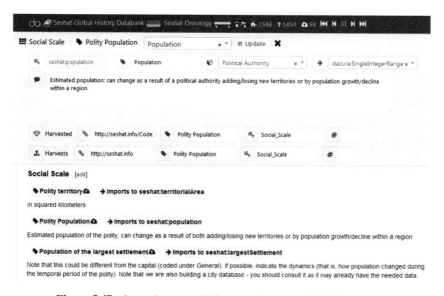

Figure 5.47 Importing a model from semi-structured HTML source.

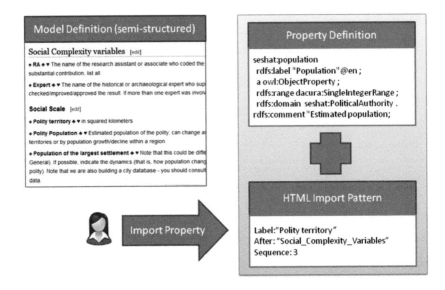

Figure 5.48 Process for associating property definitions in a model with a pattern within a semi-structured HTML page.

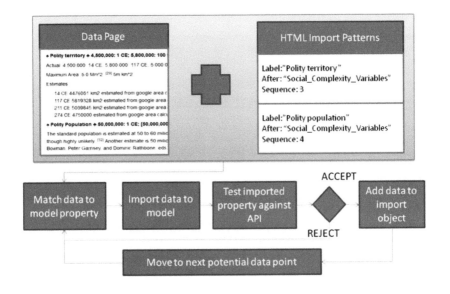

Figure 5.49 Process for using patterns to extract data from semi-structured html pages.

5.8 Model-Driven Data Curation

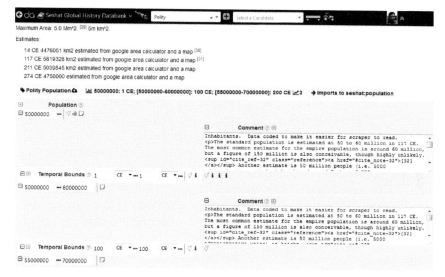

Figure 5.50 Screenshot showing results of automated importing of semi-structured HTML data into structured model.

It rewrites the Web page to show the user what the data would look like if it were imported into Dacura, as shown in Figure 5.50. This allows users to visualise and refine the mappings to ensure that as much data as possible can be imported. Once the user is happy with the mapping, they can import all the data from the entire wiki with a single click.

5.8 Model-Driven Data Curation

The model-driven data curation interfaces provide tools for the automatic generation of data-curation interfaces. These interfaces enable the creation of ontological models and the update of data, which respects these models with a high level of agility and flexibility of model.

The interface specifications, known herein as frames, are generated automatically from ontologies.

These ontologies are specified in RDF/OWL. The frames are generated by the Dacura Quality Service.

These frames are consumed by the Dacura platform, which utilises it both for its back-end management and for the more user-focussed Dacura console. Utilising the Dacura console, users can introduce new data or edit existing data from DQS via entry forms generated by javascript from the frame specification. Additionally, the model itself can be incrementally updated from the Dacura console in architect mode (for users with suitable permissions).

The software facilitates the Seshat use-case, which requires that we are able to import, track, update and delete from a large existing dataset (on a wiki) which is highly unstructured, into a highly structured format suitable for mathematical analysis of various historical trends. The software has already been utilised in modification of the ontologies developed in ALIGNED and has improved the agility of our model development, and consequently the automatically generated user interfaces.

Highly structured Linked Data often suffers from poor quality. Hence, the software helps to guarantee strong data quality standards by the structure of the user interface itself. Furthermore, it can be enhanced by quality checks after data have been constructed.

While this code is used in the Seshat use-case, its flexibility makes it broadly applicable to a wide range of data-curation uses. This would include any use case in which there needs to be model flexibility and data entry via the Web and especially collection of human or automatically facilitated collection of information from highly unstructured data sources.

The current implementation is a basis for further development, which will include enriching the user interface with additional data entry types, which enhance the user experience of data entry. This will include the ability to describe territories on maps, the inclusion of data ranges and autocomplete comboboxes for entering pre-existing objects. Additionally, richer constraints will be checked on the client side using code auto-generated from restrictions given in frame specifications.

We begin with the specification of frames which are generated by DQS. We then describe the production of the user-interface elements from these frames.

5.8.1 Dacura Quality Service Frame Generation

The Dacura Quality Service has been extended to produce frames, which constitute specifications for user interfaces derived automatically from ontologies, which are described using RDF/OWL. The service is structured as a plugin to the ClioPatria semantic Web server and providing a number of new API endpoints which allow clients to interact with the ontology.

We briefly describe the structure of frames, which are detailed as abstract datatypes using JSON. We will then describe the API which is used to obtain frames and the data associated with them for a given ontology.

All code and API endpoints documentation for DQS is available on Github.[31]

5.8.2 Frames for UserInterface Design

Frames are specified using JSON[32]. This provides a useful interchange format for Web APIs, for which there is tool support available in virtually every modern programming language.

Frames give information about an object, the classes they are associated with and which properties are accessible to them given the ontological specification. Since, in general, it is possible for the entire RDF graph to be transitively accessible to a given class, we further restrict the generation of frames to truncate the graph at any object which has been described as a dacura:Entity (that is, the given class is an owl:subclassOf dacura:Entity). This gives us a fragment of the graph which is amenable to the creation of a usable dataentry interface.

In every case, we give the domain and range of the properties associated with a given class. If the range is a class which is not a dacura:Entity type then we include the frame associated with that class. If it is a datatype, we give back sufficient information to aid in the construction of the userinterface element. This includes the datatype, which is entered along with a potential restriction on that type, which further constrains its behaviour.

5.8.3 SemiFormal Frame Specification

In Table 5.2, we demonstrate the grammar of frames in a variant of EBNF, which describes the JSON objects that are produced by the DQS framework in accordance with a given ontology.

First, we describe some of the idioms used in our EBNF, which has been modified to reflect the use of JSON as the objects of interest. This should be considered indicative of the actual format useable by software engineers who are working with the object, rather than as a strictly formal specification.

There are two primary formats that are returned for frames. One is the purely abstract empty object associated with a class for use as a template for user interfaces, and the second is a filled frame, which is a frame that fills

---

[31] Dacura Quality Service Cliopatria plugin https://github.com/GavinMendelGleason/dacura
[32] The JavaScript Object Notation (JSON) Data Interchange Format RFC 7159 http://rfc7159.net/

such an abstract object with concrete triples from the instance graph for the given class.

In Table 5.2, the Frame syntactic element provides the toplevel object which is returned in JSON format by the endpoints. The dominValue and

Table 5.2 Dacura Quality Service Frame Grammar

```
Language := "en" | ...
XSDType := "xsd:integer" | "xsd:gYear" | ...
Literal :=
   { "lang" : Language,
       "data" : "..."}
  | { "type" : XSDType,
       "data" : "..."}
OwlProperty := URI
OwlClass := URI
Op := "and" | "or" | "not" | "xor"
PropertyRestriction :=
    true
  | { "type" : Op, "operands" : [PropertyRestriction] }
  | { "mincard" : N, "valuesFrom" : OwlClass }
  | { "maxcard" : N, "valuesFrom" : OwlClass }
  | { "card" : N, "valuesFrom" : OwlClass }
  | { "hasValue" : OwlClass }
  | { "allValuesFrom" : OwlClass }
  | { "someValuesFrom" : OwlClass }
Property :=
   { "type" : "objectProperty",
      "domain" : OWLClass,
      "property" : OWLProperty,
      "range" : OWLClass,
      <"label" : Literal >, <"comment" : Literal >,
      <"domainvalue" : Value(PropertyType) >,
      <"frame" : FRAME >,
      <"restriction" : PropertyRestriction > }
  | { "type" : "datatypeProperty",
      "domain" : OWLClass,
      "property" : OWLProperty,
      "range" : OWLClass,
      <"label" : Literal >, <"comment" : Literal >,
      <"domainValue" : URI >,
      <"rangeValue" : Literal >,
      <"restriction" : PropertyRestriction > }
  | { "type" : "restriction",
       "property" : OWLProperty,
       "restriction" : PropertyRestriction }
PropertyFrame := [Property]
LogicalFrame := {"type" : Op, "operands" : [Frame]}
OneOfFrame := {"type" : "oneOf", "elements" : [URI]}
EntityFrame := {"type" : "entity", "class" : URI, <"domainValue" : URI>}
Frame := LogicalFrame | PropertyFrame | OneOfFrame | EntityFrame
```

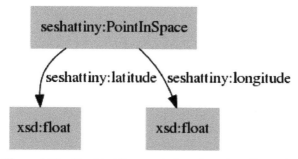

Figure 5.51 Graphical Representation of ontology fragment.

rangeValue elements are optional, and are only returned when querying for filled frames.

Schematically, Frames are used to produce empty forms with the appropriate userinterface elements for the data, while filled frames are used to create prepopulated entry forms, in the event that the data for an object is already known.

The optional "label" and "comment" fields are not essential in all cases, but are used in the automatic production of userinterface element labels and tool tips when present. Figure 5.51 shows an ontology fragment.

5.8.4 Frame API Endpoints

We briefly note here API endpoints used in the DQS for the generation and manipulation of frames.

/dacura/entity_frame
POST variables: class, schema, instance
Requires: class, schema, instance
Returns: Frame |Error

Returns the frame associated with a given entity instance, filled with its respective values. The 'class' post variable is the URI of a valid class in the schema provided by the post variable 'schema'.

/dacura/class_frame
POST variables: class, schema
Requires: class, schema
Returns: Frame |Error

Returns the frame associated with a given class. The 'class' is the URI of a valid class in the given schema.

/dacura/element_annotation
POST variables: schema, instance, property, element
Requires: schema, property, element
Returns: Frame |Error
The endpoint returns a Frame associated with a given annotation in the annotation graph given by 'instance' and associated with the data element 'element'.

6

Use Cases

Kevin Feeney[1], Christian Dirschl[2], Andreas Koller[3], James Welch[4], Dimitris Kontokostas[5], Pieter Francois[4], Sabina Łobocka[6] and Piotr Bledzki[6]

[1]Trinity College Dublin, Ireland
[2]Wolters Kluwer Germany, Germany
[3]Semantic Web Company, Austria
[4]University of Oxford, UK
[5]University of Leipzig, Germany
[6]Wolters Kluwer Poland, Poland

6.1 Wolters Kluwer – Re-Engineering a Complex Relational Database Application

6.1.1 Introduction

The publishing industry is – like many other industries – undergoing major changes. These changes are mainly based on technical developments and related habits of information consumption.[1] The world of the customers has dramatically changed and as an information service provider, Wolters Kluwer wanted to meet these changes with the best solutions for the customers and their work environment.

Wolters Kluwer has already engaged for a couple of years in new solutions to meet these challenges and to improve all processes of generating good quality content in the backend on the one hand and to deliver information and software in the frontend that facilitates the customer's life on the other hand.

One of these frontend applications is a platform called JURION[2] – an innovative legal information platform developed by Wolters Kluwer Germany (WKD) that merges and interlinks over one million documents of content and

---

[1]See e.g., this article about the information consumption in the US http://hmi.ucsd.edu/pdf/HMI_2009_ConsumerReport_Dec9_2009.pdf

[2]https://www.jurion.de/de/home/guest

data from diverse sources such as national and European legislation and court judgements, extensive internally authored content and local customer data, as well as social media and Web data (e.g., from DBpedia). In collecting and managing this data, all stages of the Data Life Cycle are present – extraction, storage, authoring, interlinking, enrichment, quality analysis, repair and publication. On top of this information processing pipeline, the JURION development teams add value through applications for personalisation, alerts, analysis, and semantic search.

The JURION use case is addressing both software life cycle and data life cycle. Therefore, their combination and integration is a key challenge within this use case. Still, currently both life cycles are highly independent from each other, which lead to a lot of errors and inefficient use of resources.

In order to address this challenge in a practical and pragmatic way, we have developed based on our daily operational experience two dedicated use case scenarios that shed a first light on the challenge and also on our view how to address it.

We have deliberately chosen one use case scenario that is triggered by the data life cycle and a second scenario triggered by the software life cycle. We also tried to describe common, yet not too complex situations, so that we could cover them in a sufficient granularity.

6.1.2 Problem Statement

JURION is an innovative legal information platform developed by Wolters Kluwer Germany that merges and interlinks over one million documents of content and data from diverse sources such as national and European legislation and court judgements, extensive internally authored content and local customer data, as well as social media and Web data (e.g., from DBpedia). In collecting and managing this data, all stages of the Data Life cycle are present – extraction, storage, authoring, interlinking, enrichment, quality analysis, repair and publication. On top of this information processing pipeline, the JURION development teams add value through applications for personalisation, alerts, analysis and semantic search. Based on the FP7 LOD2 project, parts of the Linked Data stack have been deployed in JURION to handle data complexity issues (see Figure 6.1). Currently, the software development process and data life cycle are highly independent from each other and require extensive manual management to coordinate their parallel development, leading to higher costs, quality issues and a slower time-to-market.

6.1 Wolters Kluwer – Re-Engineering a Complex Relational Database

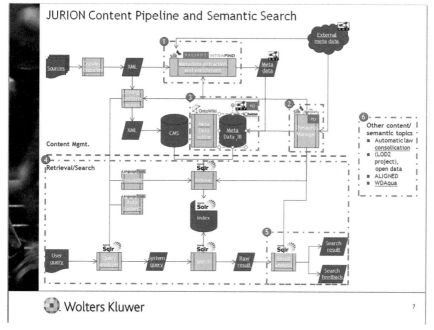

Figure 6.1 JURION Content Pipeline and Semantic Search.

By adopting the ALIGNED methodology and tools, software development and data processing pipeline maintenance will gain integrated governance mechanisms. These mechanisms will provide unified views of software and data engineering tasks enabled by linked enterprise Linked Data representations of both engineering teams. This will build on a common system specification language that produces and maintains links between data entities and code, executable code and program transformations that take account of how both systems co-evolve. The engineering process for both systems will be improved by the presence of new tools to integrate bug tracking and test results in both systems. ALIGNED methods and tools will streamline the processes for data acquisition, data processing, and data integration. These are all data curation activities that will be supported by workflows, model-driven generation of dataset-specific curation interfaces, automated data unit test generation, execution and reporting, data quality frameworks, and rule-based data integrity gateways. ALIGNED will enable JURION to address more complex business requirements that rely on tighter coupling of software and data.

6.1.3 Actors

Role	Description
CMS Expert	responsible for the technical correctness of process and data
Content Architect	responsible for the overall process and schemas
Legal Domain Expert	responsible for ensuring that legal data are correct
Legal Editor	responsible for editing legal information
Product Owner	wants the best possible product
Quality Manager	responsible for data quality assurance
Schema Expert	responsible for executing and documenting schema changes

The requirements on which the JURION use case was based are detailed in Appendix A.

Architecture

Based on the FP7 LOD2[3] project, parts of the Linked Data stack have been deployed in JURION to handle data complexity issues (see Figure 6.2). The

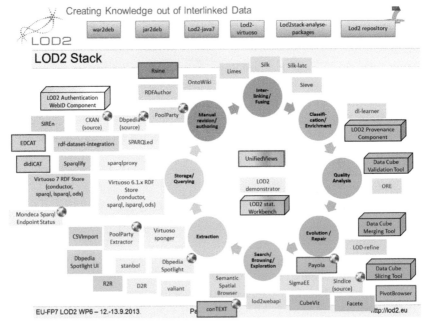

Figure 6.2 Distribution of the Linked Data stack components w.r.t. Linked Data Publishing cycle.

---

[3]http://lod2.eu/Welcome.html

6.1 Wolters Kluwer – Re-Engineering a Complex Relational Database

FP7 LOD2 project aimed at developing novel, innovative Semantic Web technologies and also at the expansion and integration of openly accessible and interlinked data on the Web. WKD acted as a use case partner for these technologies, supported the development process of semantic technologies and integrated them to support the expansion of Linked Data. WKD also published some domain specific datasets.

The software development process and data life cycle at WKD are highly independent from each other and require extensive manual management to coordinate their parallel development, leading to higher costs, quality issues and a slower time-to-market. This is why the JURION use case in ALIGNED is located both within the software engineering as well as in the data processing area (see Figure 6.3).

In the initial prototype implementation, we aimed at the creation of a stable prototypical environment, in which we can start testing and evaluating implementations to encounter the current issues. In this first phase, we concentrated mainly on the enhancement of data quality and repair processes. Based on requirements, we started to work on data transformation issues and the improvement of data quality processes in PoolParty.

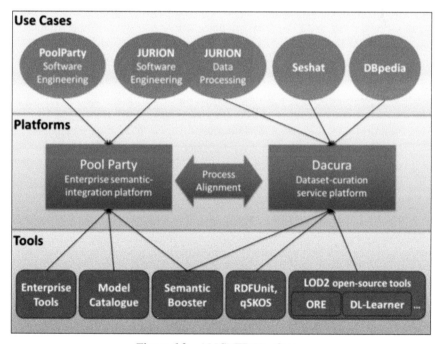

Figure 6.3 ALIGNED Use Cases.

6.1.4 Implementation

6.1.4.1 PoolParty notification extension

Development and maintenance of controlled vocabularies such as thesauri is mostly a manual and thus error-prone process. Especially in environments where multiple contributors are allowed to perform changes to the vocabulary, structural complexity increases, which makes it harder for individuals to maintain an overview. Furthermore, conflicting opinions arise and lead to inconsistent description, meaning and structure of the thesaurus' concepts. This problem is even more important when using software that allows for collaborative vocabulary development or publishing vocabularies as Linked Data.

Furthermore, maintaining an overview is not only necessary for thesaurus development, but also for curators responsible for datasets published as Linked Data on the Web serving various use cases. Users are allowed to change or add information (metadata) to existing data anytime. Therefore, errors can be introduced and hence manual review is required.

6.1.4.2 rsine notification extension

In order to address certain scenarios, the rsine[4] notification service as well as its integration into PoolParty had to be extended. The changes incorporated into rsine's code dealt with support for persistence transaction and attaching rsine to receive notifications from other PoolParty repositories than the default vocabulary repository. Regarding the former change, multiple triple changes that are written into rsine managed triple store as one transaction are now combined and treated by rsine as one changeset. This allows for easier formulation of notification subscription documents and more robust notifications. On the PoolParty side, we added integration code that forwards changes to, e.g., the custom schema repository or the user account repository to rsine, so that it is also possible to get notified on schema and user account changes. However, this is just a temporary solution as we aim to get PoolParty working with a single repository and organise all other information in separate named graphs. Once this has been accomplished, also the rsine integration code can be simplified.

6.1.4.2.1 *Results*

To cover the most important scenarios, we implemented five new rsine notification subscription documents that enable notifications for

---
[4]See https://github.com/rsine/rsine

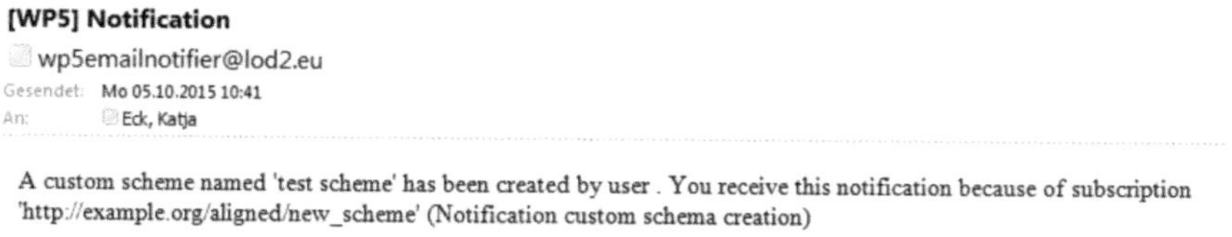

[WP5] Notification
wp5emailnotifier@lod2.eu
Gesendet: Mo 05.10.2015 10:41
An: Eck, Katja

A custom scheme named 'test scheme' has been created by user . You receive this notification because of subscription 'http://example.org/aligned/new_scheme' (Notification custom schema creation)

Figure 6.4 Notification message.

- Creation of a new custom class
- Creation of a new custom schema (see Figure 6.7)
- Deletion of a custom schema
- Creation of a new user account
- Creation of a new project

The other scenarios can be covered with similar subscriptions. Up to now, the details covered in the notification messages sent out to the users cannot cover information like

- who (username) created a custom schema or user account, and
- the name of the newly created schema

The reason for this is that (i) this information is not available in the persisted data or (ii) the repository holding the data is not available for querying through a SPARQL endpoint. Figure 6.4 shows a sample notification message.

6.1.4.3 RDFUnit for data transformation

As part of the core, CMS tasks within JURION each WKD XML document that is checked-in through internal workflow functionality and is converted to RDF based on the Portal Content Interface (PCI) ontology. The PCI ontology is a proprietary schema that describes legal documents and metadata in OWL. Due to change requests and new use cases for the RDF metadata in the ontology, the conversion logic or both the conversion logic and ontology need amendments. In these cases, we need to ensure that the RDF data that are generated from the WKD XML documents still comply with the PCI ontology for quality assurance.

Current Situation

As a gatekeeper to avoid loading flawed data into the triple store, each result of the conversion from WKD XML into PCI RDF is sent to a proprietary dedicated Validation service that inspects the input and verifies compliance with the ontology. This approach assures that the conversion results

208   *Use Cases*

are verified but comes with some major issues. The three most important ones are:

- The current service can only process larger data packages. This makes error detection on single data units quite difficult and one error blocks the whole processing pipeline
- the service is a SOAP-based Web service that works asynchronously with many independent process steps, which imposes high complexity on its usage
- it depends on other services and requires permanent network access and therefore is potentially unstable

To improve these issues, we want to implement unit test scenarios that can be run directly coupled to the conversion project development environment (this project hosts XSLT logic to convert WKD XML into PCI RDF). The tests should be run both automatically on every change in the project, but also be able to be manually triggered. Tests should be easily extendable and expressive enough to easily spot issues in the conversion process. The feedback loop should be coupled as tight as possible to the submitted change.

Implementation

To allow comparable and reproducible test results with suitable execution time, a number of WKD XML reference documents have been selected, against which the actual conversion into PCI RDF is executed and each resulting RDF dataset is verified individually.

The prototyped solution (see Figure 6.5) integrates RDFUnit as the core driver of the tests. The integration is currently based on auto generated tests, which are generated from a current version of the PCI ontology every time the test suite is run.

It also integrates seamlessly into the general development toolchain. Any change in the conversion project automatically leads to an entire build of the project including validation. As the test suite is integrated in the underlying standard test mechanisms, a developer can trigger this test chain manually on his local workstation to retrieve direct feedback at any time.

As a proof of concept RDFUnit's test results (the validation model based on the Test-Driven Data Validation Ontology[5]) linked to this test is stored into Virtuoso triplestore to enable future analysis/reviews of historical data.

---

[5]See http://rdfunit.aksw.org/ns/core

Results

Each of the test results manifests in the validation model, which is based on RDFUnit's Test-Driven Data Validation Ontology. As we currently rely on RDFUnit's auto generators, all statements are spotted that outcast rules that have been derived/interpreted from the ontology. These are especially cardinality and domain/range violations.

In any case, a summary of the test results is presented to the user. As this is always in the context of a concrete RDF-dataset (in the form of a file) one can immediately spot issues on the exact resource, which avoids unnecessary lookups and helps to identify the defective part of the conversion.

The integration of RDFUnit into the development cycle and build pipeline (see Figure 6.5) enabled the following possibilities that were entirely missing before:

- run automated tests based on the ontology
- steadily monitor project health
- capture metrics

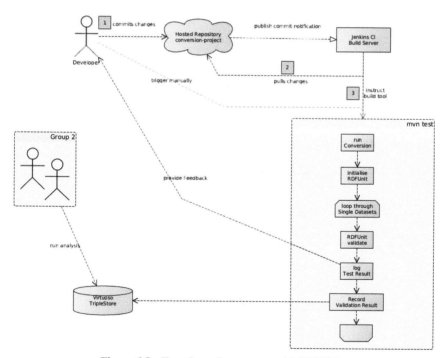

Figure 6.5 Transformation process with RDFUnit.

210 *Use Cases*

In the past, most issues aroused after the changes to the conversions have been released without proper and reliable testing – as this was only possible in manual developer tests. Moving forward, we can make sure that reproducible tests are run with each change especially before releases. Tests and tested documents can be easily extended to increase coverage of corner cases.

Figures 6.6 and 6.7 shows some of the test results, which can easily be stored and used on a regular basis in current and future QA reports.

Early and quick feedback on changes to the project are very valuable to assure that the project is in good health and existing functionality meets the defined expectations. Good coverage with automated tests prevents bugs from slipping in released functionality which may have bad side effects on other parts of the system.

RDFUnit enables possibilities but still needs a tighter integration as a library with our existing toolchain to improve reporting capabilities and make its feedback even more useful.

RDFUnit proves as being very useful and will be a fixed component of the operational tech stack within WKD JURION from now on.

We will provide further requirements to improve RDFUnit's integration into our development pipeline. At a later point in time, we will utilise RDFUnit to enable monitoring the existing data store to implement quality assurance on operational side.

```
target/test-classes/junit7523938743608749278/output/baulast_13211.meta.rdf

[ERROR] http://wolterskluwer.de/ceres/wk-
de/lexdb/181634/baulast_13211#Hinweis01bea53a31ad369b9dabd6a4704230ef
 . Cardinality of http://wolterskluwer.com/ceres/concept-v1.0/anchorId
different from 1 (is 0) for type http://wolterskluwer.com/ceres/content-
warehouse-v1.0/BlockAnchor
 . Cardinality of http://wolterskluwer.com/ceres/concept-v1.0/anchorId
different from 1 (is 0) for type http://wolterskluwer.com/ceres/concept-
v1.0/Anchor

[ERROR] http://wolterskluwer.de/ceres/wk-de/lexdb/181634/baulast_13211
 . http://wolterskluwer.com/ceres/wk-
de/referenceInformation.ChapterReference does not contain a literal value
(http://www.w3.org/1999/02/22-rdf-syntax-ns#XMLLiteral)
 . http://wolterskluwer.com/ceres/wk-
de/referenceInformation.ChapterReference has rdfs:domain different from:
http://wolterskluwer.com/ceres/ltr-v1.0/ReferenceInformation
 . http://wolterskluwer.com/ceres/wk-de/searchTuningKeyword has rdfs:domain
different from: http://wolterskluwer.com/ceres/concept-
v1.0/InformationClass
 . http://wolterskluwer.com/ceres/content-warehouse-
v1.0/isDocumentInstanceOf has rdfs:domain different from:
http://wolterskluwer.com/ceres/concept-v1.0/FileResource
```

Figure 6.6 RDFUnit results.

6.1 Wolters Kluwer – Re-Engineering a Complex Relational Database

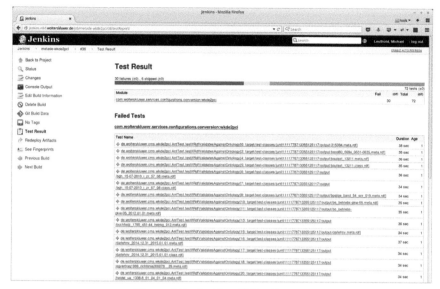

Figure 6.7 Jenkins-CI Test Report.

6.1.4.4 PoolParty external link validity

WKD document metadata and controlled vocabularies are linked to several external sources. These sources are mainly DBpedia[6] and thesauri like Eurovoc[7], Thesoz[8] or STW[9]. On a larger scale, we plan to include more of these kinds of sources to connect with additional internal and external data for the enhancement of several services. To control the process of change and to evaluate what kind of effects this can have on the quality of data, we want to control changes of Linked Data that can cause problems.

In addition to the validity of external links, we also aim to monitor the validity of internal links between different projects and datasets as also internal WK sources will need validity control.

Current Situation

Currently, we have no effective overview over the validity of linked sources. This causes, for example, frontend problems in the published vocabularies (see Figure 6.8). Currently, the only way to evaluate the quality is to analyse

---

[6] See http://de.dbpedia.org/
[7] See http://eurovoc.europa.eu/drupal/
[8] See http://www.gesis.org/en/services/research/thesauri-und-klassifikationen/social-science-thesaurus/
[9] See http://zbw.eu/stw/version/latest/about.en.html

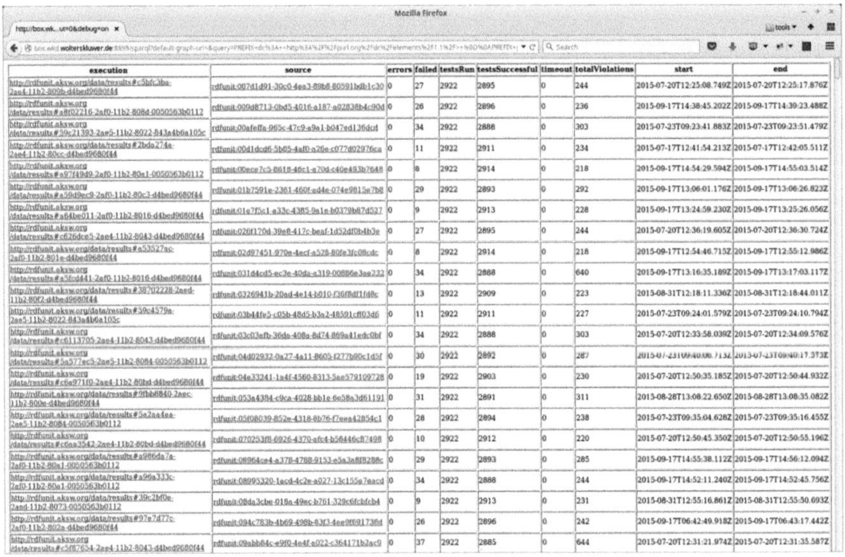

execution	source	errors	failed	testsRun	testsSuccessful	timeout	totalViolations	start	end
http://rdfunit.aksw.org/data/results#c5bfc3ba-2ae4-11b2-809b-d4bed9680f44	rdfunit:007d1d91-30c0-4ea3-89b8-80591bdb1c30	0	27	2922	2895	0	244	2015-07-20T12:25:08.749Z	2015-07-20T12:25:17.876Z
http://rdfunit.aksw.org/data/results#a8f02216-2af0-11b2-808d-0050563b0112	rdfunit:009d8713-0bd5-4016-a187-a02838b4c90d	0	26	2922	2896	0	236	2015-09-17T14:38:45.202Z	2015-09-17T14:39:23.488Z
http://rdfunit.aksw.org/data/results#59c21393-2ae5-11b2-8022-843a4b6a105c	rdfunit:00afeffa-965c-47c9-a9a1-b047ed136dcd	0	34	2922	2888	0	303	2015-07-23T09:23:41.883Z	2015-07-23T09:23:51.479Z
http://rdfunit.aksw.org/data/results#2bda274a-2ae4-11b2-80cc-d4bed9680f44	rdfunit:00d1dcd6-5b65-4af0-a26e-c077d02976ca	0	11	2922	2911	0	234	2015-07-17T12:41:54.213Z	2015-07-17T12:42:05.511Z
http://rdfunit.aksw.org/data/results#a97f49d9-2af0-11b2-80a1-0050563b0112	rdfunit:00ece7c5-8618-46c1-a70d-c40e493b7648	0	8	2922	2914	0	218	2015-09-17T14:54:29.594Z	2015-09-17T14:55:03.514Z
http://rdfunit.aksw.org/data/results#a59d9ec9-2af0-11b2-80c3-d4bed9680f44	rdfunit:01b7591e-2361-460f-ad4e-074e9815e7b8	0	29	2922	2893	0	292	2015-09-17T13:06:01.176Z	2015-09-17T13:06:26.823Z
http://rdfunit.aksw.org/data/results#a64be011-2af0-11b2-8016-d4bed9680f44	rdfunit:01e7f5c1-a33c-4385-9a1e-b0379b87d527	0	9	2922	2913	0	228	2015-09-17T13:24:59.230Z	2015-09-17T13:25:26.056Z
http://rdfunit.aksw.org/data/results#c626dce5-2ae4-11b2-8043-d4bed9680f44	rdfunit:026f170d-39e8-417c-beaf-1d52df0b4b3e	0	27	2922	2895	0	244	2015-07-20T12:36:19.605Z	2015-07-20T12:36:30.724Z
http://rdfunit.aksw.org/data/results#a53527ac-2af0-11b2-801e-d4bed9680f44	rdfunit:02d97451-970e-4ecf-a528-80fe3fc08cdc	0	8	2922	2914	0	218	2015-09-17T12:54:46.715Z	2015-09-17T12:55:12.986Z
http://rdfunit.aksw.org/data/results#a5fcd441-2af0-11b2-8016-d4bed9680f44	rdfunit:031d4cd5-ec3e-40da-a319-00886e3ae232	0	34	2922	2888	0	640	2015-09-17T13:16:35.189Z	2015-09-17T13:17:03.117Z
http://rdfunit.aksw.org/data/results#38702228-2aed-11b2-80f2-d4bed9680f44	rdfunit:0326941b-20ad-4e14-b010-f36f8df1f48c	0	13	2922	2909	0	223	2015-08-31T12:18:11.336Z	2015-08-31T12:18:44.011Z
http://rdfunit.aksw.org/data/results#59c4579a-2ae5-11b2-8022-843a4b6a105c	rdfunit:03b44fe5-c05b-48d5-b3a2-48591cff03d6	0	11	2922	2911	0	227	2015-07-23T09:24:01.579Z	2015-07-23T09:24:10.794Z
http://rdfunit.aksw.org/data/results#c6113705-2ae4-11b2-8043-d4bed9680f44	rdfunit:03c03afb-36da-408a-8d74-869a41edc0bf	0	34	2922	2888	0	303	2015-07-20T12:33:58.039Z	2015-07-20T12:34:09.576Z
http://rdfunit.aksw.org/data/results#5a577ec5-2ae5-11b2-8084-0050563b0112	rdfunit:04d02932-0a27-4a11-8605-f277b90c1d5f	0	30	2922	2892	0	287	2015-07-23T09:40:00.713Z	2015-07-23T09:40:17.373Z
http://rdfunit.aksw.org/data/results#c6a971f0-2ae4-11b2-80bd-d4bed9680f44	rdfunit:04e33241-1a4f-4560-8313-5ae579109728	0	19	2922	2903	0	230	2015-07-20T12:50:35.185Z	2015-07-20T12:50:44.932Z
http://rdfunit.aksw.org/data/results#9fbb8840-2aec-11b2-800e-d4bed9680f44	rdfunit:053a4384-c9ca-4028-bb1e-6e58a3d61191	0	31	2922	2891	0	311	2015-08-28T13:08:22.650Z	2015-08-28T13:08:35.082Z
http://rdfunit.aksw.org/data/results#5a2aa4ea-2ae5-11b2-8084-0050563b0112	rdfunit:05f08039-852e-4318-8b76-f7eea42854c1	0	28	2922	2894	0	238	2015-07-23T09:35:04.628Z	2015-07-23T09:35:16.455Z
http://rdfunit.aksw.org/data/results#c6aa3542-2ae4-11b2-80bd-d4bed9680f44	rdfunit:070253f8-6926-4370-afc4-b58446c87498	0	10	2922	2912	0	220	2015-07-20T12:50:45.350Z	2015-07-20T12:50:55.196Z
http://rdfunit.aksw.org/data/results#a986da7a-2af0-11b2-80a1-0050563b0112	rdfunit:08964ce4-a378-4788-9153-e5a3a8f8288c	0	29	2922	2893	0	285	2015-09-17T14:55:38.112Z	2015-09-17T14:56:12.094Z
http://rdfunit.aksw.org/data/results#a96a333c-2af0-11b2-80a1-0050563b0112	rdfunit:08995320-1acd-4c2e-a027-13c155a7aacd	0	34	2922	2888	0	244	2015-09-17T14:52:11.240Z	2015-09-17T14:52:45.756Z
http://rdfunit.aksw.org/data/results#39c2bf0e-2aed-11b2-8073-0050563b0112	rdfunit:08da3cbe-018a-49ec-b761-329c6fcbfcb4	0	9	2922	2913	0	231	2015-08-31T12:55:16.861Z	2015-08-31T12:55:50.693Z
http://rdfunit.aksw.org/data/results#97e7d77c-2af0-11b2-802a-d4bed9680f44	rdfunit:094c783b-4b69-498b-83f3-4ee9f691738d	0	26	2922	2896	0	242	2015-09-17T06:42:49.918Z	2015-09-17T06:43:17.442Z
http://rdfunit.aksw.org/data/results#c5f87654-2ae4-11b2-8043-d4bed9680f44	rdfunit:09abb84c-e9f0-4e4f-a022-c364171b2ac9	0	37	2922	2885	0	644	2015-07-20T12:31:21.974Z	2015-07-20T12:31:35.587Z

Figure 6.8 Validation Data stored for Analysis.

the frontend representations of the linked sources or to follow a link to detect a missing source. There is in general no process in place to control the validity of external sources. Figure 6.9 shows a sample defect.

Implementation

To check the validity of external links, we use the same technique as qSKOS.[10] All URIs used in the vocabulary that do not point to the local host are dereferenced and the remote server's response is checked. If the HTTP status code is 200, the link is considered valid. In case redirects occur, they are followed properly. All other responses are to be classified as invalid.

Results

URI checking can be invoked from the PoolParty user interface in the current experimental version. The result overview (see Figure 6.10) shows the URIs of the violated links and the total number of checked links as well as the number of violated links. The quality manager can use these links to change or delete the respective relations.

However, since each URI gets resolved and duplicate URIs are not omitted, this process can take a lot of time. In future versions of PoolParty, we

[10]See https://github.com/cmader/qSKOS

6.1 Wolters Kluwer – Re-Engineering a Complex Relational Database   213

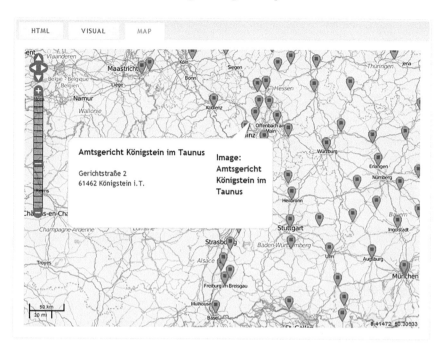

Figure 6.9 Example defect: the Image file of the external source does not exist anymore[11].

---

[11] See frontend http://vocabulary.wolterskluwer.de/court/10592.html

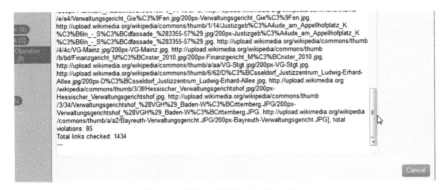

Figure 6.10   Validation Results.

will investigate ways of running these kinds of checks in the background and notify users on the results.

6.1.4.5 Statistical overview

As we are integrating more and more controlled vocabularies and custom schemas in the metadata management tool PoolParty, we are in need of solutions that give an overview of existing relations between projects and external data and schemas. Besides, the number of user roles is growing so that we need a solution that enables a best overview for a number of different users with different purposes. By different queries and enhancements, we want to get an impression about the relations between projects and the usage of specific custom schemas.

Current Situation

Connections of projects and schemas are not easily traceable. Owners of vocabularies need to document everything so that others can also understand the projects and its relations and possibilities. Without this documentation, it is hard to analyse the different projects. Within the tool, the user can only analyse the individual concepts for relations to investigate any relations with schemas. For linking to other projects it is possible to get a list of links. This list does not provide the number of links and specific numbers for different kinds of linking. These figures need to be searched manually.

Implementation

We currently implemented two different kinds of statistical metrics and integrated them into the PoolParty UI (i) checking for external links validity

6.1 Wolters Kluwer – Re-Engineering a Complex Relational Database

and (ii) links to other PoolParty projects on the server. These metrics differ in the methodology they are evaluated. Checking the validity of external links cannot be done using SPARQL and requires external tool support (e.g., Java code, see section 6 on external link validity). Reporting links to PoolParty projects can be achieved in a similar way than checking for data consistency violations. Each statistical property can be formulated as a SPARQL query, which is executed on the relevant project data, i.e., the current project data and metadata as well as all linked project data and metadata.

Results

The checking of project relations can be invoked from the PoolParty user interface in the current experimental version. The results (see Figure 6.11) show the kind of used relations, the frequency of these relations, the detailed list of linked resources and the total number of linked resources.

This way users can check how and to which extent projects are related to each other and they get an overview of used relations.

6.1.5 Evaluation

The Jurion Use case is split into two sections within the ALIGNED project: (1) the Jurion platform, and (2) the Jurion IPG tool. The developments concerning the Jurion platform took place in the first half of the project, based on the respective categories of measurement and will be repeatedly described here for completeness.

For the prototype of the JURION platform use case, we focussed on the data development processes.

The ALIGNED tools that were used for this prototype are RDFUnit and PoolParty. We had four major features for the initial prototype.

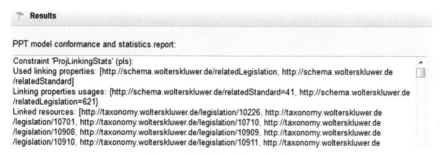

Figure 6.11 Statistical checks.

Use Cases

- RDFUnit for Data Transformation
- Notification Service in PoolParty
- Project Linking Statistics in PoolParty
- ELV in PoolParty

The methods of collection are divided into three categories, namely productivity, quality, and agility, as follows in Figure 6.12.

Tasks	Comment	Productivity (Prototype testing)	Quality (Prototype Testing, expert evaluation/ interviews)	Agility (expert evaluation)
RDF Transformation	quality test of data transformed from XML to RDF	Time Measurement for Quality Checks Time Measurement for Error Detection Need for Manual Interaction	number of detected error categories test coverage expert evaluation	Time to include new constraints/adapt the testing to new requirements
Notification	notification about predefined changes	Number of Scenarios Time Measurements Usefulness	Notification completeness expert evaluation	Time to include new constraints/adapt the testing to new requirements User roles that can modify Notifications Time to configure a new Notification Integration of a customized Notification Configuration Time to configure new requirements
Statistics	Statistics about relations of projects	time detected links	usability aspects result consistency	Detection Issues Integration of Statistics Time to configure new Requirements Extension
External links	quality of external links	checked links violations time	usability aspects expert evaluation/inter correctness of results	Scope of External Link Checks Integration of Internal Link Checks Time to configure new Requirements Extension

Figure 6.12 JURION: Overview.

The evaluation of the prototype showed clearly that during the Jurion prototype development, we have achieved our aim to improve the productivity and quality of data processes within the data life cycle. With the presented features, these improvements could be made visible. Performance and quality/error rates of the test results were satisfactory. Nonetheless, evaluation outcomes suggest further improvements are possible, especially with regard to usability, performance, integration of functionalities and required details that are not yet fully optimised.

6.1.5.1 Productivity

In summary, the productivity of data processes is clearly improved by the Jurion prototype. The data transformation service enables a testing that points directly to the detected error source and improves the bug fixing process this way. The notification service provides notifications as soon as an action is executed. This is a helpful tool to ensure quality analysis and data monitoring. Nonetheless, there needs to be a solution that helps to send the notification precisely where it is needed to avoid spamming. The statistics and ELV functionalities can help to save much time by replacing time-consuming manual work with efficient data overviews.

6.1.5.2 Quality

Concerning the quality of the prototype functionalities, the results are very satisfying. For notifications and ELV, there are only few issues. For the data transformation with RDFUnit and the statistics part, there needs to be further investigation to enable comprehensive and extensive data testing results. Usability issues need to be tackled in all the features for a better operational implementation. As this is only an initial prototype, usability was less of a priority.

6.1.5.3 Agility

The testers' feedback for agility of features is quite positive. The agility of RDFUnit is seen as satisfying as the automated service allows the implementation of new requirements easily. With regard to notification, adaptations are dependent on the specific notification use case and the respectively available data. In the same way, the agility of statistics feature is highly dependent on the availability of required underlying data.

ELV has a reasonable agility and is planned to be done by an external application to address performance issues.

The evaluation of the Jurion tasks was done in an early phase of the project, based on an earlier evaluation approach. We will analyse one on of the tasks based on the latest suggested method to show the adaptability of the test results for this approach. Task 4 ELV service serves as a good example for this analysis.

6.1.5.4 Measuring overall value

JURION is a legal information platform that merges and interlinks over one million documents of content and data from diverse sources such as national and European legislation and court judgements, extensive internally authored content and local customer data, as well as social media and Web data (e.g., from DBpedia). The JURION development teams add value through applications for personalisation, alerts, analysis and semantic search. Revenue is generated by customers paying for the platform content and related services.

PoolParty serves as the metadata management tool of controlled vocabularies that are used for specific search functionalities and the development of further functionalities in applications. The ELV is a PoolParty functionality.

ELV is a new feature that evaluates the links to external sources and informs the user in case the sources are not available anymore. Previously, it was only possible to check the links manually in random samples. Therefore, it provides a fast and efficient curation service to guarantee an error-free linking to external sources. A measure of value could be curation cost of maintaining a given quality of service as measured by revenue. The saving of time needed for the error detection is the most important parameter for this calculation.

6.1.5.5 Data quality dimensions and thresholds

Data accuracy, completeness and consistency are essential for this task. Jurion Customers pay for the curated information and related services so that high-data quality is a major requirement. Data accuracy was analysed in the evaluation by analysing the errors – 100% of the found errors have been data inconsistencies. In average, 81% have been outdated links, we were looking for. Nineteen per cent have been unexpected inconsistencies that exceeded our expectations. Completeness was checked via mapping the errors that were found manually, against the system results. All manually detected issues have been detected by the links.

6.1.5.6 Model agility
As the functionality is embedded in Poolparty, the assessment of the Poolparty use case is also valid in this case. With regard to the functionality, the configuration of new requirements for the ELV is possible. Determining which URIs should be resolved can be done either with the methods SPARQL provides or within the Java resolution algorithm. In each case, the effort for change is low, allowing for agile reaction on changed requirements. However, changes to the current configuration require recompilation and redeployment of PoolParty.

6.1.5.7 Data agility
As the functionality is embedded in Poolparty, the assessment of the Poolparty use case is also valid in this case. Based on the pattern to detect URI patterns for links to be checked, the solution can also be used for (or constrained to) "internal" links. Therefore, appropriate methods must be evaluated.

6.1.6 JURION IPG

6.1.6.1 Introduction
The Jurion IPG system is a commercial intelligence system, providing a means for business contractors to perform due-diligence queries, serving historical data about companies and their relationships with other companies, responsible individuals, and business documents. As a reliable provider of credibility and financial information for over five million entities, the integrity and consistency of the data is of vital importance, and increasingly hard to manage at scale. In this use case, we are deploying the ALIGNED tools to find problems in the existing data and to improve the integrity of data submitted in the future. ALIGNED tools are also helping increase the scope of the data, by enabling the linking of data stored within the system to external related datasets.

Figure 6.13 shows the flow of content through components of the system. Source data are manually imported or acquired through crawling non-formatted data sources, and pushed into a relational data store. Metadata is extracted and enriched, before being entered into a separate RDF data store.

220 Use Cases

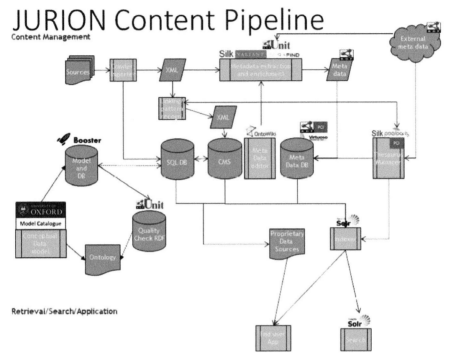

Figure 6.13 JURION Content Pipeline, showing ALIGNED tools integrated with existing functionality and datasets.

The schema for the relational data store is versioned and updated through an instance of the Model Catalogue; and data integrity is maintained through Semantic Booster-generated stored procedures. This relational data may also be viewed in an RDF format, where the RDFUnit tool may be used for further data validation. External metadata is managed through use of the SWC's PoolParty thesaurus manager and linked with the RDF representation of the core dataset. Existing end user interfaces to the data will be supplemented with an administrative interface automatically generated by the Semantic Booster tool.

New software integration points can be found where ALIGNED tools interact or communicate. In particular, such interactions occur between the Model Catalogue and Semantic Booster, where Booster models are generated from the Model Catalogue, and where metadata in the catalogue is used to supplement the end user interfaces. Further integration is between Booster and RDFUnit, where the D2RQ tool is used to help convert relational

6.1 Wolters Kluwer – Re-Engineering a Complex Relational Database

data into RDF: the configuration for this may again be parameterised by a transformation of the model in the Catalogue. The ALIGNED vocabularies are used to standardise these interactions.

Use Case

Figure 6.14 shows the problem space and more specifically the complexity of the JURION IPG system. The utility of the JURION IPG system is dependent on the maintenance and evolution of a large, semantically consistent dataset. Huge amounts of daily processed data originally from pdf sources; and maintenance through a proprietary, obsolete CMS makes the IPG case extremely suited as an ALIGNED use case. Business value of the system is dependent on the maintenance and evolution of a large, semantically consistent dataset. The overall goal is to ensure the quality of the system used to enter and maintain the data and to improve the value by linking to external datasets. To provide this solution by implementing ALIGNED tools, we used in parallel two approaches including Semantic Booster first and Dacura afterwards. For the purpose of the Jurion IPG use case, we have chosen to concentrate on a number of key critical concerns:

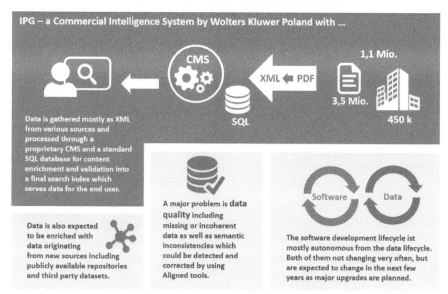

Figure 6.14 IPG problem statement.

- The use of Semantic Booster will allow a wider range of semantic integrity constraints and business rules to be applied to the data upon entry, ensuring high-quality data. The automatic data-migration tools provided with Booster will minimise the impact of upgrading and evolving the underlying data model whilst maintaining data consistency.
- The administrator interface in the IPG system currently requires manual development each time the database changes; increasing the cost of evolving the data store. The model-driven Booster default interface can be used: either in its entirety, or components reused to save development effort. (Figure 6.14)
- The existing data store is currently stored in a relational format. Whilst Booster can help enforce a range of integrity constraints, there are some consistency checks which would be more reliably performed using RDF reasoning; some additional constraints may be enforced in a less severe manner: not enforced globally but treated on a case-by-case basis. We will use the existing D2RQ[12] tool to convert data stored within a Booster database into RDF format, making it available to the RDFUnit testing tool. D2RQ is a platform and language for accessing standard relational data, as that found in Booster, as triples. It is the basis from which the R2RML[13] W3C standard was developed. In D2RQ, each element of a Linked Data schema can be mapped to data from a relational database, using standard SQL queries, embodied in a mapping file in the D2RQ formalism. The additional testing and monitoring this enables will also provide insight into productivity and quality gains through use of the ALIGNED tools.
- Semantic integrity of the data can be compromised by a lack of understanding of the model. Here the Model Catalogue can be used to provide accurate descriptions of data fields, including those from linked external data sources. Such descriptions can aid correct data entry, and permit additional reuse of the data within the organisation. The Catalogue will also serve as a provider of models to the generated tools, and an environment where new versions of the data model can be created and evolved.

In the ALIGNED use case, the IPG domain model is edited and versioned within the Model Catalogue: Figure 6.15 shows a screenshot including a subset of the model. The model can be used as the foundation for a model in Semantic Booster, but the Catalogue is also able to generate

---

[12]http://d2rq.org/
[13]https://www.w3.org/TR/r2rml/

6.1 Wolters Kluwer – Re-Engineering a Complex Relational Database   223

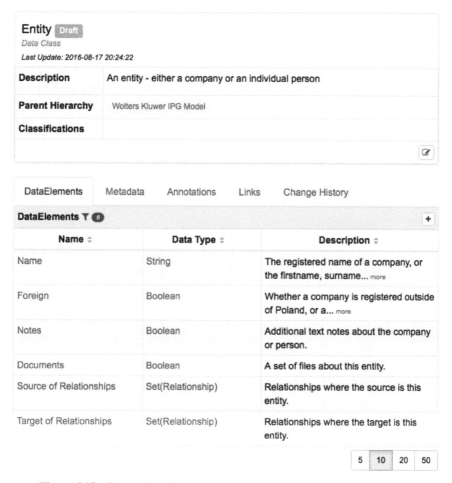

Figure 6.15 Screenshot of a subset of the IPG model in the Model Catalogue.

documentation files, data interchange specifications (such as in XML), and other useful system components.

The Booster model may be further edited within the Eclipse-based IDE (Figure 6.16) to extend concepts with further business rules and update methods. The Booster generation system is then used to generate a database with stored procedures for updates, a programmatic API, and a Web-based administrative interface (Figure 6.17).

The data within the Booster system can be extracted in RDF format using the D2RQ tool. These RDF data are now suitable for linking to external datasets, or further reasoning. The RDFUnit tool can be used for

224 Use Cases

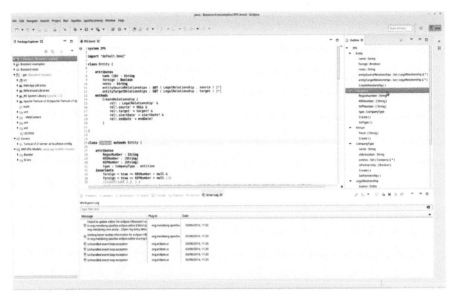

Figure 6.16 The Eclipse-based Booster tool.

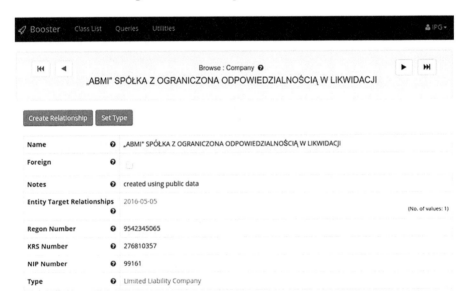

Figure 6.17 Screenshot of the Booster administrator interface for the JURION IPG system.

performing extra validity checks on these data (Figure 6.18) – checks that might be hard to describe or perform within a relational framework, or properties concerning relationships with external data.

6.1 Wolters Kluwer – Re-Engineering a Complex Relational Database

```
47  <urn:uuid:8d2adb48-2b2c-11b2-80c3-dc0ea1faf74d/8d2adb57-2b2c-11b2-80c3-dc0ea1faf74d>
43     a                         rut:TestCaseResult , rut:StatusTestCaseResult , rut:AggregatedTestResult ;
44     dcterms:date              "2016-07-11T07:58:58.296Z"^^xsd:dateTime ;
45     dcterms:description       "https://w3id.org/ipg#hasLegalRelationshipKind does not contain a literal value (htt
46     rut:resultCount           0 ;
47     rut:resultPrevalence      -1 ;
48     rut:resultStatus          rut:ResultStatusSuccess ;
49     rut:testCase              rutt:aligned.cs.ox.ac.uk_schemas_ipg.generated.ttl-OBJFUNC-39400effb8c242f0cf91e3b62
50     rut:testCaseLogLevel      rlog:ERROR ;
51     prov:wasGeneratedBy       <urn:uuid:8d2adb48-2b2c-11b2-80c3-dc0ea1faf74d> .
52
53 <urn:uuid:8d2adb48-2b2c-11b2-80c3-dc0ea1faf74d/8d2adb4d-2b2c-11b2-80c3-dc0ea1faf74d>
54     a                         rut:TestCaseResult , rut:AggregatedTestResult , rut:StatusTestCaseResult ;
55     dcterms:date              "2016-07-11T07:58:58.378Z"^^xsd:dateTime ;
56     dcterms:description       "https://w3id.org/ipg#hasKRSNumber has rdfs:domain different from: https://w3id.org/
57     rut:resultCount           0 ;
58     rut:resultPrevalence      -1 ;
59     rut:resultStatus          rut:ResultStatusSuccess ;
60     rut:testCase              rutt:aligned.cs.ox.ac.uk_schemas_ipg.generated.ttl-RDFSDOMAIN-4f6f956f9f7ec5ed20640f
61     rut:testCaseLogLevel      rlog:ERROR ;
62     prov:wasGeneratedBy       <urn:uuid:8d2adb48-2b2c-11b2-80c3-dc0ea1faf74d> .
63
64 <urn:uuid:8d2adb48-2b2c-11b2-80c3-dc0ea1faf74d/8d2adb70-2b2c-11b2-80c3-dc0ea1faf74d>
65     a                         rut:AggregatedTestResult , rut:StatusTestCaseResult , rut:TestCaseResult ;
66     dcterms:date              "2016-07-11T07:58:58.28Z"^^xsd:dateTime ;
67     dcterms:description       "https://w3id.org/ipg#hasEndDate does not contain a literal value (http://www.w3.org
68     rut:resultCount           0 ;
69     rut:resultPrevalence      -1 ;
70     rut:resultStatus          rut:ResultStatusSuccess ;
71     rut:testCase              rutt:aligned.cs.ox.ac.uk_schemas_ipg.generated.ttl-OBJFUNC-86dafa592722657289d4b04f6
72     rut:testCaseLogLevel      rlog:ERROR ;
73     prov:wasGeneratedBy       <urn:uuid:8d2adb48-2b2c-11b2-80c3-dc0ea1faf74d> .
74
75 <urn:uuid:8d2adb48-2b2c-11b2-80c3-dc0ea1faf74d/8d2adb4f-2b2c-11b2-80c3-dc0ea1faf74d>
76     a                         rut:StatusTestCaseResult , rut:TestCaseResult , rut:AggregatedTestResult ;
77     dcterms:date              "2016-07-11T07:58:58.403Z"^^xsd:dateTime ;
78     dcterms:description       "https://w3id.org/ipg#legalRelationshipType has different range from: https://w3id.o
79     rut:resultCount           0 ;
80     rut:resultPrevalence      -1 ;
81     rut:resultStatus          rut:ResultStatusSuccess ;
82     rut:testCase              rutt:aligned.cs.ox.ac.uk_schemas_ipg.generated.ttl-RDFSRANGE-807576591b580938ed3a7f5
83     rut:testCaseLogLevel      rlog:ERROR ;
84     prov:wasGeneratedBy       <urn:uuid:8d2adb48-2b2c-11b2-80c3-dc0ea1faf74d> .
```

Figure 6.18 Results of using the RDFUnit tool against data from a Semantic Booster database.

Dacura provides several services to software engineers developing software that utilises the data curated by the system. These include reliable access to data models, change notifications and the automatic production of simpler formats, which are more familiar to developers.

In order to verify the effectiveness of both approaches – Semantic Booster and Dacura – we created a list of unsolvable issues existing in the current Jurion IPG system. Based on the results of an evaluation, we will be able to determine from a business point of view, which approach suits best for a real business use case.

6.1.6.2 Architecture

Figure 6.20 shows the architecture of the platform that was constructed to support the Jurion IPG use case. The IPG system is shown on the left. It consists of a CMS and an SQL database, upon which a suite of Business Intelligence services have been developed – some of which access the database directly and some of which use the API provided by the CMS. In this scenario, we compare two alternative approaches to solving the IPG problems.

Use Cases

Category	ID	Problem
Information is not there or is wrong	1	The lack of the trustee in bankruptcy proceedings
Information is not there or is wrong	2	Member of management board in companies where management board doesn't exist
Information is not there or is wrong	3	Member of management board without function in board e.g. board chairman, vice chairman
Information is not there or is wrong	4	Commercial proxy without type of proxy e.g. joint commercial representation
Information is not there or is wrong	5	Proxy without type of proxy
Information is not there or is wrong	6	Lack of additional information about way of appointment of a trustee
Information is not there or is wrong	7	Limited partner without limited liability amount
Information is not there or is wrong	8	Are there multiple shareholders if company is labeled as „Sole Shareholder"
Information is not there or is wrong	9	Information about the suspension of a member of the management board - only YES or NO
Information is wrong	10	The same person in management board and as commercial proxy
Information is wrong	11	Member of management board is a member of supervision or a commercial proxy, a official receiver, a trustee
Information is wrong	12	Receiver is a member of supervision or a trustee
Information is wrong	13	Official receiver is a member of management board or a member of supervision ,a trustee, an appointed person
Information is wrong	14	Trustee is a receiver or an official receiver, a member of management board
Information is wrong	15	There should be at least one person (natural or legal) in representation (management board, partners, trustee)or receiver / official receiver at any moment in time.
Information is not there or is wrong	16	Partners without information like amount of shares
Information is not there or is wrong	17	Do value of partners shares at every moment in time is equal or lower than capital value.
Information is not there or is wrong	18	Lack of amount of capital value in joint stock company and limited liability company
Information is not there or is wrong	19	Lack of information about way of formation of a company only information about circumstances of formation
Information is not there or is wrong	20	Lack of information about circumstances of formation of a company only information about way of formation
Information is not there or is wrong	21	Did a company publish multiple annual reports.
Information is not there or is wrong	22	Lack of the post office in company address when is not the same like the place where headquarter is
Information is not there or is wrong	23	Lack of a date of validation of expunging company from the court registry
Information is not there or is wrong	24	Lack of title of organ - supervisory board
Information is not there or is wrong	25	PESEL No. with less digits than 11 when first digit is 0
Information is not there or is wrong	26	Email and web page address with space
Information is not there or is wrong	27	Lack of @ in email address
Information is not there or is wrong	28	@ in webpage address
Information exists, but is hardly understandable	29	Complexity of the data model in the table describing attributes for company and relationship – the table szczegol_instyt_watrosc
Information exists, but I can't do anything with that – no process for consuming info	30	Information from the legal notice about a ban on economic activity and ban to be a member in representation (management board), supervision (supervisory board or audit committee)
Process exists, but isn't working or is too slow so solution is unknown	31	Find out relationship at specific moment in time between company and company, company and person, person and person
Information exists, but is difficult to get	32	Cycle loop - find out if company A is owner of company B, than company B is owner of company C where C is owner of company A

Figure 6.19 Jurion IPG unsolvable issues.

6.1 Wolters Kluwer – Re-Engineering a Complex Relational Database

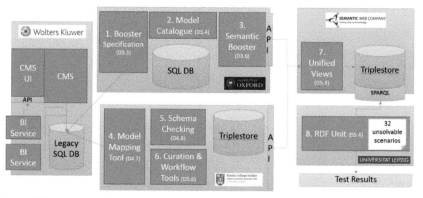

Figure 6.20 Jurion IPG use-case architecture showing integration across all major project tools and partners.

The first approach is provided by a configuration of the Oxford MDE approach. A booster specification is created (1) which generates SQL statements to extract the data from the legacy SQL DB and saves it in a format that can be managed by the Model Catalogue tool (2), the booster specification should ensure that these extracted data are correct by construction according to the booster specification. Then, these data are made available as RDF via Semantic Booster.

The second approach is provided by a configuration of Dacura services developed at Trinity College Dublin. The Model mapping tool (4) transforms the SQL schema of the legacy DB into an OWL ontology which is then used by the schema checking tool (5) to ensure that all data conform to the model. The curation and workflow tools (6) allow data managers to change the model and migrate the data and manage the process.

In order to properly compare the results of the two approaches, the RDF that they produce must be mapped to a common model – the UnifiedViews tool (7) provides this service and saves the resulting data to a triplestore. Finally, RDF Unit is used to test the output against the 32 unsolvable scenarios shown in Figure 6.19 to evaluate the success of the competing results. As RDFUnit supports arbitrary SPARQL queries, it is possible (although sometimes inconvenient) to encode all the evaluations as RDFUnit tests.

6.1.6.3 Tools and features

The tools and features used in the JURION IPG system are detailed in Figure 6.21.

228 Use Cases

Software	Type	RESTful API	Triple Store	SPARQL	Linked Data	Shared Ontologies supported
PoolParty	Platform	X	X	X	X	RDF(S), PROV, SKOS
Model Catalogue	Platform	X				RDF(S), OWL, PROV
RDFUnit	Command line tool		X	X	X	RDF(S), PROV, DQV, DataID, SHACL, RUT,
Semantic Booster	Command line Tool	X	X			RDF(S), OWL, PROV

Figure 6.21 Integration Paradigms and vocabularies supported by ALIGNED tools and platforms.

6.1.6.4 Implementation

Modifying Seshat Schema: Dacura provides tools to allow users to edit and modify ontologies on the fly (Figure 6.9). Using the Dacura browser plugin, users can browse the current Seshat code book and create properties and objects, adding them to the Seshat ontology and allowing researchers to collect information on these newly added properties.

Data complexity: Wolters Kluwer has managed and is still managing a tremendous business transformation process from a publishing house to a global information service provider (Figure 6.22). This development requires that high value-added services like IPG are also transformed from a traditional monolithic technical environment to a modular, flexible and sustainable infrastructure. Due to its data complexity and data quality issues (e.g., the main added value lies in the complex relationship model), ALIGNED tools can heavily support this transformation process.

Semantic Booster and the Model Catalogue. Semantic Booster has its strengths in the automatic model and software code creation process. It has also strong quality constraints so that no invalid data get into the transformation process. This approach was augmented by using RDFUnit for further data quality checks and which is the prerequisite to connect external open datasets to the IPG application in an easy and sustainable way (see Figure 6.23).

A booster specification is created which creates a model from the SQL database, along with formal constraints, which ensure that the data remain correct by construction. The Model Catalogue tool is then used to manage this data model. Semantic booster is used to make these data available as RDF via an API.

6.1 Wolters Kluwer – Re-Engineering a Complex Relational Database

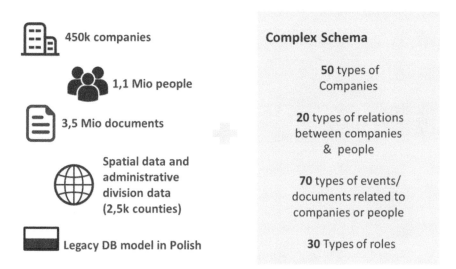

Figure 6.22 Complexity of the Jurion IPG use case.

How Aligned Tools are integrated

 We use the **Model Catalogue** to collaboratively manage a subset of the IPG data model. Future Releases will broaden the scope of the model.

Semantic Booster is used to generate a working implementation of the model in a SQL database with API and GUI. This tool also enforces a range of data integrity constraints and business rules improving quality already.

In addition, Semantic Booster exposes the data as RDF triples which is then used by **RDF Unit** to run advanced data quality checks. Having the Data in RDF opens us further for linking and integration into existing semantic web tools and data sets.

Figure 6.23 Integrating Semantic Booster and the Model Catalogue.

230 Use Cases

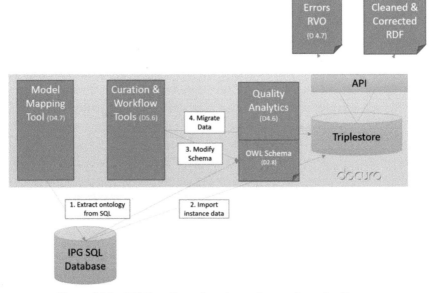

Figure 6.24 IPG Data Error detection and correction using Dacura.

Dacura provides an alternative method of achieving the same results (Figure 6.24). Firstly, the model mapper tool is used to generate an OWL ontology from the IPG SQL table structure (Figure 6.25). This ontology is deployed as the schema for the graph into which the instance data are imported.

Dacura's curation tools provide user interfaces which enable the data manager to view and modify the data and to analyse it for validation errors (Figure 6.26). The manager can use these tools to change the schema to include complex constraints on data quality. The results are provided as a cleaned, schema conformant RDF dataset and a list of errors expressed using ALIGNED's RVO ontology.

Unified Views: in order to ensure that the results of the validation processes carried out by Dacura and Semantic Booster can be evaluated, they must be mapped to comparable schema for testing. The Unified Views tool is used to manage this mapping and transformation and to save the transformed data to a triple store.

RDFUnit: Each of the unsolvable issues is encoded as RDFUnit scripts, which run SPARQL queries against the final data to check whether the issues are still present in the data. These queries are run against both the data

6.1 Wolters Kluwer – Re-Engineering a Complex Relational Database

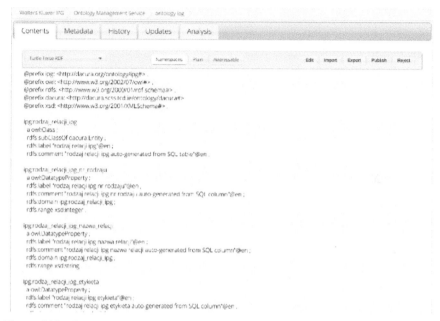

Figure 6.25 Ontology generated from IPG SQL database by Dacura's Model Mapper Tool.

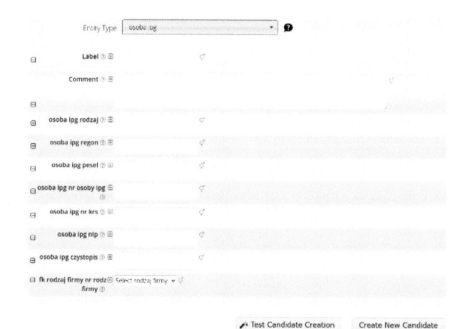

Figure 6.26 Using Dacura's curation tools to analyse the IPG data model.

produced by Dacura and by Semantic Booster. In situations where issues still remain, RDFUnit can be used to fix some of these outstanding issues.

Conclusion: The platform produces a result set which describes the errors that have been found in the data which can be passed back to the DB administrators for correction.

6.1.6.5 Evaluation
Measuring Overall Value

The IPG database supports a variety of business services for customers which provide revenue to Wolters Kluwer – a very clear measure of the value provided by the system. The most important metrics in this case are: firstly, the curation cost of maintaining a given quality of service as measured by revenue; and secondly, the cost of improving the overall quality of service to provide more value and increase revenue, for example, by adding new features and new business services to the system. As the scale of the system – both in terms of the size of the database and the complexity of the services consuming the data – has increased, the curation costs have increased to such a stage that the cost of improving Quality of Service may be greater than the increase in business value that will accrue.

Data Quality Dimensions and Thresholds

Accuracy of certain information is very important in a commercial intelligence system. For example, accurate identification, accurate contact information and accurate shareholder and other relationship information are all significant in terms of the overall value provided to customers by the system. Users will tolerate some errors, but there is a threshold at which they will lose confidence in the value provided by the system.

Model Agility

The existing data model in the IPG SQL database have evolved to a stage where it is difficult to understand. Database access is heavily optimised and any changes to the structure require modification of caching and other optimisations. In addition, if we change any part of the structure of the database, we will likely break existing programs which use that part and it is very expensive to change the code of existing programs. Rather than attempting to modify the existing data model, a path to migrating to a new and easier to understand data model is required. In this case, the task is to

6.1 Wolters Kluwer – Re-Engineering a Complex Relational Database

create a new data model that is easier to understand than the existing model without increasing the overall complexity of the system.

Data Agility

Once again, scale is the obstacle to overcome in enabling data agility in IPG. The overall system is highly optimised, and it requires significant effort to integrate any changes into existing infrastructure, testing, and so on, before we can safely ensure that the new service will not negatively affect the overall QoS of the system. Particularly problematic are queries that contain many complex joins and programs that make repeated round trips to the database, for example by executing a query in a loop, as they can put high load on the server and decrease the QoS across the entire system. To evaluate the overall system's data agility, the task is to create a new application, based on existing data, which identifies illegal relationships between people and companies.

Task 1: Curation

Identify and correct 10 different types of significant errors in the database.

Before ALIGNED

- Write queries to extract all the fields to be analysed from the database
- Write and maintain functions to analyse the extracted fields for the various types of errors identified and apply auto-correction where possible.
- Develop and maintain a system which allows users to view and correct identified errors
- Operate this system until the error rate has fallen below the desired threshold.
- Write queries to insert the corrected values back into the database.
- Either of:
  - Trace the source of errors back to the programs that produced them, fix and redeploy the programs
  - Periodically test the Database and rerun the process if quality levels have fallen below minimum thresholds.

Task 2: Model Agility

Create a new, easier to understand model and deploy it so that all new services can use the new model while existing services still use the old model, with state being shared between them, without increasing the overall complexity of the system, increasing the data curation costs or reducing agility.

Before ALIGNED

What makes this problem hard is that, with current technologies (SQL), this requires a complete refactoring of the database and all the services that use it and the solution, at best, slightly improves the problem by slightly increasing the scale limit at which agility drops to zero. If we allow the existing data model to remain untouched, so that we do not have to change the code of existing services, and use middleware or an ORM architecture to create a new model for new data and link it to the existing model through code, we do not solve the problem at all – it actually makes it worse by introducing another layer of code that must be understood and maintained in order to understand or change the data or model. Thus, the only real alternative, using current technologies, is a complete refactoring of the database and re-engineering of the entire system and all of its services to use the new database schema.

Carry out complete system refactoring, with new, simplified schema, existing data mapped to that schema and all software services updated to use new model.

Task 3: Data Agility

Implement a new application which uses the IPG data to identify several different types of 'illegal' relationships between people and companies.

Before ALIGNED

- Write, test and maintain complex recursive SQL queries using CTE or CONNECT BY syntax to identify relevant instances.
- Write and maintain program to execute queries against database and return results
- Write and maintain program to browse and display result

6.1.6.6 Experimental evaluation

We used ALIGNED tools to address 31 unsolvable problems identified by IPG. This required us to complete the following tasks:

- Create a semantic model to represent the entities and relationships referenced in the IPG unsolvable problems.
- Create a mapping from the IPG SQL schema to the ontology

- Import the data to the ontological version and analyse it using the 22 quality constraints represented statically in the ontology
- Run a set of queries against the knowledge base to identify 8 of the 31 problems that could be expressed as graph queries.

We measured the time and effort required to complete these steps and identify all constraint violations in the SQL data, covering 30 of the 31 problems. The vast bulk of the effort was expended on scaling our reasoner and storage engine and toolchains to handle the scale of the data, as, when transformed into semantic representations, the IPG database amounted to tens of billions of triples.

We established a pipeline which lifts IPG's SQL schema to an OWL ontology and then converts SQL row data into triples. We then transform the data triples using our Dacura mapping tool into an ideal schema. At this point, the instance data are checked for consistency with the ontology.

The results of the consistency check are reported as a file of JSON objects which elaborate the problem and its source. After running the consistency check, we found an initial set of 2,103,583 errors. A high proportion of a sample of these errors are genuine and have been verified ($>95\%$), but since work on the project is ongoing, the number of false positives has not been completely determined.

The ontology design phase took around two days of expert ontology designer time. Since most of the importation is automatic, the process took around one day of developer time. Mapping of triples into the ideal schema took an additional one day. The total development time is then around one working week of effort.

Since IPG's original use case described "unsolvable" problems, i.e., problems deemed too difficult to solve given the state of the current software and database setup, setting a baseline is somewhat difficult.

However, these problems are not genuinely unsolvable in abstract but simply too time intensive and expensive to solve. We can estimate the time that it would take in excess of one month of developer time, and very likely 6 months in order to find the $\sim$2 million errors. This gives a range of speedups from using our methodology of between 400% and 2400%.

6.2 Seshat – Collecting and Curating High-Value Datasets with the Dacura Platform

The Seshat: Global History Databank is an international initiative of humanities and social science scholars to build an open repository of expert-curated historical time-series data. The Seshat project began by selecting a sample of 30 areas from around the world. For each area, they recorded all societies that had controlled it throughout history, and answered over a thousand questions about each – describing its population, technology, religion, infrastructure, and so on. This made it possible to answer a wide range of questions about each of them – describing its population, technology, religion, infrastructure, and so on. The Seshat has been designed to test theories about the evolution of social complexity, from the point of view of historians and anthropologists. The databank extracts data from a combination of databases, Linked Data, websites, academic publications, and human experts.

A special code book defined the full list of questions, and researchers added data to the system by creating a copy of the code book page for each society, and adding data points using a special syntax that encoded uncertainty, disagreement, and temporal scope, along with comments and citations in relation to domain-specific provenance information. In the initial stages of the Seshat project, a wiki was used to collect the data. The system amassed over 200,000 data points on hundreds of civilisations, but whilst the unstructured wiki data store allowed great flexibility at the start of the project, it did not scale to the number of contributors, data users, data points, or the complexity of the data.

The Seshat evolved to encompass new areas that were not originally anticipated. In particular, this involved recording societies from the prehistoric past, which required a collection of archaeological data. It soon became obvious that many Seshat variables were unsuitable for capturing this part of human past. There was also a lack of relevant proxies that would allow translation of archaeological evidence into coding templates. Accordingly, the Archaeological Seshat code book was designed and developed in order to fill in the gap, and the data were collected independently.

A wiki-based approach, used in Seshat for the data collection task, posed numerous problems, in particular for the verification of data correctness, and the extraction of data in usable forms. As the dataset grew and the focus moved from collection to integration and analysis, several other significant problems emerged. The fundamental problem is that a wiki is designed for human presentation and editing of data. To a machine, it is semi-structured, which lacks any type information and the meaning of the elements depends on their context within a jumble of HTML. Without any support for validation, errors proliferated.

The limitations of the wiki also impacted agility. As the Seshat code book was rapidly evolving, any changes needed to be manually copied to all existing data pages was a costly and error-prone task. There was also no easy way to express spatial data through the wiki, so these data were stored in separate GISs. This solution also offered no support for publication, while the scraping tool could extract the raw datapoints, citations, and comments were also important but were encoded in totally unstructured HTML.

Productivity suffered as increasing resources had to be devoted to curation and cleaning. Some of the corrections were not copied back to the wiki, and spreadsheets became the authoritative source for some sections of the data. Moreover, there was no way of incorporating third-party data into Seshat dataset.

6.2.1 Use Case

6.2.1.1 Problem statement

A group of researchers, distributed geographically and across multiple teams and disciplines are collaborating on the compilation of the Seshat dataset describing human social evolution since Neolithic times. The goal is to record geo-temporal time-series datasets describing how hundreds of variables describing social complexity changed with time and place. The Seshat researchers are currently using a wiki and a polity-based template, which includes a simple syntax for encoding machine-readable variable values, to collect the data. The Seshat researchers can currently be roughly broken down into three roles:harvesters – typically RAs who are paid to input data to fill the datasets on a per-polity basis;experts – scholars with specific expertise in particular geo-temporal slices of human history, their role is to correct, interpret and validate the data for their particular areas of expertise;architects – the core Seshat editorial team, who are responsible for designing and modifying the dataset Schema. The high-level goal of ALIGNED in this use case is to produce tools for the Seshat researchers which will increase productivity and data quality and improve the availability of data for analysis.

Actors

Role	Description
Harvester	non-technical users who add and update data
Editor	moderate, correct and manage the data in the system over time
Expert	use domain-specific knowledge to analyse and interlink data in the system
Architect	make changes to the schema and manage transitions between schema versions

238 Use Cases

The requirements on which the Seshat use case was based are detailed in Appendix A.

6.2.2 Architecture

Figure 6.27 shows the architecture of the system that was developed to support the Seshat use case trials, highlighting the places where common ALIGNED integration paradigms and ontologies were exploited. The system demonstrates integration between three of the project's major suites of tools, and three of the project's use cases.

The full suite of Dacura tools form the core of the system, importing the data, ensuring it meets consistency requirements, automatically producing user-interfaces and curation tools to allow the expert contributors to use the system without any knowledge of the underlying semantic technologies being used, and finally publishing the data and making it available to software engineers.

The Model Catalogue tool was used to help develop and manage the ontologies used by the system – it supports OWL models and provides a RESTful API to support easy integration with third-party tools and incorporating into complex workflows.

The Unified Views tool, developed as part of the SWC's PoolParty semantic suite was used to manage the integration of datasets from third-party datasets into the Seshat dataset. In this case, the DBpedia knowledge base was the data source being exploited.

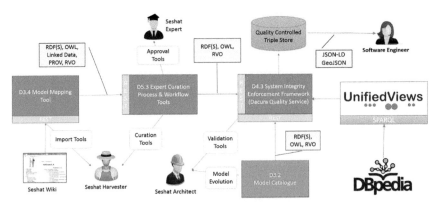

Figure 6.27 Seshat Use Case Trial System Architecture, showing the tools provided to different Seshat users, the use of ALIGNED integration standards and interoperation paradigms.

6.2 Seshat – Collecting and Curating High-Value Datasets

The final platform thus directly incorporated the research outputs of six of the seven research groups involved in ALIGNED and demonstrated integration across three of the use cases – DBpedia provided data and the PoolParty use case provided tools and expertise in establishing the data import pipelines.

In April 2017, this platform was used to prepare and publish the first public release of data from Seshat, which in an attractive and well-structured format for appraisal by other researchers – particularly focussed on scientific reviewers who needed to evaluate the data on which several of the major Seshat publications were based. Since then, the project's major focus has been made to deploy the system in a software engineering context, which has involved making the RDF/OWL data stored within the system available in simpler forms, such as GeoJSON and JSON-LD available to software engineers.

The platform has been constructed to support the following Seshat data curation tasks:

- Importing the large volume of wiki data that they have accumulated in a semi-structured form, into a structured, rich semantic format according to a pre-defined model, that is amenable to statistical analysis and automated quality control.
- Analysing the data to identify a large number of new constraint violations – e.g., datatype constraints, referential integrity constraints, cardinality constraints. In the current Seshat data collection workflow, such problems only show up at data-analysis stage and it requires a very significant manual effort to amend them at that late stage.
- Providing approval pipelines and workflow tools to allow moderators to inspect and correct problems identified in the data and to give them the agility to be able to use lower-skilled data collectors with higher error rates without sacrificing overall quality.
- Providing model rapid prototyping tools to allow our archaeologist partners to experiment with the definition of large new segments of the Seshat schema to allow them to define semantic mappings between entities at different levels of abstraction and time-depth. This supported the accumulation of archaeological evidence and extended the timedepth of the Seshat ontology which was initially conceived primarily to investigate societies that were historically known.
- Importing data from third-party datasets such as DBpedia and Pleiades historical gazette and integrating it with the existing Seshat data.

Software	Feature	Used for
Model Catalogue	Model Definition User Interfaces	Model Prototyping
Model Catalogue	Model Export to OWL	Model Integrity Enforcement
Dacura	Model Mapping Tool	Importing wiki data
Dacura	Real-time Instance Data Validation	Testing imported data
Dacura	Model Generated User Interfaces	Correcting imported data
Dacura	Curation Workflow Tools	Update approval queues
PoolParty	Unified Views ETL	Import third party data

Figure 6.28 Features of the ALIGNED tools used to support the Seshat trials.

6.2.2.1 Tools and features

Figure 6.28 shows which features of the ALIGNED software tools have been deployed in order to support these scenarios.

6.2.3 Implementation

6.2.3.1 Dacura data curation platform

The Dacura Linked Data curation platform1 is developed at Trinity College Dublin. Dacura provides support for dataset capture, curation, and publication. The major components of Dacura in the context of the Seshat use case are shown in Figure 6.29.

In the initial prototype developed for Seshat four of the components from Figure 6.29 are used: (1) the wiki data entry/validation tools (top left in the figure) which are user-facing data curation widgets; (2) the schema management tools which include the Model Catalogue tool in the demo; (3) the data quality controls (lower middle of the figure) which perform schema and data integrity checks and act as a data quality gatekeeper for the RDF triple store; and (4) the data export tool or wiki scraper which can transform Seshat data into the TSV file dumps required by statistical analysts within Seshat.

6.2.3.2 General description

Dacura provides tool support to improve the efficiency and accuracy of Seshat's data collection processes.

- The wiki data entry/validation tools make data entry easier for Seshat researchers. This also assists in collecting more complex data and data validation at the point of entry.
- The schema management tools check that OWL-based schemas are consistent and correct as they grow.

6.2 Seshat – Collecting and Curating High-Value Datasets

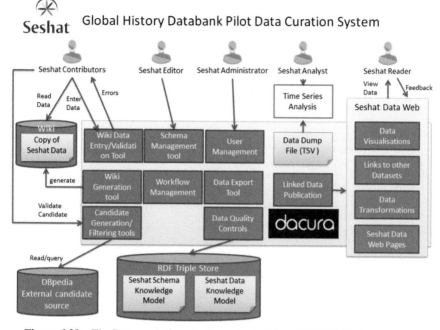

Figure 6.29 The Dacura platform in the context of the ALIGNED Seshat use case.

- The data quality controls ensure that both data entered through widgets and data already entered in the Seshat wiki is checked for conformity with the Seshat schema before it is added to the triplestore.
- The data export tool allows multi-format data publication. It also allows Seshat administrators to get a first look at how their dataset is growing and evolving.
- The Seshat OWL ontology developed for this demonstrator and used by our tools enables more structured information to be captured than the original Seshat.

6.2.3.3 Detailed process

This section describes the use of each of the components developed for the demonstrator system, along with a screenshot of the components in use where applicable.

Dacura data entry validation tools (Figure 6.30) are embedded in the Seshat wiki, allowing researchers to validate previously entered data and add new variables to the dataset. Researchers can validate or enter data directly

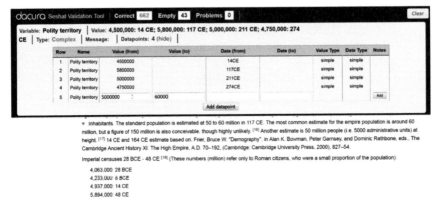

Figure 6.30 Screenshot of TCD's Seshat Data Entry/Validation tool in Demonstrator System.

from the wiki page. These tools reduce the complexity of entering data in the wiki, as the need for complicated syntax is reduced and any errors in data will be immediately revealed. A version of this tool that supports validation of data entered into the wiki has already been deployed in the live Seshat system.

Modifying Seshat Schema: Dacura provides tools to allow users to edit and modify ontologies on the fly (Figure 6.31). Using the Dacura browser plugin, users can browse the current Seshat code book and create properties

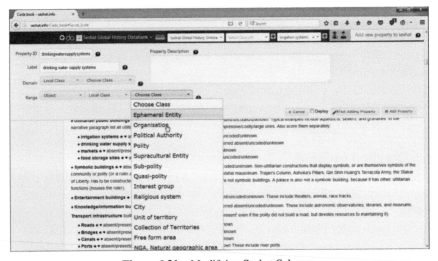

Figure 6.31 Modifying Seshat Schema.

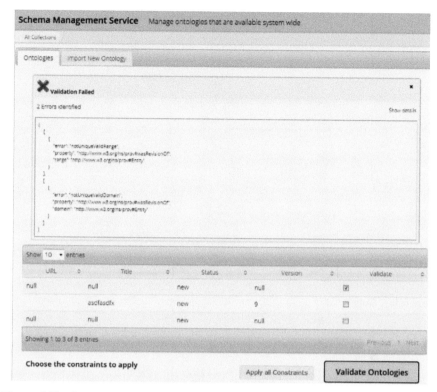

Figure 6.32 Screenshot of TCD's Schema Management component using the prototype integrity enforcement framework in the Demonstrator System.

and objects, adding them to the Seshat ontology and allowing researchers to collect information on these newly added properties.

Triplestore integrity enforcement (Figure 6.32) is a key feature of Dacura. Preventing data that are not in accordance with the schema or preventing a malformed schema from entering the triplestore ensures that all data are of high quality. This reduces the need for Seshat researchers to spend time correcting errors in the dataset. Dacura checks that imported vocabularies are consistent before allowing them to enter the triple store and constrain instance data.

The schema management components (Figure 6.33) in Dacura allow changes to the schema to be analysed to ensure that the schema remains consistent. Checks are performed on schemas before they are used in the data store, highlighting errors and potential issues for attention. A range of

244   *Use Cases*

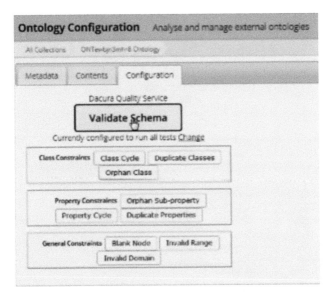

Figure 6.33   Screenshot of TCD's Schema Validation Service in Demonstrator System.

checks are performed using a constraint-based interpretation of the Seshat OWL ontology.

Finally, the wiki export component (Figure 6.34) extracts the historical data entered in the Seshat wiki, parses them, and produces a TSV of these values ordered by date. This allows Seshat administrators to perform analyses without needing to manually extract the values from the large and constantly growing dataset. It also produces error reporting, allowing researchers to identify errors in the dataset and see how the data are evolving.

The introduction of the Dacura data validation component into the live Seshat data collection process has reduced the rate of errors in the Seshat wiki. Despite a large increase in the size of the wiki of 29% (from 56,160 to 72,252 data points) between March and June 2015, the absolute number of errors has decreased by 19%. The rate of errors per variable has decreased by 42%, from 0.035 errors per variable to 0.02 errors per variable. This shows the positive impact of deploying data quality/data curation tools on the Seshat workflows. This trend is shown in Figure 6.35.

Managing complex workflows: The Dacura approval queue allows dataset administrators to monitor added data for quality and completeness (Figure 6.36). Administrators can approve, deny, publish and unpublish the Linked Data objects submitted by Seshat researchers.

6.2 Seshat – Collecting and Curating High-Value Datasets 245

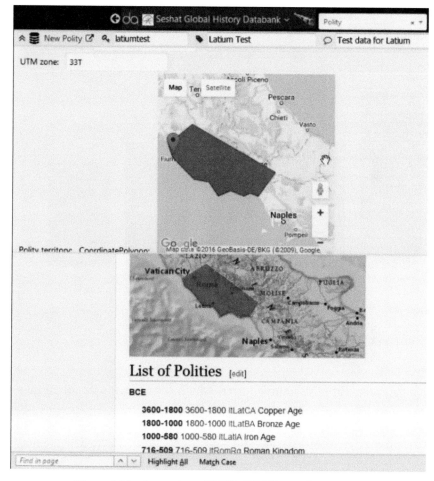

Figure 6.34 Screenshot of TCD's Wiki Export Component.

Importing third-party datasets: The Unified Views tool (Figure 6.37) allows data to be imported via SPARQL from third-party datasets, in this case, DBpedia is used as a source of data. Unified Views allows the establishment of processing workflows to automate the importation of such data.

Publication: The Dacura system allows the generation and publication of processed and curated data in easy-to-use forms. Casual users can browse Web pages (Figure 6.38 which lay out the information in a simple and structured manner). Seshat team members looking to perform analysis on

246 *Use Cases*

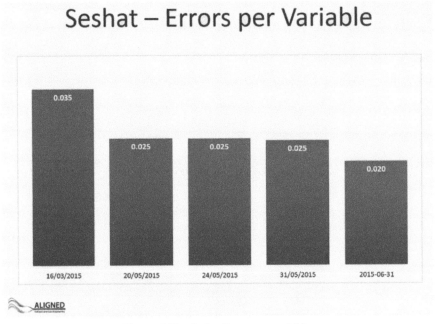

Figure 6.35 Seshat Errors per variable.

the data can access it in structured forms, which can be easily imported into analysis software.

Software Engineering support: The Dacura system provides several services to software engineers developing software that utilises the data curated by the system (Figure 6.39). These include reliable access to version controlled data models at a well-known URL, change notifications and the automatic production of simpler formats which are more familiar to traditional Web-developers. In this case, a GeoJSON stream is automatically made available describing all the features in the dataset that have a geographical location associated with them.

6.2.4 Overview of the Model Catalogue

When dealing with large, complex datasets, it is important to have tools to help collaborators understand what each data point means, how it has been collected, and how groups of data points may be interrelated. Typically, a large number of tools are used for this kind of metadata management:

6.2 Seshat – Collecting and Curating High-Value Datasets

Figure 6.36 Managing Complex Workflows.

data dictionaries for storing information about variables and allowed values; data manuals or procedures for describing the intended meaning of data points; specifications of forms describing how data are to be collected; or diagrams describing relationships between groups of data points. The Model Catalogue toolkit is being developed at Oxford University for the purposes of collaborative editing and sharing of such documents within a common framework. Based on previous work on international standards for metadata registration and previously explored in the context of clinical research, the tool is now being developed and extended to support the Seshat use case.

The tools have been built with a model-driven software development process in mind: programmatic interfaces allow communication between the

248 Use Cases

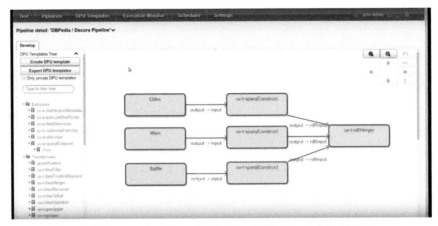

Figure 6.37 Importing data to Seshat from DBpedia with Unified Views.

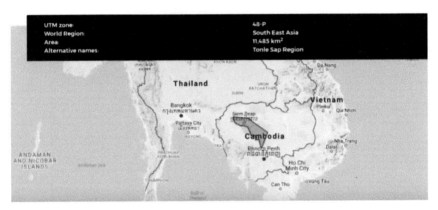

Figure 6.38 Publication.

catalogue and other systems; a number of export tools have been written to automatically produce or configure software artefacts such as databases or data messaging schemas from data model descriptions stored in the catalogue.

The extended tool is initially designed to support two key use cases – for the Seshat editors and data managers to cooperate in the incremental evolution and description of the Seshat data model or code book; and for researchers interested in using the Seshat data to understand which data points have been collected and their meaning and provenance. To support

6.2 Seshat – Collecting and Curating High-Value Datasets

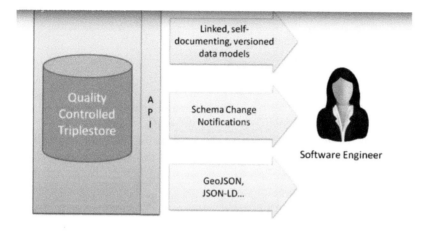

Figure 6.39 Services to support software engineering.

Figure 6.40 The Model Catalogue user interface showing a section of the code book.

the first use case, the toolkit provides facilities for automatic import of existing documents and software artefacts to initialise data models, and uses careful structuring to minimise the amount of user input required. The Web-based editing environment (see Figure 6.40) promotes collaborative editing of modes, with processes for publication and versioning. To support the second use case, the Web interface allows exploration of data models, and the creation of user-friendly reports or exports. Data points can be linked to provide additional meaning or context and can be compared to understand differences between different datasets, or a single dataset over time.

6.2.4.1 Model catalogue in the demonstrator system

This section describes the Model Catalogue components deployed in the demonstrator system. The section starts with a discussion of the Seshat use cases addressed by the components and then provides a description of the model curation processes supported by the Model Catalogue component. Finally, there is a subsection presenting some initial results from the deployment of the components.

General description

The Seshat databank is a complex dataset comprising more than 1,000 variables categorised into approximately 100 groups or classes. A code book describes each group and variable: a description about the intended meaning, or semantics, of the data points collected against it; a selection of possible values that the data point may take; links to related or similar variables in another category.

Furthermore, this set of variables has been evolving and expanding; in 3 years, more than 300 revisions to the code book have been made: variables have been added, removed, or extended; descriptions have been enhanced; the permitted range of values may have items added or removed.

This code book holds the key to understanding how data points should be collected and stored and how potential users of the data can make sense of what is made available. Until this point, the Seshat code book has been encoded into a wiki page. This has served the purpose required, but in order to scale, a new approach may be required. The Model Catalogue built by OxSE is intended to support:

- Collaborative editing of the code book as a curated data model
- Annotating variables and linking to related datasets or standards
- Versioning and publication life cycles for change management
- Informed reuse of data collected, through discovery and exploration and comparison of data models and variables
- Automated import and export of models into other data formats
- Generation or configuration of software artefacts within an iterative development approach

The Model Catalogue prototype makes progress towards schema evolution. The "Expert" and "Architect" user roles can modify the schema, allowing view-only access to the schema for "Harvester" and "Editor" user roles. There is also some progress towards addressing expert interpretation, where experts can us the Model Catalogue to understand and modify the code book to improve support for capturing complexity in the Seshat databank.

Detailed Process

The Model Catalogue stores concrete models such as dataset descriptions, form designs, database schemas, and so on. alongside an abstract representation common to all models. Each Data Model contains a number of Data Classes, which in turn may contain sub-classes and Data Elements, in our case representing the variables of the Seshat code book. This structure provides easy interoperability between models, and allows a simple tree-view for viewing and exploring models, such as in the screenshot shown in Figure 6.41. This view is useful for those exploring the structure of a data model: users interested in requesting or working with items of data, or editors wishing to make changes to the structure.

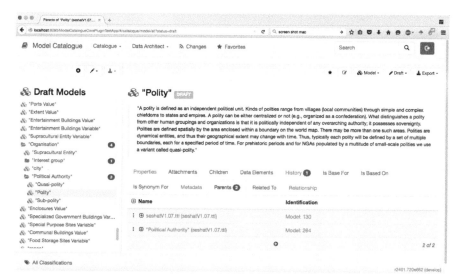

Figure 6.41 Screenshot of the Model Catalogue Web interface, showing the 'tree view' and a section of the Seshat code book.

252 Use Cases

Data models and their components may be linked with a number of different relationships describing the type of similarity between them. A plugin for discourse has been integrated: this allows users to comment on parts of a data model in a familiar fashion. Comments may include links to other data elements, or mention of other users, who can be prompted to respond. Attachments can be added too: links to websites, or file attachments giving more information about the meaning of a variable.

For prospective users of the data, it is important to understand how data collected against different versions of the code book may be related. Figure 6.42 shows a screen written for this purpose: highlighting differences in descriptions, sub-components (for models and classes), and datatypes (for data elements). Sophisticated search functionality allows data elements to be found within all models; all finalised models, or within the currently displayed model.

Model importers allow existing structures to be imported into the catalogue without manual transcription. One such importer has been written to ingest the OWL description of the Seshat code book, extracted by the TCD team for use in Dacura. Although written with Seshat in mind, this component may be re-configured for use in the other ALIGNED use cases during the next phase of the project.

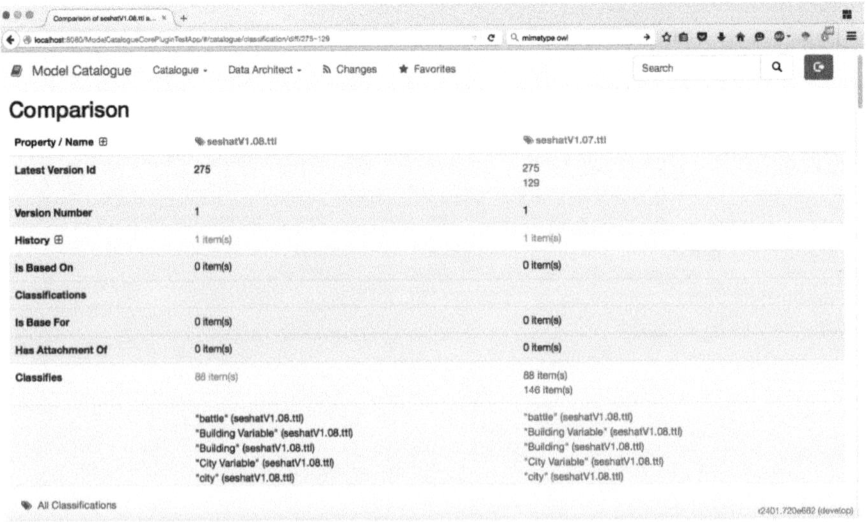

Figure 6.42 Screenshot of the Model Catalogue Web interface showing the comparison between two versions of the Seshat code book.

6.2.5 Seshat Trial Platform Evaluation

6.2.5.1 Measuring overall value

Ultimately, the most important metric for the Seshat project is the cost of going from a hypothesis to a published scientific paper presenting an empirical evaluation of that hypothesis. Everything else is a means to that end. The nature of the variables chosen and the data collected are explicitly designed in order to enable computational analysis of particular hypotheses against historical evidence. The validity and significance of the analysis is then validated by the peer-review process of the world's top scientific journals [1, 2]. The best proxy for the value delivered by the overall system is the number of papers that are published in top-tier scientific journals.

6.2.5.2 Data quality dimensions and thresholds

seshat has unusually high Data Quality requirements across a large number of separate dimensions. The Seshat researchers need to be able to analyse the data statistically, which imposes high thresholds for syntactical accuracy and structural integrity. Furthermore, the historical accuracy of the data is also extremely important – as the Seshat researchers want to be able to identify patterns in long term historical processes in order to make predictions – these predictions can only be as good as the accuracy of their data. Because it is often the case that historical facts can only be known probabilistically, uncertainty must be incorporated into the data in such a way that statistical analysis can still be applied. Furthermore, it is the norm in top-tier scientific journals that datasets are scrutinised closely by reviewers. Because Seshat is pioneering new data-driven methods in the social sciences, it is thus particularly important that published datasets are robust in the face of expert scrutiny, as any significant errors risk undermining the credibility of the approach and not just the individual publication.

In terms of data agility, the most important requirements for Seshat are the ability to make data, of the required quality, available for analysis with the programs used by Seshat's data-analysts (R, Mathematica) and to make the data available for inspection for academic reviewers in whatever format will make the best impression upon them.

In terms of model agility, the most important requirements for Seshat is the ability to make changes to the 'code book' to reflect the input of new experts and experience. The Seshat code book has changed many times over the last few years as the community has grown and more expert opinion has been incorporated into the selection of variables and proxies to collect.

254   Use Cases

The Seshat data collection effort started some 2 years before the start of ALIGNED and has continued for almost 5 years at this stage. There are currently two major papers which are under review for publication in major top-tier scientific journals (both resubmitted with changes, as advised by the editorial committees). All the data that went into these publications were collected, curated, cleaned, and analysed using methods that pre-date the introduction of ALIGNED technology, with one exception – the publication of data for the reviewers. This provides us with a very good baseline measure with which we can evaluate the impact of ALIGNED technology: the total cost of producing these two papers – including all the cost that went into producing the final publication and associated datasets.

We can break it down into the following tasks:

1. Running expert workshops to identify interesting hypotheses and suitable proxy variables to form the dataset's schema.
2. Employing and training research assistants to fill in data on a wiki for each historical society of interest
3. Soliciting reviews from volunteer experts to validate the data entered in 2.
4. Developing and maintaining programs to extract data from the wiki, identify syntactical errors, make it available as a CSV for analysis and then transform it into the format required for each analysis and then, finally, perform the analysis and produce the results.
5. Employing and training research assistants to correct data entered in 2 in response to errors identified in 4, and reviews received in 3, to ensure that it remained accurate and true to the schema over time, as the schema changed to reflect the outputs of 1.
6. Developing and maintaining a program for publishing the data for appraisal by the general public and reviewers.

All these tasks were necessary for the production and publication of the journal papers and their associated datasets and must be included in the consideration of the system's productivity before the introduction of ALIGNED tools and methods. Furthermore, by looking at the cost of achieving required data quality levels as the schema was changed over this period and the cost of making data available in new ways, we can gain a reasonable estimate of the likely future productivity of the system. In order to demonstrate the impact of our tools and methods, therefore, we would ideally repeat the process from start to finish with our tools, starting from a new set of hypotheses and compare the overall cost of the two.

6.2 Seshat – Collecting and Curating High-Value Datasets

In December 2018, at a Seshat workshop in Oxford, in association with the ALIGNED general meeting, we began such an evaluation. In this study, we apply our tools and methods to the entirety of this process – from modelling tools to support 1, curation interfaces to support 2 and 3, data manipulation tools to support 4 and 6, error detection and correction tools to support 5. We have identified a set of new hypotheses that we wish to test and we will measure the total cost of transforming these hypotheses into published scientific papers. However, this is an irreducibly time-consuming task due to the nature of the domain. In particular, experts are a scarce resource and it inevitably takes considerable time to elicit all the domain knowledge necessary from the relevant experts for any particular historical question. Therefore, the full results of this comparison will not be known until the entire process is complete, which will be beyond the timeframe of ALIGNED.

However, what we can do in a short space of time is to measure the impact of our innovations on steps 4, 5, and 6 of the process and compare it with the existing system. The Seshat wiki contains several sections of data that have been collected by RAs and approved by experts but have not yet been cleaned, analysed or published. We choose one of these sections, comparable in terms of size and complexity to the datasets that were used to publish the first two papers. Starting from this point, we extract, clean, analyse and publish the data to the same level of quality as was achieved with the already published datasets and we measure the total cost in doing so, in terms of both time and money. This comparison will provide us with a minimum valuation of the benefits of our technology to the system's curation costs. As these are all necessary steps, any productivity gains in this section will be realised generally.

In order to measure the likely future value and productivity that the system will exhibit, we need to estimate the likely cost of changing the schema and repurposing the data for new uses. For schema changes, we can simply use those changes that were introduced during the compilation of the published papers, measure how much it cost to achieve the required data quality levels after those changes and repeat the experiment with the new system. For data agility, we can measure the cost of making the data available for review with our tools and estimate the amount of effort that this would have required had our tools not been available – as this was the one part of the process of publishing the initial two Seshat papers that our tools were responsible for.

Together these three measures provide a very comprehensive way of evaluating and comparing the performance of the two alternative systems of interest – Seshat before and after ALIGNED tools were introduced. By looking at the likely evolution of scale of the system, we can provide reasonable estimates not only about the current value provided by our tools and methods, but the future value that they will provide.

Thus, in this document, we evaluate the system that was introduced by ALIGNED against the existing wiki-based system in terms of the cost of three particular tasks:

Curation

Extract, clean, and analyse a new section of wiki data to test a hypothesis so that the quality of the data and analysis is considered ready for publication.

Data Agility

Publish the data in a form that allows reviewers to evaluate all the data that went into the analysis, complete with sources, citations, uncertainty, and disagreement.

The major task that required data agility in Seshat was the publication of data for reviewers. The requirements were that the reviewers should be able to browse the dataset, view every individual datapoint with citations and indicators of uncertainty and disagreement, and that they, or any member of the public, could provide feedback on each individual datapoint. This was deemed important as a way of signalling to the research community that expert feedback and corrections were invited and taken seriously. This task was never competed with pre-ALIGNED methods, ALIGNED tools were entirely responsible for the dataset that was published to support the first major Seshat publication in April 2017. However, it is possible to provide a reasonably accurate estimate of how much it would cost using a standard software engineering approach:

- Develop and maintain custom program to transform Seshat extracted CSV into interactive website.

Model Agility

Change the schema by adding, removing, and changing the definition of variables, adding complex relationships between entities and returning the system to the quality level it had before the schema was changed.

For all schema changes, we need to carry out the following tasks:
1. Update Seshat code book wiki page
2. Copy updates to every wiki page that uses that section of the code book.
3. Update, test and deploy publication program to reflect changes in schema
4. Update and test script to transform CSV into analysis-ready format

Experimental Deployment

On 21 December, 2017, the first major paper was published utilising the full power of the Seshat data in the Proceedings of the National Academy of Sciences.[14] ALIGNED tools were used to model, harvest, correct, improve, enrich, and publish the full data describing social complexity for 416 different historical polities that were used for the analysis. The analysis which formed the basis of the paper was carried out using the old – before ALIGNED – process. Upon publication, the published data were fully migrated to the Dacura platform and the processes where compared to evaluate the relative productivity, agility and quality – and the overall cost – of the process with and without the ALIGNED tools.

On 17 January, 2018, all of Seshat's social complexity data were imported into a semantic model from the Seshat wiki. Figure 6.43 presents a summary of the results. The results are grouped into two groups – the WS30 group which represented the data that had been used to make the analysis and the Macrostates group which had not yet been analysed and therefore represented a more 'raw' version of the data with less effort expended in data quality control and analysis. The two groups were used to evaluate the impact of our tools when starting from different stages of an existing workflow.

---

[14]Turchin, P., T.E. Currie, H. Whitehouse, P. François, K. Feeney, D. Mullins, D. Hoyer, C. Collins, S. Grohmann, P.E. Savage, G. Mendel-Gleason, E. Turner, A. Dupeyron, E. Cioni, J. Reddish, J. Levine, J. Jordan, E. Brandl, A. Williams, R. Cesaretti, M. Krueger, A. Cecceralli, J. Figliulo-Rosswurm, P. Peregrine, A. Marciniak, J. Preiser-Kapeller, N. Kradin, A. Korotayev, A. Palmisano, D. Baker, J. Bidmead, P. Bol, D. Christian, C. Cook, A. Covey, G. Feinman, Á. D. Júlíusson, A. Kristinsson, J. Miksic, R. Mostern, C. Petrie, P. Rudiak-Gould, B. ter Haar, V. Wallace, V. Mair, L. Xie, J. Baines, E. Bridges, J.G. Manning, B. Lockhart, P.-J. Tuan, A. Bogaard, and C. Spencer. 2017. "Quantitative historical analyses uncover a single dimension of complexity that structures global variation in human social organization." Proceedings of the National Academy of Sciences of the United States of America. doi: 10.1073/pnas.1708800115. http://www.pnas.org/content/early/2017/12/20/1708800115.full.

#	Process Before	Service Provision Costs
1	Use validation tool to identify and fix syntax errors on each wiki page	Develop and maintain custom validation tool code Operating validation tool and fixing errors identified
2	Use scraper tool to extract data from wiki pages into CSV	Develop and maintain custom scraper tool service
3	Use script to transform CSV into required format for analysis	Develop and maintain transformation scripts
4	Analyse data and identify any remaining errors in the data.	Operate statistical analysis tool and inspect the results to identify anomalies, outliers, missing values and errors which might impinge upon the accuracy of the result.
5	If possible, correct all errors in CSV, complete the analysis, and copy corrections back to wiki from CSV	Identify and record all errors in the data and carry out known corrections. Manually copy all the corrections back to the wiki.
6	If impossible, deploy RAs to correct the data in the wiki manually and return to step 1.	Identify and record all errors in the data. Collect the corrections in the wiki. Carry out another iteration of the process.

Figure 6.43 Seshat: Comparison.

	WS30	Macrostates	Total
Polities	416	54	470
Variables	301,288	33,100	334,388
Non-empty variables	162,200 (54%)	10,779 (33%)	172,979
Imported variables	27,693 (17% of nonempty)	2,487 (23%)	30,180
Triples	995,580	75,716	1,001,296
Syntax Errors Detected	4	2	6
Semantic Errors Detected	218	117	335
Semantic Errors Corrected	86 (39%)	18 (15%)	104
Semantic Errors Remaining	132	99	231
Entity References	658	28	686
Correctly Imported	214 (33%)	5	219
Incorrectly Imported	444	23	467

These results then fed into our curation workflow, where provenance information was used to identify errors that had been detected but not corrected for manual correction. By far the biggest problem detection was in correctly identifying entity interlinks due to the lack of a consistent naming convention. Significantly, it was possible to identify 218 new errors that had evaded the human analysts and it was possible to automatically correct 39%. Nine days of RA labour was expended on fixing the problems identified, completed on January 29th 2018. The publication of the full dataset is scheduled for the first week of February 2018, and the full evaluation results will be published in a paper that is under preparation.

6.3 Managing Data for the NHS

6.3.1 Introduction

Oxford University Software Engineering researchers have been involved in four separate projects involving health research data, which have made extensive use of the Model Catalogue and components of Semantic Booster.

In the first application, the National Institute for Health Research (NIHR) commissioned the Health Data Finder – an online tool for discovering national healthcare datasets (Figure 6.33). These datasets primarily contain routine hospital data for audit and economic reasons, but may be made available to researchers in academia and industry with appropriate governance approval. The datasets are maintained by a number of separate organisations, and so data users wishing to discover data and request access may have to make a number of requests, often with inconsistent results (Figure 6.32).

In the second application, the NIHR Health Informatics Collaborative, five of the largest teaching and research hospital trusts in the country have been asked to share routine clinical data in five therapeutic areas. Each trust maintains data to differing standards and semantics, and rather than unifying data to a lowest common denominator, sites are asked to build their own data warehouses for a federated data store. Users of the data can make a request to the hospitals, and data can be linked and unified on a per-usage basis, taking the research purpose into account. This allows hospitals to maintain ownership of their data and ensures data quality is as high as possible for any given research study. The model catalogue is used to document national data standards in each of the therapeutic areas, alongside local differences for each hospital trust. Models in the catalogue are used as the source for the generation of MS-Word documentation, and for data transfer specifications in the form of XML Schema.

A third application of the catalogue, and related technologies, has been made in the UK 100,000 Genomes Project. As in the Health Informatics Collaborative project, the catalogue has been used for the collaborative, iterative development of models for sample tracking, cancer and rare disease data models, and the generation of non-technical documentation, XML Schema, and also Case Report Forms, compatible with a commonly used clinical trials management system. In the pilot phase of the project, the models in the catalogue were also used to generate relational databases, sufficient for storing data collected according to the specification.

The fourth application of the ALIGNED tools is in the construction of a data warehouse for Oxford University Hospitals Foundation Trust. This instance of the catalogue acts as a detailed asset register for the hospital, detailing field-level metadata about databases and spreadsheets of patient data around the hospital, as well as describing dataflows and message-passing between systems, and specifications for audit and research datasets. It is planned that these models can be used as part of a data-science platform for the trust: allowing clinical researchers to request data, and be automatically guided through governance processes, as well as provided with the data presented in a secure environment.

6.3.2 Use Case

6.3.2.1 Quality

In all four applications, reuse of existing data without detailed documentation can be problematic: researchers are unable to make good use of the data without understanding its semantics: linkage between datasets may be inaccurate; transformation of data into different formats may be incorrect; interpretation of statistical results is error prone. In the Health Data Finder example, such data reuse is minimal: researchers do not know what data may be available to them; different providers may return inconsistent results on data governance, and data must be re-interpreted each time, which may result in costly errors.

In similar projects preceding the Health Informatics Collaborative and 100,000 Genomes projects, collecting comparable data from multiple hospitals has proven difficult. Precise specifications have been hard to produce, mechanisms for data capture and transfer have been manually programmed, often by non-technical domain experts, and inconsistencies have resulted in data that are often incomplete, incomparable, or completely unusable.

6.3.2.2 Agility

The quality and accuracy of data documentation is difficult to maintain during an iterative process. In all the health data research projects, datasets are continually evolving, data specifications are continually being improved. Without careful version management and automation, it is very easy for the documentation to get left behind.

Similarly, software artefacts must keep pace with the changes in requirements: changes to the data or the software specifications must invoke updates to the XML schema, database schema, or Case Report Forms. Manual coding slows the iteration process, which in turn can result in outdated or inaccurate specifications.

Productivity

Domain experts find it difficult to provide documentation or simple modelling because of the technicalities involved: XML schema and Case Report Forms require specialist technical knowledge: domain expertise is often left out, or modelling is undertaken poorly.

Implementing efficient database structures requires a lot of repetitive works: implementation of a domain class will involve a familiar pattern of tables, association tables, keys, and indexes. Such work is time-consuming and error prone, yet ripe for automation.

Data scientists looking to reuse health data currently spend a lot of time searching for usable datasets, often requiring long periods of interaction where inventories and documentation are not available online. Applying for governance, asking technical questions, and retrieving data in a suitable format often require further time and energy. Interpretation and curation of the data is a typically manual task, which may be repeated and reproduced by every scientist receiving a data extract.

6.3.3 Architecture

In each project, ALIGNED technologies are being used in slightly different ways.

In the NIHR Health Data Finder, the model catalogue is the central resource, holding the master copy of models and documentation. A REST-based API provides services used by the front-end website that provides shopping-cart and dataset overview functionality. Metadata is imported into the catalogue by means of a bespoke spreadsheet-based format, which is suitable for domain experts and data curators to populate.

In the NIHR Health Informatics Collaborative, each site hosts its own instance of the model catalogue, documenting their own data landscape: a data warehouse, source patient record systems, research systems and local data flows. A central installation of the catalogue contains the shared data specifications, along with local variations, and relevant national specification.

Local catalogue installations can automatically import the latest version of the central models, and the central catalogue is used to generate XML schemas for use by all partners.

In the UK, 100,000 Genomes Project, the architecture of the pilot is of particular interest: information is provided by the hospitals in the form of XML, matching a schema generated by the Model Catalogue, or manually through online Case Report Forms, hosted in a system called OpenClinica. Information is extracted via an ETL process from OpenClinica, and combined with a shredded form of XML, and stored in a matching relational database, generated by a component of Semantic Booster.

Finally, the architecture of the OUH data warehouse follows a similar pattern to the right-hand-side of Figure 6.44. Almost 100 local databases and data specifications are modelled within the catalogue, along with the design for the main data warehouse. The catalogue is used to document field-level metadata, summary metadata, and dataflows, and this information will be used in the construction of research data extracts and for generating hospital auditing and service improvement metrics.

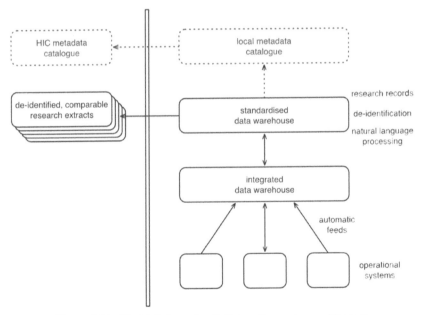

Figure 6.44 Health Informatics Collaborative system architecture.

6.3.4 Implementation
6.3.4.1 Model catalogue

Whether you have a small spreadsheet, or a large federated data warehouse, the key to making the most of your data is understanding its semantics. In order to share your data with others, to reuse it for a different purpose, or to link it to other data stored elsewhere, you have to know what it means. At its simplest, this is just knowing a datatype, and having a description of how those data have been collected. But you may also know how the data have been curated, where it was created in the context of business processes, or how it relates to recognised standards. To do this at scale requires automation: tools that can do the hard work for you and allow the rest to be done collaboratively. Our metadata catalogue tool provides a common framework in which to store descriptions of data alongside data standards, terminologies and dictionaries, providing common reference points by which to describe data.

The catalogue is able to automatically import models – structured descriptions – from relational databases, XML schema, spreadsheets, and UML diagrams. A collaborative editing environment allows the iterative development of models in a clean, simple fashion: just suitable for domain experts to really focus on the things that are important. The catalogue facilitates reuse of data models: parts of one model can be dropped into another. This will make it easy to reuse data in the future and can help to proliferate data standards. Describing data is made easy: links to existing descriptions are automatically suggested; classes of data can be described in a single place, and creating new versions of models maintains any semantics already expressed. Finally, the models can then be exported in a variety of different formats: as relational databases for storing data, or as XML schema for data transport, or as forms for collecting data from scratch. Models for software engineering tools can also be generated – for example by generating specifications for our Semantic Booster tool, we enable the complete, automatic generation of working information systems. In this way, the catalogue can be used as an IDE for an agile, model-driven approach to software- and data engineering. Figure 6.45 shows the catalogue interface.

6.3.4.2 NIHR health informatics collaborative

We now illustrate the advantage of the catalogue with three case studies. In the first, the Oxford team have led the coordination of the Health Informatics Collaborative – a project funded by the National Institute for Health Research

Figure 6.45 The front page of the catalogue interface.

to promote the sharing of healthcare data in the UK. Five of the largest research hospitals in the country – across London, Oxford, and Cambridge, were asked to share routine clinical data on five therapeutic areas: in critical care, ovarian cancer, acute coronary syndromes, hepatitis and renal transplantation. Clinicians at the hospitals were asked to collaborate on the definition of a new dataset, suitable for addressing a wide range of research issues within each clinical specialty, and the hospitals were asked to share anonymised data matching these data specifications. The metadata catalogue provided tools for collaborative editing of dataset specifications, maintaining older versions for reference. XML schema were generated for data transfer between the sites, and the catalogue was able to generate Excel spreadsheets for documentation.

One of the main problems with data sharing amongst healthcare providers in the UK is that each site may record their data points differently. Here the catalogue provided another useful feature – allowing each site to document their own variations, and details of any transformations required to translate data from one format into another. Each of the five data models had, on average 250 data points of interest, and we were able to map relationships between the NHS's own data dictionary, as well as existing standards and audits in each area. The project has created combined datasets for the first time in these areas of clinical interest, enabling new research and, in some cases, better treatment. Figure 6.46 shows the project in action.

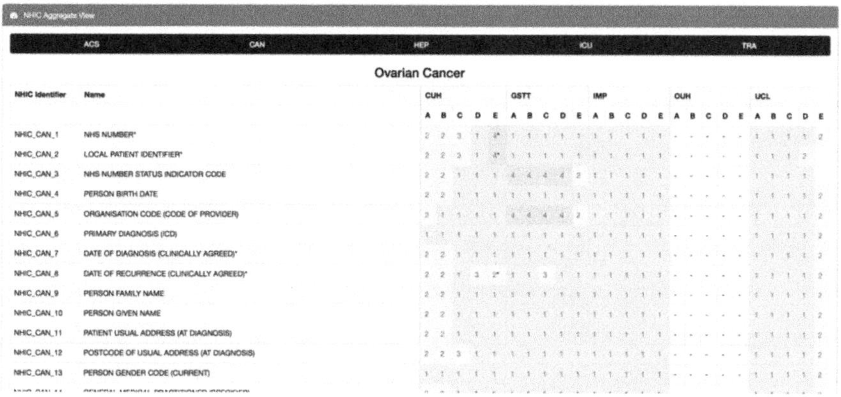

Figure 6.46 Data comparison in the Health Informatics Collaborative.

UK 100,000 Genomes Project

In the last couple of years, Oxford has also been involved in a large genetics programme – the UK 100,000 Genome Project. The project was set up to revolutionise personalised medicine in the UK, starting with the whole genome sequencing of NHS patients with key forms of cancer, as well as patients and family groups with rare inherited diseases. Again, the catalogue was used by the scientists to develop new datasets for routine clinical data, and brand new, bespoke models for each of nearly 200 rare diseases. The metadata catalogue was again used to generate XML schemas for data transfer, but also for electronic case report forms, compatible with a widely used clinical trials management software. These forms were built to include terms from existing medical ontologies, including the Human Phenotype Ontology, and SNOMED CT. For the pilot studies, the catalogue was also used to build databases, used to store the clinical and sample-tracking data on submission. These databases were entirely generated by the data model: a change to the model in the catalogue resulted in a new schema for the database, along with an appropriate data upgrade. Figure 6.47 shows the catalogue.

The project is now halfway to completion and would not have succeeded without the catalogue's provision of a central data model. The national Genomics Medicine Centres rely on the catalogue as the specification for prospective data collection, and those interpreting the data rely on its descriptions to make sense of the data collected. Initial results include confirmed diagnoses for patients with unspecified rare-diseases, and the refinement of lab processes for processing DNA samples at scale.

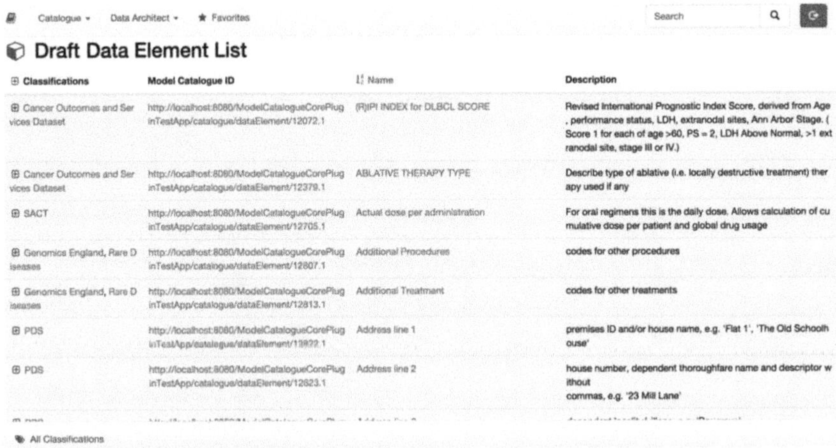

Figure 6.47 Data elements in the UK 100,000 Genomes Project catalogue.

NIHR Health Data Finder

A final example of where the catalogue has been providing benefit is the UK's Health Data Finder. This instance of the metadata catalogue, commissioned by the National Institute for Health Research, provides a portal for healthcare researchers in industry and academia, allowing them to discover national datasets. These datasets, collected at scale across the whole health service, are primarily collected by a number of different bodies for commissioning or audit purposes, but are of great value because of their size. There was no easy way to inform potential users exactly what those datasets contained, and the process for requesting data was time-consuming and prone to error. The catalogue now provides element-by-element descriptions for over 3,000 data points, across more than 20 datasets. It stores summary metadata and usage information, sufficient for researchers to understand whether the data will help them answer a particular question before starting to request any of the valuable data. We are currently streamlining the process for requesting data, by using the catalogue as a 'shopping cart', allowing researchers to select a set of data points to request, and generating queries to return those data points once sufficient governance checks have been made. The shopping cart is shown in Figure 6.48.

The catalogue has provided a number of benefits to healthcare projects across the UK, but is continuing to be developed and extended. Figure 6.49 shows the catalogue in the Health Data Finder. We are increasing the range of models that can be imported into the catalogue, and we are continuously

6.3 Managing Data for the NHS 267

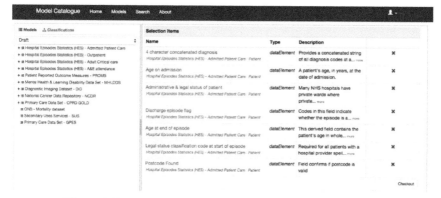

Figure 6.48 An example shopping cart in the Health Data Finder.

Figure 6.49 The model catalogue in the Health Data Finder.

improving the usability for non-technical domain experts – including graphical editing tools, automated search and suggestion, and new visualisations. We are especially interested in using these models as the basis for MDE, and so plugins are being written to generate or configure software components so that reuse of models can really instigate reuse of data. Figure 6.50 shows an example of catalogue metadata. The catalogue has been extensively used in the domain of healthcare, but is fundamentally nonspecific to any particular domain – our work with the ALIGNED partners is helping us prove the

268   *Use Cases*

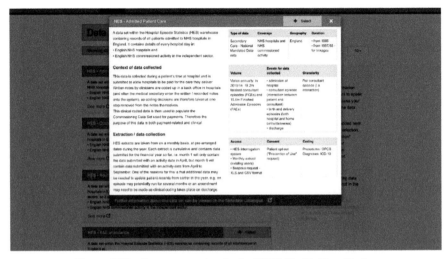

Figure 6.50   Dataset metadata in the NIHR Health Data Finder.

technology in other domains. Our experiences with the tool show that it can be invaluable for software engineers and data engineers alike.

6.3.5 Evaluation

In the Health Data use case, the Model Catalogue has been deployed in four main projects: The UK 100,000 Genomes project, the NIHR Health Data Finder, the NIHR Health Informatics Collaborative (HIC), and the Oxford Biomedical Research Centre's data warehousing activity. In all four projects, the catalogue has provided functionality that was not previously available, or automated tasks that were previously undertaken by hand. The utility of the ALIGNED tools can be measured by their usage: if the tools are used frequently, then they provide a valuable service.

Across the four projects, the model catalogue was primarily used for two separate use cases: firstly the management and documentation of existing data assets – allowing potential data users to search and discover datasets of interest; secondly the collaborative development and publication of new data standards – reusing existing definitions where available. The Health Data Finder and Oxford BRC Data Warehouse projects are primarily focussed on the cataloguing of existing datasets or databases, and the Health Informatics Collaborative and UK 100,000 Genomes Project are primarily concerned with the development and publication of new data standards in a number of medical therapeutic areas.

For each of the measures: "Productivity", "Quality" and "Agility", we will consider each of the four projects and assess the impact made in these areas. In most cases, quantitative measures of improvement are not easily obtained: the Health Data use case was a late addition to the project and the relevant baseline measures were not taken; however, the use of the catalogue in all four cases does not replace any existing functionality or tool provision – originally any software or data engineering tasks were carried out by hand or not at all.

6.3.5.1 Productivity

The NIHR Health Data Finder was set-up to be a single portal for researchers – both academic and in industry – to find out about existing national audit datasets that can be requested for research purposes. Before the introduction of the Model Catalogue, this could be a painful process: the datasets are held by one of a number of public health bodies: NHS Digital, Public Health England, the Clinical Practice Research Datalink, the National Institute for Health Research, and the Medical Research Council. Each maintained their own documentation for the datasets, usually stored in non-computable formats, and, in general, not made publicly available. If a research data user required data, they would have to first find out whether such data existed, and telephone a help desk to ask any questions about the data; detailed questions could take weeks to be answered. Requesting the data would require a different governance process for each provider, and data would be provided in different formats by each provider. All data are anonymised before being conferred: if data from multiple providers were required to be linked before anonymisation, this would increase the complexity of this largely manual process.

The Model Catalogue provides a solution to some of these problems, and forms part of a greater plan to streamline all data requests. The catalogue provides a single portal where all datasets are described, datapoint-by-datapoint, with information about the scope, coverage and completeness for each dataset. Information pertaining to 'frequently asked questions' is stored alongside each data element, and adherence to national standards is recorded. As well as advanced functions for browsing and searching, the catalogue provides a 'shopping cart' function which allows users to compile requests made up from multiple datasets.

The time saved by the use of the catalogue tool is hard to quantify, as each request is different. However, the site has been used more than 4,200 times in 2 years since its launch, with an average of six visitors per day. Of these, approximately 40% are returning visitors, indicating some degree of success on their first visit. The average 'session' duration for all visitors is well over 3 min, suggesting that a lot of users are taking the time to browse and explore. Although the number of visitors has dropped since the first launch of the site, the numbers remain stable.

In the Oxford BRC Data Warehousing project, a team of developers are building a large warehouse of patient data, extracts of which will be made available to local researchers for specific purposes. In order to maintain an asset register and to provide documentation to potential users, every data source and data flow is being documented. Before the introduction of the catalogue, this documentation would have been maintained in a series of spreadsheets and shared (perhaps in a source control system) to allow collaboration. The catalogue provides plugins that automate the transcription of database metadata, and descriptions can be collaboratively edited via the online interface – a vast improvement to productivity. There are currently 12 developers and data engineers using the catalogue – some on a daily basis – and allowing access to Oxford University researchers is planned in 2018.

In the Health Informatics Collaborative, and in the UK 100,000 Genomes project, a key output is the development of new data standards – to facilitate the transfer of clinical data from a number of different hospitals to a centralised location. In such projects, collaboration is required from a range of different people: those with clinical expertise to assess the availability of data; those involved in research to assess the requirements for each data point, and technical people at each hospital who can assess the feasibility of providing data. Previously such collaboration may not have happened or taken place via email and teleconferences; with the use of the Model Catalogue such collaboration is much easier, and can reduce the number of iterations required to reach a viable data specification.

In the UK 100,000 Genomes project, complex models for Cancer and over 200 Rare Diseases have been developed and published, iterating through a number of intermediate versions. In the NIHR HIC project, models for five therapeutic areas have been developed: originally using spreadsheets and email; latterly using the catalogue. The catalogue has reduced the amount of communication required and simplified the task of development and documentation of the model; a further five new therapeutic areas are to be addressed with new models in the NIHR HIC project, during the first quarter of 2018.

6.3 Managing Data for the NHS

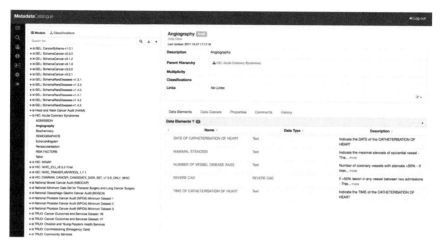

Figure 6.51 Screenshot from the NIHR HIC Model Catalogue.

6.3.5.2 Quality

In the UK 100,000 Genomes and NIHR HIC projects, as well-documented data standards, key outputs are software components to allow the storage and transfer of data according to the standard. Without automation, it would be very easy for mistakes to be made in the development of tools such as XML schema or database schema: differences between the standard and the tools could result in data not being transmitted or stored correctly. The plugins developed for the Model Catalogue allow these components to be generated automatically. In the early stages of the HIC project, when such a manual process was in place, discrepancies arose frequently, and this caused delays and frustration as errors had to be corrected, new standards or tools re-tested, published and distributed. The introduction of the catalogue has seen a complete reduction in these errors, and also reduces development effort (Figure 6.51). In the UK 100,000 Genomes project, further components were required to configure off-the-shelf software, and suitable plugins were developed to ensure that these also remained consistent with the standards.

In the Oxford BRC Data Warehousing project, data quality can be improved by allowing those entering the data to see the descriptions of the intended values – so they know how to complete fields correctly – or to see the data already submitted – in order to fix any problems with existing data. In its current state, running metrics on the existing data has identified a number of potential issues with the data and other local reporting, and so the Model

Catalogue has become a useful tool for the reporting and discussion of these issues.

In the Health Data Finder project, an improvement has been made, not in the quality of the actual data, but in the linking and usage of the data. With detailed descriptions of every data point, researchers are better able to make decisions on how to use the data – in many cases preventing mistakes in analysis, or, where previously the semantics of data points were unknown, preventing researchers having to collect new data from scratch to ensure its validity for the particular purpose.

6.3.5.3 Agility

One of the key advantages of the catalogue product is the ability to create new versions of a model with ease, ensuring that all participants can be kept up-to-date, and by using plugins to generate software components, updates to a data model can be reflected in changes to the related software much more quickly. In the NIHR HIC project, this is an essential requirement: the XML schema required for transferring data between sites can be made available as soon as the new data model is finalised – giving technical staff the maximum amount of time to adapt to the new model. Previously, delays in the generation of XSDs (and subsequent fixing of any errors), could delay the timely collection of data. A similar improvement has been made in the UK 100,000 Genomes project, where without the use of the Model Catalogue, manual approaches to collaborative model evolution, publication and software development would result in a much slower turn-around time.

In the Health Data Finder and Oxford BRC Data Warehousing projects, the key notion of agility is in the time taken to update the documentation in response to a new version of the database schema. Again, the plugins have proven invaluable in this respect: the importer plugins can automatically import the new structures, and existing descriptions can be copied, meaning that minimal effort is required from domain experts.

6.4 Integrating Semantic Datasets into Enterprise Information Systems with PoolParty

6.4.1 Introduction

PoolParty Semantic Suite is the SWC's platform for enterprise information integration based on Linked Data principles. Since it was created, the product has evolved to include entity extraction from unstructured information. To align product development with ongoing technology trends, market monitoring and trend scouting features have been incorporated. Atlassian Confluence is used to support the requirement engineering process while Atlassian JIRA is used for issue tracking, including an external system for customers.

The developers of the SWC's software have numerous sources of information that is relevant to their product development role – bugs, feature requests, usage information, and so on. They would like to ensure that the information relevant to any particular development task is made available to the relevant developers in as timely, well-structured and meaningful way as possible, regardless of the source. Customers of PoolParty would like to integrate a variety of models, schemata, ontologies and vocabularies into their PoolParty knowledge bases. In many cases, they do not have a deep understanding of semantic technologies and would benefit from as much assistance as possible in understanding what they need to do to integrate their models into PoolParty.

To support and document the development process, SWC operates installations of Atlassian Confluence and JIRA. Confluence is used for drafting, specifying and discussing new features and requirements in a text-based format which is only structured visually with headings and paragraphs. Most requirements captured in Confluence follow a defined structure: they declare the high-level goal (or summary), which is a description of the functionality the application should provide so that the requirement is met. The requirements document breaks down this description into multiple "user stories" which are detailed descriptions of how the application should behave from a user perspective. They also add preconditions, acceptance criteria and test scenarios so that the responsible developer can identify what changes need to be performed and infer JIRA tickets for each of them. A requirements document also defines various stakeholders, i.e., people and their responsibilities and roles they fulfil in the course of processing the requirement.

JIRA defines a data schema to hold the details of each ticket, like type, description, priority, or assignee. On the most general level, tickets (also sometimes called issues in this section) are assigned to various "spaces". A space is used to classify issues by project (e.g., LOD2 or ALIGNED), product (PoolParty Thesaurus Manager PPT or PoolParty Extractor PPX) or general kind (ideas, which are "nice-to-have" features or improvements for which it is not yet decided if and how they will be implemented). Each ticket can only be assigned to one space and the space, to some degree, also influences the properties that can be assigned to a ticket. For instance, valid types that can be assigned to a ticket are, e.g., "bug", "task", "epic" or "story" in the PPT space while "epic" or "story" cannot be assigned to tickets in the PoolParty Support space. Besides the affected software components, status, resolutions methods and much more, also metadata is attached to the ticket like creation and last-updated date. The properties mentioned above which are relevant for querying in the ALIGNED use case(s) are modelled in the Design Intent Ontology (DIO) by OxSE, which is used for publishing the data held by Confluence and JIRA as RDF.

6.4.2 Problem Statement

The developers of the SWC's software have numerous sources of information that is relevant to their product development role – bugs, feature requests, usage information, and so on. They would like to ensure that the information relevant to any particular development task is made available to the relevant developers in as timely, well-structured and meaningful way as possible, regardless of the source. Customers of PoolParty would like to integrate a variety of models, schemata, ontologies and vocabularies into their PoolParty knowledge bases. In many cases, they do not have a deep understanding of semantic technologies and would benefit from as much assistance as possible in understanding what they need to do to integrate their models into PoolParty.

6.4.2.1 Actors

Role	Description
PPT Developer	performs software development work on the PoolParty platform
PPT User	uses PoolParty
PPT Taxonomy Developer	responsible for developing taxonomies for PoolParty
PPT Admin	responsible for administering PoolParty services
Requirements Engineer	responsible for defining and maintaining software requirements
SWC System Administrator	responsible for administering SWC assets

The requirements on which the PoolParty use case was based are detailed in Appendix A.

6.4.3 Architecture

Figure 6.52 shows the different roles (orange figures), tools (green rectangles), repositories (cylinders), and files (parallelograms) involved in the PoolParty architecture and workflows. On the left side, the diagram describes

6.4 Integrating Semantic Datasets into Enterprise Information Systems

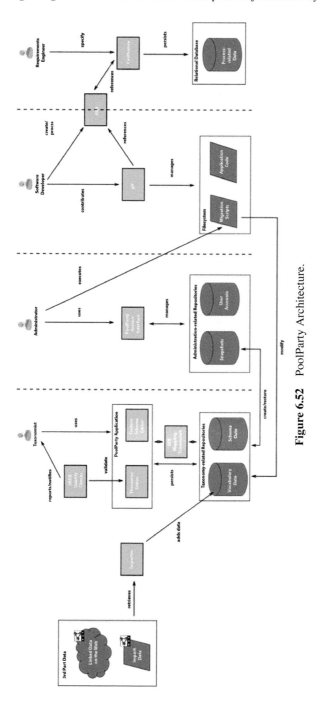

Figure 6.52 PoolParty Architecture.

the direct interaction of the customers (taxonomists) with the PoolParty application. The taxonomist creates controlled vocabularies using both the thesaurus editor and the custom schema editor. Currently, there are two components where data consistency needs to be satisfied: (i) when persisting vocabulary and schema data to the underlying triple store using a custom SWC-developed RDF Mapping Framework and (ii) when changes to the controlled vocabulary are performed which violate certain quality criteria (validated by the SKOS Quality Checks component). The RDF mapping framework converts instances of annotated Java domain classes into an RDF representation and vice versa. However, data consistency violations between the triple store(s) and the application via the RDF mapping framework can occur because in the application code, the framework is sometimes bypassed and data changes are written directly to the store. Furthermore, changes to the domain classes may require migration scripts which can easily be forgotten to develop and run. Also note that the data importer component which retrieves data either from the LOD cloud or from imported files currently persists these data directly to the triple store, which in many cases violates data consistency requirements.

6.4.4 Implementation

The demonstrator system consists of four components, which we shortly outline in the following paragraphs:

- Consistency violation detector
- RDFUnit test generator
- PoolParty integration
- Notification adaptations

6.4.4.1 Consistency violation detector

We implemented the consistency violation detector as a separate component that can be either invoked on the command line or integrated into PoolParty as a library. It takes as input the id(s) of the consistency violation check(s) it should detect as well as an arbitrary number of RDF files that contain all necessary data for performing the check(s). All these RDF data are then added to a local in-memory OpenRDF repository, together with the RDF definition of the SKOS data schema. All but one of the identified consistency violations can be detected by using SPARQL queries over the provided

6.4 Integrating Semantic Datasets into Enterprise Information Systems

RDF input files. The one constraint where SPARQL queries do not suffice is the validation of external links. This is done by a Java algorithm that dereferences all URIs (that do not reference localhost) and checks if the HTTP response code indicates an error (i.e., other than 200).

6.4.4.2 RDFUnit test generator

Test cases for RDFUnit are expressed in RDF as resources of, e.g., type http://rdfunit.aksw.org/ns/core#ManualTestCase. Our demonstrator system can generate the test cases for RDFUnit automatically, based on the SPARQL queries we defined for each data consistency check. Currently, four of the 16 data consistency violation checks can be automatically converted to RDFUnit tests. The RDFUnit test cases can then be executed on the in-memory repository mentioned above and a HTML report page is generated by RDFUnit which shows the results (success or failure) of each test case.

6.4.4.3 PoolParty integration

For the demonstrator, we integrated the consistency violation detector into PoolParty's data import functionality. The current implementation checks for violations of any of the 16 identified data consistency constraints. Therefore, it first collects the data of all linked projects, the project metadata, and custom schema data and passes it to the consistency violation detector. The generated textual report is then displayed to the user, along with the option to view the HTML page that has been generated by the RDFUnit test run.

6.4.4.4 Notification adaptations

We improved the rsine notification system1 which has been originally developed in the course of the LOD2 project (see Section 5, Improved Notifications, or the project's GitHub page for additional information) to:

- Be transaction-aware: Due to improvements on how PoolParty invokes data changes, rsine can persist them as a transaction. This enables us to write easier and more powerful notification subscriptions.
- Support of project management, custom schema, and user repositories.

Until now, only changes to the taxonomy project repository were communicated to the notification service. We changed that so that it is also possible to subscribe for changes to the project management, custom schema and user management repositories to, e.g., receive notifications on creation of new projects, new custom classes, or new PoolParty user accounts.

6.4.4.5 RDFUnit

RDFUnit is integrated in PoolParty RDF Validation for performing constraint checks. The checks are defined as RDFUnit test cases using RDF. These test cases can also be run by RDFUnit independently of PoolParty on external data. For each of the constraint checks, there is an RDFUnit test case which is based on a SHACL constraint or a SPARQL query that identifies resources that cause violations.

UnifedViews is an ETL tool for RDF data developed as part of the PoolParty semantic suite. Using this tool, we extracted data from Atlassian Confluence and JIRA and transformed it into RDF using a DPU developed for ALIGNED. The transformed data are annotated with the PoolParty Knowledge.

Graph using the extractor DPU and finally similarity scores are calculated based on the annotated data.

The Issue Integration feature is integrated in PoolParty product, which allows user to automatically create JIRA support tickets whenever an internal server error occurred in the application.

Similarity scores are calculated on development artefacts using the annotations of the PoolParty Knowledge Graph Thesaurus as a basis. Two algorithms are implemented that represent a lexical and a graph-based approach to similarity.

Graph Search, a faceted search application and part of the PoolParty product, is used to analyse the development artefacts. We integrated similarity retrieval into GraphSearch to find duplicate bugs and relations between issues.

6.4.4.6 Validation on import
General Description

Currently, users can import any RDF data into a PoolParty thesaurus project. In the best case, invalid data just lingers in the triple store where PoolParty stores all the data it operates on and consumes memory or hard disk space. However, these data also can cause problematic behaviour such as inconsistency in the user interface and a corrupt data model, manifesting in fatal exceptions in the PoolParty Thesaurus Editor. We can identify three different PoolParty functionalities where data consistency is required:

- Basic internal operations: The thesaurus editor expects certain properties for the various controlled vocabulary resources, such as concepts or concept schemes

- Schema-specific: SKOS or other data schemas impose custom restrictions on the data or encourage conformance with best practices that are not formally stated
- Reasoning: PoolParty asserts and expects class membership information to controlled vocabulary resources and interprets them with constraint semantics.

Addressed Challenge

The main challenge is to match the imported data, which follow the open world assumption with the local data model required by PoolParty. This is basically a challenge each application that consumes open data from the Web faces. Because these data are very volatile, efficient methods have to be in place that allow transition of data scraped from the Web into a meaningful local representation that can be further processed by the application's business logic.

Identify Sample Set of Data Consistency Violations

We can break down this challenge to a set of sub-goals we want to solve in the course of the ALIGNED project:

- Provide full coverage of data consistency constraints
- Identify repair strategies
- Invoke repair strategies and fix constraint violations either automatically or based on user input

Proposed Approach

We plan to support the import use case with a two-step semi-automatic scenario: in the first step, the imported data must be checked against PoolParty's internal data model and requirements on the data and any non-conformance must be reported. In a second step, users should have the option to adjust the imported data in order to fulfil PoolParty Thesaurus Editor's requirements. Based on the kind of data consistency violation, various repair strategies may be invoked. Some violations can be fixed automatically and some require additional input from the user. It should also be possible to fix similar kinds of consistency violations in one go so that it is possible to deal with a large number of violations.

Identified Data Consistency Constraints

For demonstrating the problem domain and working towards the implementation of an approach that tackles the addressed challenge, we first focussed on the sub-goals 1 and 3. We extracted 16 data consistency constraints, i.e., requirements for RDF datasets so that they match the internal PoolParty Thesaurus Editor (PPT) data model. Violations of these constraints can vary in severity: some constraints must never be violated (ERROR), some can be tolerated (WARNING) and some are just of informative value (INFO). For each of the identified consistency constraint, we propose one or more repair strategies that describe possible ways to fix the dataset.

We implemented a tool (usable both at the command line as well as a library for integration into existing applications) that checks provided RDF data against violation of these constraints. A current development branch of PoolParty makes use of this tool and displays a report if constraint violations on imported data were detected. Four of the consistency constraints listed above have also been formulated as RDFUnit test case and can thus be integrated into existing test suites.

ID	Constraint	Description	Severity	Resolution Strategies
br	Bi-directional Relations	If a resource A is related to a resource B by a property p and if p has an inverse property p', then the statement that B is related to A by p' must also be manifested in the data.	ERROR	Add complementary statement Remove relation
cd	Concept Deletion	In order for PPT to recognise deleted concepts, these concepts must be marked with owl:deprecated, must not have asserted any type information, and must contain information in the history graph for being properly displayed in the application.	ERROR	Remove other facts that are not asserted by owl:deprecated Remove "owl:deprecated true" fact
cta	Concept Type Assertion	Concepts must have the type skos:Concept asserted because no RDFS inferencing is performed in PPT.	ERROR	Add (infer) missing type declarations
dcl	dcterms Creator Literal	Using URIs for dcterms:creator to describe skos:Concepts and skos:ConceptSchemes in PPT leads to error message	ERROR	Convert provided creator agent to literal Replace with some default literal

6.4 Integrating Semantic Datasets into Enterprise Information Systems

Continued

ID	Constraint	Description	Severity	Resolution Strategies
dta	Direct Type Assertion	Concepts having asserted a class using swcs:appliedType must also be instances of this class.	ERROR	Add missing type statement Remove resource
elv	External Link Validity	Outgoing links from a thesaurus to another dataset on the Web may not be resolvable anymore.	INFO	Prompt user for replacing URI with "valid" link Apply resolution strategy suggested when dereferencing the URI Remove affected statement
hc	Hierarchical Consistency	Each resource of type skos:Concept must have a resource linked by skos:broader or skos:topConceptOf in the vocabulary namespace. Each resource of type skos:Concept must have at least one path (via skos:broader/skos:topConceptOf) to a resource of type skos:ConceptScheme in the vocabulary namespace.	ERROR	Prompt for parent resource Add to some existing default parent resource Remove (do not create or ignore) concept
lam	Label Ambiguities	Identical concept labels may indicate duplicate concepts.	WARNING	Remove Label from one concept (prompt user for which one) Merge Concepts Add descriptive note (prompt user for text input)
lav	Label Availability	Resources of type skos:Concept must have assigned exactly one Literal in the default language, using the predicate skos:prefLabel. Resources of type skos:ConceptScheme must have assigned exactly one Literal in the default language, using the predicate rdfs:label.	ERROR	Auto-generate label (based on URI, timestamp, increment, from parent/related...)

(*Continued*)

282   *Use Cases*

Continued

ID	Constraint	Description	Severity	Resolution Strategies
lpc	Linked Project Consistency	If two PPT projects are linked to each other, each of the referenced resources must exist.	WARNING	Remove Link Restore Data (i.e., create new local concept with deleted concept's label)
sc	Schema Compatibility	Detect statements using resources from namespaces that are not included in the default PoolParty schemas or in schemas that are available as custom schemas. Such statements would not be visible within PoolParty and may lead to unwanted side effects.	WARNING	Enable relevant schemas in PPT Ignore statements
sdr	Schema Domain Range Match	Domain and range axioms on a property are interpreted as constraints – that is, a property with specified domains (using swcs:domain) A and B can only be used in triples with resources that are instances of A or B. Likewise for swcs:range.	ERROR	Apply missing type(s): either one (prompt user which one) or all Remove relation
tpc	Type Propagation Collections	All concepts that are members of a collection which is instance of a class (using the property swcs:appliedType) also are instances of this class.	ERROR	Assert missing Types Remove from collection
tph	Type Propagation Hierarchical	Concepts that are part of a hierarchy (using skos:broader properties) and one of the parents (e.g., a resource being an instance of skos:ConceptScheme and skos:Concept) have either a type asserted (using swcs:appliedType for skos:ConceptSchemes) or propagated (using swcs:propagateType) must also be instances of this class.	ERROR	Assert missing types Remove from hierarchy

6.4 Integrating Semantic Datasets into Enterprise Information Systems

Continued

ID	Constraint	Description	Severity	Resolution Strategies
upl	Unique Preferred Labels	A concept must have at most one preferred label per language tag (SKOS integrity constraint)	ERROR	Remove one preferred label (prompt user which one) Add disambiguation information as notes (prompt user to supply them) Add disambiguation information as parenthesis to label (bad practice) Extract new concept (prompt for broader or insert as sibling) Remove concept
ut	Unsatisfied Type	Concepts must either be instances of skos:Concepts or instances of classes that are assigned directly or by type propagation.	ERROR	Remove type assertions Remove affected resources Import type as custom class

Detailed Process Description

In the following, we show how import data validation is implemented in a proof-of-concept branch of PoolParty:

(1) Accessing the RDF Data import functionality of PoolParty: the newly adapted import dialog provides an option for checking the imported data for conformance against the consistency constraints (Figure 6.53).

284 Use Cases

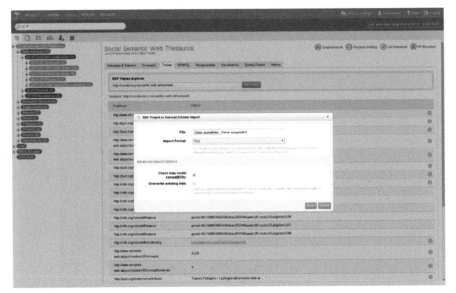

Figure 6.53 Import dialogue.

(2) Report on the resources that violate certain consistency constraints (Figure 6.54).

Figure 6.54 Consistency constraint violations as reported by RDFUnit.

6.4.5 Results

By providing a demo implementation of an import validator, we found that RDF datasets can be checked against the identified data consistency constraints, either by using SPARQL or by a hybrid approach, processing a subgraph generated by SPARQL with custom Java algorithms. Based on the query results, reports containing the resources that violate consistency constraints are created. We also found that the consistency constraints

6.4 Integrating Semantic Datasets into Enterprise Information Systems

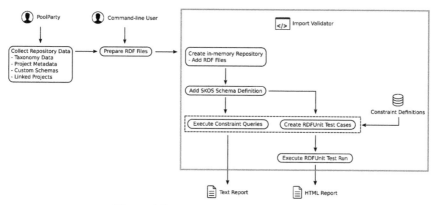

Figure 6.55 High level technical overview.

which can be solely expressed using SPARQL (i.e., no custom Java algorithms for validation are needed), can also be expressed as test cases for RDFUnit.

For identification of the above described consistency constraints, we analysed the algorithms PoolParty uses internally for creating, processing and persisting a controlled vocabulary. While this is efficient for getting an initial set of constraints, we cannot retrieve a complete set of consistency constraints which covers all error cases this way. The reason is that a formal model of the data that PoolParty operates on does not yet exist. Therefore, the consistency constraint checks must be manually crafted in SPARQL, independent from the algorithms creating or accessing the data. As a consequence, the checks constitute an additional entity that must be maintained in sync with changes to the application logic.

Figure 6.55 illustrates the workflow for checking RDF data for conformance to the PoolParty data model. The PoolParty integration collects data from various sources necessary to evaluate potential consistency violations. Alternatively, command line users of the Import Validator can prepare RDF files and pass them to the Import Validator Component (green frame), which applies the constraint definitions to this data and outputs a textual or HTML report that contains violation information.

6.4.5.1 RDF constraints check

Figure 6.56 shows the constraints checks integrated using RDFUnit. When importing data into a PoolParty project, the constraint checks are performed,

Use Cases

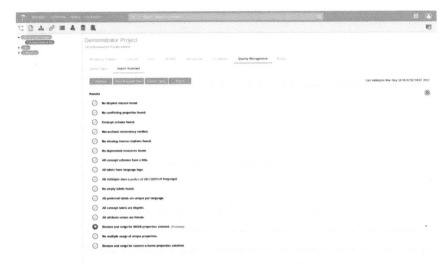

Figure 6.56 RDF validation conformance checks.

and a result list is presented to the user who outlines all the violations that have been detected. For the first release of PoolParty containing the RDF Validation, we defined a minimal set of 13 constraints so that imported data are required to conform with PoolParty to operate normally. In later releases, this set will be extended by quality checks to assist with data modelling. The declarative approach taken for defining the constraints ensures easy maintainability and extendability of the RDF Validation for future releases.

The user can browse through the detected constraint violations, select a repair strategy for each of them and apply the repair to the data (Figure 6.57). When all violations are repaired and conformance is achieved, the user can transfer the imported data into the project without the risk of application failures caused by inconsistent data.

6.4.5.2 RDF validation

The first part is the RDF Validation, which is integrated into the PoolParty product to support consistency within the application's data storage.

Data within PoolParty projects have to follow conformance rules for the application to work correctly. Usually, data are modified by the application itself, and the conformance is therefore given naturally. However, it is possible for users to import arbitrary RDF into the project. These data have to

6.4 Integrating Semantic Datasets into Enterprise Information Systems

Figure 6.57 Repair strategy for the constraint check.

be checked and eventually corrected to conform to the PoolParty application. To ensure this conformance, the import component of PoolParty was extended with an RDF Validation component. It is responsible for checking the imported data based on a set of defined constraints and reporting the results. The user is then given the opportunity to correct the import by using one or more presented repair strategies that will manipulate the data so they satisfy the constraints. Bulk repair options are also given for constraints where it is appropriate to do so. Furthermore, general quality checks can be done on the data that do not interfere with PoolParty's operations, but represent data modelling problems and would therefore be of interest to the Taxonomist.

For performing the constraint checks, the RDF Validation has integrated RDFUnit. The checks are defined as RDFUnit test cases using RDF. These test cases can also be run by using RDFUnit only and therefore can be used independently of PoolParty on arbitrary data. Also, the maintainability of the constraint checks is high because of the declarative approach of the test case definition using RDF. Changing the checks does not require changes to the application's code. The repair strategies and other metadata are also defined as RDF and extend the RDFUnit test cases for an integrated representation of validation and repair. For each of the constraint checks, there is an RDFUnit test case, which is based on a SPARQL query that identifies the resource that causes the violation. Each check also defines repair strategies that can be applied to fix the violation. The information needed for the repair strategies to determine changes that have to be done can be retrieved using a constraint specific query that returns the context of the violation as RDF statements. The combination of constraint, context and repair strategies is represented as an extension of the RDFUnit test case. The component implementing the test cases is designed to be independent of PoolParty and can be used separately. PoolParty integrates it to present the RDF Validation as an application feature.

288 Use Cases

When importing data into a PoolParty project, the constraint checks are performed, and a result list is presented to the user that outlines all the violations that have been detected. The user can browse through these violations, select a repair strategy for each of them, and apply the repair on the data. When all violations are repaired and conformance is achieved, the user can transfer the imported data into the project without the risk of application failures caused by inconsistent data.

For the first release of PoolParty containing the RDF Validation, we defined a minimal set of 13 constraints that are mandatory to conform with PoolParty to operate normally. In later releases, this set will be extended by quality checks to assist with data modelling. The declarative approach taken for defining the constraints ensures easy maintainability and extensibility of the RDF Validation for future releases.

RDF Validation: The user imports an RDF file into a project. A list of constraint violations is shown and explained. Constraint violation details are opened and the constraint details are shown. The repair strategy is executed. Another constraint violation is shown and repaired. Afterwards, all the violations have been resolved. It is explained that a save import is now possible (Figure 6.58).

Figure 6.58 RDF Validation Screenshot.

6.4.5.3 Improved notifications
General Description

During the LOD2 project, SWC developed rsine,[15] a publish/subscribe system that allows users to register for data changes in an RDF triple store. In the demonstration system, we reuse and adapt rsine to work with a current version of PoolParty and extend it to support additional notification types as required by Wolters Kluwer.

Addressed Challenge

The LOD2 technology stack[16] consists of multiple tools that cover the whole Linked Data life cycle. It encompasses, among others, storage, quality analysis and exploration utilities that target problem domains that also affect PoolParty. LOD2 project partners needed a way to better integrate their solutions, and being notified on data changes between stack components was one of the project goals. Therefore, the notification systems were required to be:

- easily integratable into existing stack components, and
- flexible enough to support notifications which can be adjusted to meet the component's purpose and data model.

Approach

Rsine runs as a stand-alone server and can be controlled by a REST-like interface. It can be configured against a Managed RDF Store (accessible by a SPARQL endpoint), which holds all data a LOD2 stack component works on. Addition and deletion of triples to this managed store are detected by the Change Handler. It forwards these changes to the Changeset Service, which enriches them with additional metadata such as timestamps using a standard ontology[17] and persists them into an in-memory Changeset Store.

We currently support two different types of change handlers:

- Integration with the managed store: an external component, e.g., a Virtuoso VAD extension[18] or transaction log parser[19] detects triple changes in the underlying Virtuoso triple store.

---

[15] https://github.com/rsine/rsine
[16] http://stack.lod2.eu/blog/
[17] http://vocab.org/changeset/schema.html
[18] https://github.com/rsine/rsineVad
[19] https://github.com/GeoKnow/trx_parser

290 Use Cases

- Integration with the stack component: The stack component (e.g., Pool-Party) is responsible for announcing all data changes to rsine using API calls.

To subscribe for notifications, users can submit Subscription Documents to the rsine server using the API. These are RDF documents, containing information about

- The change patterns (as SPARQL query) the user should be notified about,
- A notification message,
- Additional information the notification message should contain (fetched from the managed store using SPARQL), and
- Contact information (e.g., email address, log file) where the notification should go to.

A complete description of the information a change document should contain can be viewed at the project's GitHub page. Figure 6.59 shows the architecture of the notification system.

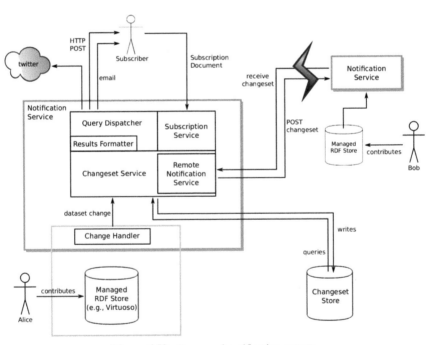

Figure 6.59 Improved notification system.

6.4 Integrating Semantic Datasets into Enterprise Information Systems   291

Improvements

For the demonstration system, we revised and improved rsine to meet the new notification requirements of WKD. We adjusted a current version of PoolParty, with the following change handler improvements:

- Dataset changes are now transaction aware: if a PoolParty action (e.g., creation of a document) creates or removes multiple triples at once, these are combined and stored as a single RDF changeset representation. This feature required us to adjust the rsine API and break compatibility with older rsine versions
- Support for other repositories than the current vocabulary repository. The change handler can now communicate data changes introduced to the project metadata repository, the custom schema repository and the user accounts repository

Notifications

These changes allow us to express and implement new types of notifications that were not possible with other rsine PoolParty integrations before. On project creation, for example, a notification containing the user who created the project and the project's name can be disseminated. Notification can also be done when creating, changing or deleting classes, attributes or properties of custom schemas in PoolParty or creating and deleting PoolParty user accounts. In the following, we provide an abbreviated example (we omitted the prefix declarations) subscription document that logs a message if a new user has been created.

```
<http://example.org/aligned/new\_account> a rsine:Subscription;
     rsine:query [
         dcterms:description "Notification user account creation";

         spin:text "SELECT ?userName (GROUP\_CONCAT(?auth; separator=', ')
AS ?auths) WHERE \{
                 ?cs a cs:ChangeSet;
                    cs:createdDate ?csdate;
                    cs:addition ?userAdd;
                    cs:addition ?userAuth;
                    cs:addition ?userInfo.

                 ?userAdd rdf:subject ?user;
                    rdf:predicate rdf:type;
                    rdf:object swcu:User.

                 ?userAuth rdf:subject ?user;
                    rdf:predicate swcu:grantedAuthority;
                    rdf:object ?auth.
```

292   Use Cases

```
            ?userInfo rdf:subject ?user;
                rdf:predicate swcu:username;
                rdf:object ?userName.

            FILTER (?csdate
>'QUERY_LAST_ISSUED'^^<http://www.w3.org/2001/XMLSchema#dateTime>)
            }
            GROUP BY ?userName HAVING (STRLEN(?auths) > 0)
            ";

    rsine:formatter [
        a rsine:vtlFormatter;
        rsine:message "A new user named
'$\bindingSet.getValue('userName').getLabel()' with the roles
'$\bindingSet.getValue('auths').getLabel()' has been created.";
        ]
    ];

    rsine:notifier [
        a rsine:loggingNotifier;
    ].
```

Detailed Process Description

The command java -jar ./rsine-cmd.jar starts the rsine notification server, accepting notification subscription documents on port 2221

The notification subscription document can be registered at the server using the command: curl -X POST -d @"create_user_account_subscription.ttl" –header "Content-Type: text/turtle" http://localhost:2221/register

Rsine detects the event and adds a notification to the log: 13:17:26.258 [qtp524197922-12] INFO e.l.r.d.n.logging.LoggingNotifier – A new user named 'aligneduser' with the roles 'PoolPartyUser, PoolPartyAdmin, Public' has been created.

Note that notifications can also be configured to be sent to an email address by adding this snippet to the notification subscription document:

```
rsine:notifier [
    a rsine:emailNotifier;
    foaf:mbox <mailto:c.mader@myhost.at>
];
```

Results

We found that the new notification subscription documents, covering project metadata, custom schema changes or user account management, were easy

6.4 Integrating Semantic Datasets into Enterprise Information Systems

to implement on the rsine side. However, we had to put more effort into adapting the Change Handler components, which are part of the newly created PoolParty-ALIGNED branch, to support the new notification types. The reason for this is that PoolParty internally organises vocabulary data (i.e., the SKOS representation and some metadata of a taxonomy project), project metadata, custom schemas and user account information in different RDF repositories, and only the vocabulary data can be accessed by a SPARQL endpoint. This has two major consequences:

- We had to integrate the Change Handler code into PoolParty's data persistence logic for each repository,
- We currently cannot cover all information that should be contained in the notification messages. For example, if adding a new class to a custom schema, the notification message can only contain the name of the new class, not the schema name it has been added to or the user who created it.

However, future releases of PoolParty will only use a single repository for the data described above and organise it into different named graphs. This will allow us to efficiently query the data and also to formulate more powerful queries, aggregating knowledge of each named graph. Therefore, PoolParty's rsine integrate will also profit from disseminating more detailed and useful notification messages to the subscribers.

6.4.5.4 Unified governance

The Unified Governance tool is used to harvest data from the tools used in the PoolParty development life cycle. The data are transformed into RDF and integrated using ALIGNED vocabularies to create unified views for supporting the development process.

The Unified Governance tool will support three use cases for the trials:

- Search over the integrated RDF software development data
- Computing similarity for software development artefacts based on a combined graph-based and text-based approach
- Statistical analysis of the software development process

The sources of development data used for the tool are Atlassian Confluence and Atlassian JIRA.

Atlassian Confluence is used for requirements engineering, organising ideas from team members, providing documentation of research projects, and publishing of technical information. Atlassian JIRA is used for issue management as part of the SCRUM-based software development process. It is also used as a ticketing system for customers to report issues. Both of these tools are used for integrated software development process, but they are not integrated with each other. This has to be done by humans as part of the process to synchronise the information. It includes manual linking of requirements in Confluence to JIRA issues and linking duplicate JIRA issues together. Generally, an integrated and interlinked view of requirements and development artefacts is needed. With the Unified Governance tool, this can be achieved automatically.

The tool retrieves the information from both Confluence and JIRA and transforms it into RDF based on the ALIGNED metamodel vocabularies DIO and DIO-PP. This has to be done on a regular basis to have up-to-date information to work on. Therefore, we use Unified Views, an Extract Transform Load (ETL) tool supporting RDF data processing, for periodic retrieval and transformation of the development data. Having integrated the data as RDF, we can query it using SPARQL. The queries can make use of the underlying metamodels to improve the results. Furthermore, SPARQL-based applications can be put on top of the triple store to support querying, filtering and facetted search.

During the integration process, the generated RDF data are annotated with concepts using a PoolParty Thesaurus. These concepts can support search applications. They can also be used as a basis for computing similarities between artefacts based on the hierarchical graph structure of the Thesaurus. A graph-based approach can leverage the underlying knowledge model and provide semantic similarity for the development artefacts. We decided to use a combined method of text-based and graph-based similarity to benefit from both approaches and improve the results. The results of the similarity computation can be applied to several tasks. First, we can automatically identify developments issues that correspond to requirements and semi-automatically link them. Second, we can identify similar requirements and ideas that should be organised together, but appear distributed in the system. Third, we can identify duplicate issues in JIRA, which is important to prevent the duplicate reporting of bugs. We can identify duplicates before an issue is submitted, inform the user about it and eventually prevent the creation of the issue.

The RDF data of development artefacts can be used for a statistical analysis of the development process. The results can then be used to apply improvements. They can be used as a reference basis for future time estimations of efforts. Flaws in the development process can be identified by analysing performance decreases. Development efforts and reported bugs can

6.4 Integrating Semantic Datasets into Enterprise Information Systems

be analysed for deviations from the expected values. Statistical data will be visually presented in form of diagrams as part of the search application.

Unified Governance: A UnifiedViews pipeline is used to extract data from Atlassian Confluence and JIRA (Figure 6.60). The data are transformed into RDF and integrated using ALIGNED vocabularies to create a unified view on the development data for supporting the development process. The transformed data are then annotated with concepts from the PoolParty Knowledge Graph. Similarity between development artefacts is calculated using a lexical and a graph-based approach in combination.

Unified Governance Search: The facetted and autocomplete search application on top of the Unified Governance data is explained in detail (Figure 6.61).

Unified Governance Similarity: The similarity computation as part of the search application and the use cases are explained.

Unified Governance Statistics: The statistical analysis and the visualisation are explained.

Issue Integration: Data inconsistencies in PoolParty can be caused by application error or can be caused by user by importing the data in PoolParty without doing constraint checks. These types of inconsistencies which cannot be handled by PoolParty can be reported by using the Issue integration feature (Figure 6.62). It allows users to configure a JIRA instance and report the issue automatically to PoolParty support (Figure 6.63). The log file is also automatically attached to the issue.

Graph Search: The faceted search which is used for managing development artefacts. It also provides a recommender UI where users can see the similarity between different issues and requirements. By using this recommender, users can find duplicate bugs, similar stories, and the requirements, which are associated with specific bugs.

Users can search for issues and see the details about it (Figure 6.64). GraphSearch provides a selection of similarity algorithms that were integrated for this use case to calculate similarities between development artefacts (Figure 6.65).

6.4.6 Evaluation

6.4.6.1 Measuring overall value

PoolParty is a software product provided to customers on premise or as a cloud service. Although the value can be measured by commercial success, the improvements done to both the application's features and the development process cannot be easily quantified. Data curation for PoolParty during

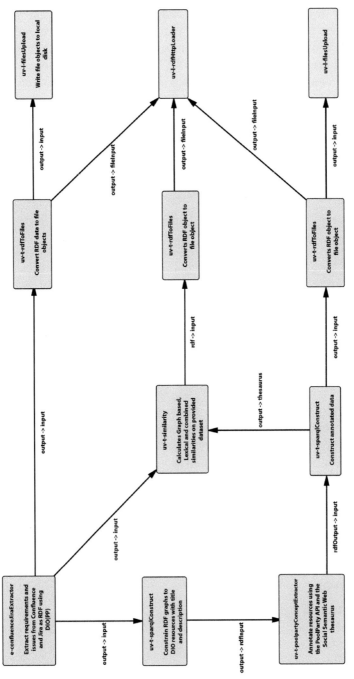

Figure 6.60 UnifiedViews pipeline for PoolParty use case.

6.4 Integrating Semantic Datasets into Enterprise Information Systems

Figure 6.61 Unified Governance Search.

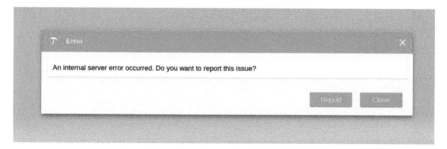

Figure 6.62 Issue Integration reporting dialogue.

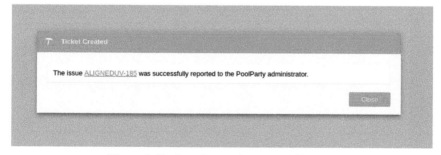

Figure 6.63 Issue Integration created dialogue.

298 Use Cases

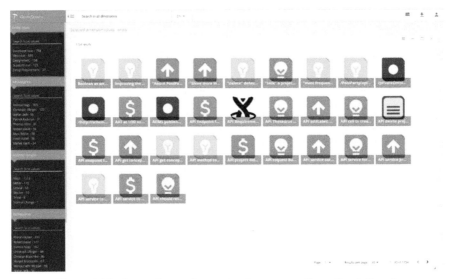

Figure 6.64 Semantic search over development artefact – Graph Search.

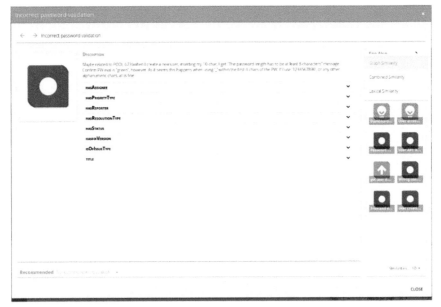

Figure 6.65 Details view of specific issue with the option to select similarity algorithm – PP Recommender.

6.4 Integrating Semantic Datasets into Enterprise Information Systems 299

ALIGNED provides identification and repair of data problems to customers. Curation reduces the need for consultant assistance, but also makes PoolParty more flexible regarding stability by allowing more freedom on data imports. This should raise the acceptance of the product, but it is hard to quantify. Value for the development process can be measured by the reduction of time efforts.

6.4.6.2 Data quality dimensions and thresholds

Data quality requirements for PoolParty are covered by two classes of validation. First, the Import Assistant feature ensures that imported data do not conflict or violate the functionality of the application and therefore provides a stability guarantee for users regardless of the data they import. Second, quality checks can be applied to the data to discover design flaws in the data modelling. These do not cause problems for the application, but might be unwanted by the data engineers. In ALIGNED, we focus on the Import Assistant and analyse the time saved by automatic identification and repair done by the user in contrast to a manual repair by a SWC consultant done directly on the data. Value is represented first by providing a feature for the user to do the repairs without assistance and second by the reduced time efforts needed to resolve data issues.

Data Agility

The Import Assistant provides users with the possibility to safely import any RDF data they want to use. The validation checks are based on stability requirements on the application. They detect all problems that would cause PoolParty to fail, and as a result, any data can be imported safely. Regarding the Unified Governance data, which is also represented as RDF, we can change or extend the set with additional information and adapt the software components using configuration rather than having a need to change the actual implementation. The pipeline processing can be configured within the pipeline steps and the Graph Search faceted search and similarity application uses ontologies to provide both the search interface and the data representation. Also, the Integrated Issue Reporting automatically provides metadata and logging information in the case of application failure.

Model Agility

If new features for PoolParty are implemented or existing features are changed, there might be the need for additional or modified validation checks.

Using a standard-based declarative approach for these allows development to add, change, test and reuse them more easily. The starting point was SPARQL-based checks, and we moved on to using SHACL shapes for validating the import data. The Unified Governance data model can be easily configured using ontologies and therefore provides adaption for the application to a changed dataset. In addition, Graph Search provides a plugin architecture to add new functionality regarding similarity and recommendations. Plugins are automatically loaded and provided via user interface for users to work with.

For PoolParty, the evaluation will compare productivity, usability and data quality, as well the connection between data development life cycle and the software development life cycle to the results in the previous validation deliveries. SWC is evaluating the PoolParty trial for import validation by using the Import Assistant to ensure data consistency. Improvements to the software developing process by using the Unified Governance methodology and tool chain, including the Integrated Issue Reporting, are evaluated regarding time efforts.

6.4.6.3 Evaluation tasks
Curation

Import data into a PoolParty project, detect the problems and repair them.

Process

- Import an RDF dataset into a PoolParty project
- Test the PoolParty Thesaurus Manager to detect problems
- Problems may show up immediately or when a specific resource is addressed
- Repair the problems manually using SPARQL

Challenges

Detection of problems is difficult. It may be that they are discovered during later work. There is no systematic checking other than manual testing, which is a lot of effort and not possible for big Thesauri.

Detection, but especially repair, requires detailed knowledge of the data model and also the PoolParty application. It is unlikely that the average user has this knowledge and can fix the issues.

6.4 Integrating Semantic Datasets into Enterprise Information Systems

Data Agility

Use the development data for managing the development process by organising issues to detect duplicates and find similar issues so that requirements and stories can be viewed in relation.

Process

- Duplicates are often created when two different users use a new or modified feature and report a bug. We can only detect duplicates manually at a later time.
- Requirements are manually linked to Jira issues on creation.
- Search has to be done separately in Confluence and Jira.
- In the case of an application failure, a Jira issue is created manually. Log files have to be requested from the customer.

Challenges

- Duplicates are created and detected at a later time. We cannot prevent the duplicate reporting.
- Finding stories and bugs for a specific requirement is a lot of effort. Using linking between Confluence and Jira is done manually.
- We cannot do an integrated search over the whole development data.
- Requesting log files from the customer in the case of application failure increases the time until issues are solved.

Model Agility

Create a unified view on the complete development data.

Process

Development data are managed in separate applications without integration using a common basis.

Challenges

- The development data are distributed in several systems.
- There is no integrated semantic description for the different parts.
- The representation is not standards based. Publishing is difficult.
- Integration and processing is proprietary.

- There is no option to change the processing using a declarative approach like ontologies.

6.5 Data Validation at DBpedia

6.5.1 Introduction

DBpedia is the centre of the current web of data. It publishes authoritative RDF-based datasets that are used as a common point of reference for interlinking and enriching most of the structured data on the Web today. It relies on an automated data extraction framework to generate open RDF data from Wikipedia documents, published in the form of file dumps, Linked Data and SPARQL (SPARQL Protocol and RDF Query Language) hosting on the Linked Data Stack.

DBpedia is a large-scale extraction project of unstructured and semi-structured data from different Wikipedia language editions to RDF. This extraction is achieved from a modular extraction framework that is customised to handle multilingualism and structural differences between different Wikipedia language editions. The latest DBpedia release (v. 2014) generated three billion facts from 125 localised versions. As Wikipedia evolves over time, the code should be able to adapt to these changes. However, identifying errors at this data scale becomes very hard, and validation workflows must be established that will ensure the quality of the extracted data. The high-level goal of ALIGNED in this use case is to produce tools for the DBpedia community, which will increase the coverage and precision of the provided DBpedia data stack.

The latest DBpedia release contains around 23,000 files from more than 100 Wikipedia language editions. At the moment, we provide a download folder for each language and detailed description only for the English dataset. We want to extend the current approach and provide a machine readable representation for the whole release and, besides dataset links to additionally provide descriptions for all datasets and languages, license and contact information using DataID (dataid.dbpedia.org).The DataID will be autogenerated by a script that will iterate over all release folders and using a pattern-based approach will assign metadata for each dataset.

6.5.2 Problem Statement

DBpedia is a large-scale extraction project of unstructured and semi-structured data from different Wikipedia language editions to RDF. This extraction is achieved from a modular extraction framework that is customised to handle multilinguality and structural differences between different Wikipedia language editions. The latest DBpedia release (v. 2014) generated three billion facts from 125 localised versions. As Wikipedia evolves over time, the code should be able to adapt to these changes. However, identifying errors at this data scale becomes very hard, and validation workflows must be established that will ensure the quality of the extracted data. The high-level goal of ALIGNED in this use case is to produce tools for the DBpedia community, which will increase the coverage and precision of the provided DBpedia data stack.

6.5.2.1 Actors

Role	Description
Extractors	DBpedia team members who run the extraction process for a given DBpedia release
Extraction Agents	Software agents that perform the extraction such as DBpedia live
Mapping editors	Community members who edit the DBpedia mapping wiki
Ontology Editors	DBpedia foundation members that edit the DBpedia ontology
Release managers	DBpedia team members that are responsible for the actions leading to a given release of DBpedia
Developers	DBpedia team members who write code for the extraction tools
Users	Users of DBpedia

The requirements on which the DBpedia use case wasbased are detailed in Appendix A.

6.5.3 Architecture

Figure 6.66 depicts the DBpedia use case trial architecture, showing the ALIGNED tools used in different stages of the DBpedia data workflow. With the DBpedia trial, we want to showcase both the reuse of ALIGNED tools as well as different integration points. For the data validation trial, we focus on validating instance data, mappings to RDF and links to other datasets. Link validation is performed with the SUMMR Mapping tool that reports results in the ALIGNED Metamodel version 2, especially the DLO and the DBpedia use case specific ontology. This means that the RDF logs produced

304 *Use Cases*

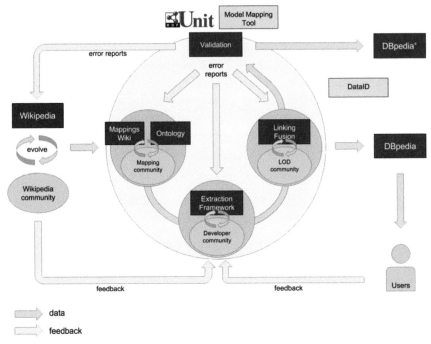

Figure 6.66 DBpedia Use Case Trial System Architecture, showing the ALIGNED tools used in different stages of the DBpedia data workflow.

by SUMMR can be consumed by the ALIGNED Unified Governance Tools in an ALIGNED tool chain. The other two validation scenarios are based on RDFUnit and use the SHACL violation reporting vocabulary as an integration point. DBpedia instance data are validated with a simple RDFUnit setup, while the DBpedia infobox-to-ontology mappings by using RML as an intermediate format. Regarding data dissemination, we use the DataID ontology as an integration point and automate the generation of the download page or a DBpedia release as well as the provision of a DBpedia release in a triple store through Docker.

6.5.4 Tools and Features

Figure 6.67 shows which features of the ALIGNED software tools are deployed in order to support these scenarios.

Software	Feature	Used For
RDFUnit	Data validation	Instance data validation
RDFUnit	Data validation	Mapping Validation
RDFUnit	Data validation	Ontology Validation
RDFUnit	Data validation	Link Validation (for metadata)
SUMMR Mapping Tool	Link Validation	DBpedia Interlink validation
DataID	Dataset description	Automatic generation of a release download page
DataID	Dataset description	Automatic generation of a triple store with data from a release using docker

Figure 6.67 ALIGNED Tools and Features used in the DBpedia trial platform.

6.5.5 Implementation

Figure 6.68 depicts an RDFUnit validation report of a DBpedia release. The report is provided as an RDF file that adheres to the RUT ontology, as well as an HTML export that is human readable. The report provides a high-level overview at the top with basic provenance metadata and statistics and continues with detailed error counts per constraint.

The mapping validation report (Figure 6.69) uses RDFUnit in the background but performs more sophisticated validation processing and reporting. The complete workflow is described in "Assessing and Refining Mappings to RDF to Improve Dataset Quality". The end user report is tailored for the mapping editors where they can select mappings errors based on language, infobox, DBpedia property, or DBpedia class.

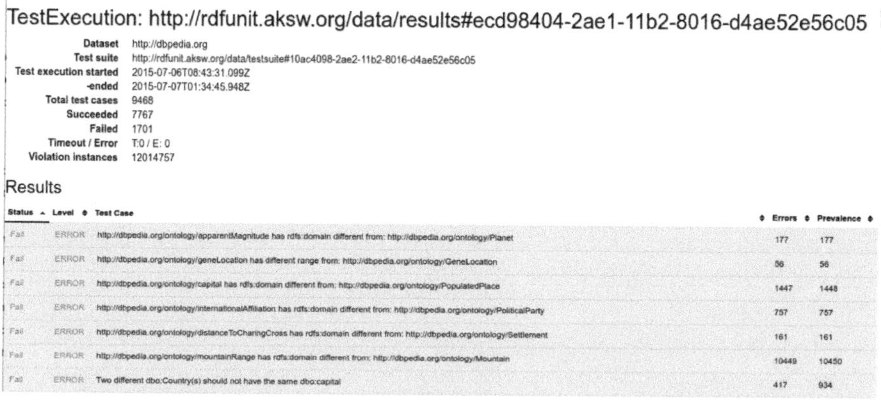

Figure 6.68 Instance data validation report with RDFUnit.

306 Use Cases

language	mapping	predicate	expected	existing
en	Infobox_sea (edit)	areaOfCatchment (edit)	Lake (edit hierarchy)	Sea (edit hierarchy)
en	Infobox_rail (edit)	voltageOfElectrification (edit)	RouteOfTransportation (edit hierarchy)	PublicTransitSystem (edit hierarchy)
en	Infobox_rail (edit)	typeOfElectrification (edit)	RouteOfTransportation (edit hierarchy)	PublicTransitSystem (edit hierarchy)
en	Infobox_public_transit (edit)	voltageOfElectrification (edit)	RouteOfTransportation (edit hierarchy)	PublicTransitSystem (edit hierarchy)
en	Infobox_public_transit (edit)	typeOfElectrification (edit)	RouteOfTransportation (edit hierarchy)	PublicTransitSystem (edit hierarchy)
en	Infobox_public_transit (edit)	numberOfStations (edit)	RouteOfTransportation (edit hierarchy)	PublicTransitSystem (edit hierarchy)
en	Football_box (edit)	numberOfGoals (edit)	CareerStation (edit hierarchy)	Event (edit hierarchy)
en	Infobox_Geopolitical_organization (edit)	officialLanguage (edit)	PopulatedPlace (edit hierarchy)	Organisation (edit hierarchy)
en	Infobox_train (edit)	yearOfConstruction (edit)	Place (edit hierarchy)	Train (edit hierarchy)
en	Infobox_television_season (edit)	numberOfEpisodes (edit)	TelevisionShow (edit hierarchy)	TelevisionSeason (edit hierarchy)
en	Infobox_diocese (edit)	numberOfMembers (edit)	Legislature (edit hierarchy)	Diocese (edit hierarchy)
en	Infobox_canal (edit)	riverBranchOf (edit)	River (edit hierarchy)	Canal (edit hierarchy)
en	Infobox_bus_transit (edit)	numberOfStations (edit)	RouteOfTransportation (edit hierarchy)	PublicTransitSystem (edit hierarchy)
en	Infobox_bay (edit)	areaOfCatchment (edit)	Lake (edit hierarchy)	Bay (edit hierarchy)
en	Infobox_academic_conference (edit)	frequencyOfPublication (edit)	PeriodicalLiterature (edit hierarchy)	AcademicConference (edit hierarchy)

Figure 6.69   Mapping validation report with RDFUnit and RML.

Starting in October 2017, DBpedia replaced the old Mappings Wiki with a new Mappings UI, based on GitHub and RML mappings. The validation report shown in Figure 6.70 has been superseded by this interface, since wrong mappings are detected automatically on commit.

The graphical interface based on the DBpedia-Links repository provides an overview of all outgoing links to other datasets and points out any inconsistency in a given linkset (Figure 6.71). DBpedia employs multiple validation methods for link validation, including the SUMMR Mapping tool (see below).

The SUMMR Mapping tool (Figure 6.72) performs an interlinking validation for all the external links in a DBpedia release. After processing the link, the tool outputs a log file and splits the links into valid and invalid. The invalid links are discarded, and only the valid ones become part of the DBpedia release.

Gathering extensive metadata throughout all extraction steps is not only helping to produce exhaustive dataset metadata (in form of DataID documents), but also allows for highly expressive logs and convenient summary reports (as shown in Figure 6.73).

The generation of a DBpedia release download page was a tedious task. We use DataID as a core release metadata component and created a flexible user interface that people can use to identify and filter specific DBpedia datasets. Figure 6.74 depicts the download page of the 2015-2010 release.

6.5 Data Validation at DBpedia   307

Figure 6.70   The new Mappings UI (using RDFUnit for validating mappings).

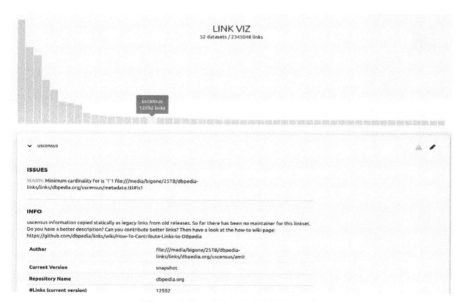

Figure 6.71   DBpedia Link Viz tool.

308 Use Cases

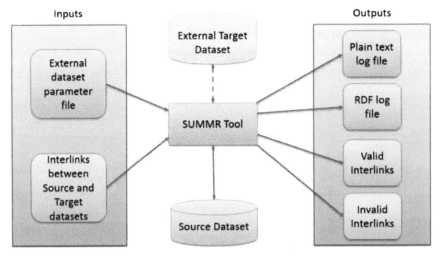

Figure 6.72 SUMMR Mapping tool.

Figure 6.73 Active extraction monitoring (here: extraction summaries forwarded to Slack).

The dockerised DBpedia (Figure 6.75) automates the digestion of a DBpedia release by downloading the datasets of the user's preferred language and loading these datasets on a Virtuoso triple store server. We use docker as the underlying technology that has recently became a very common means

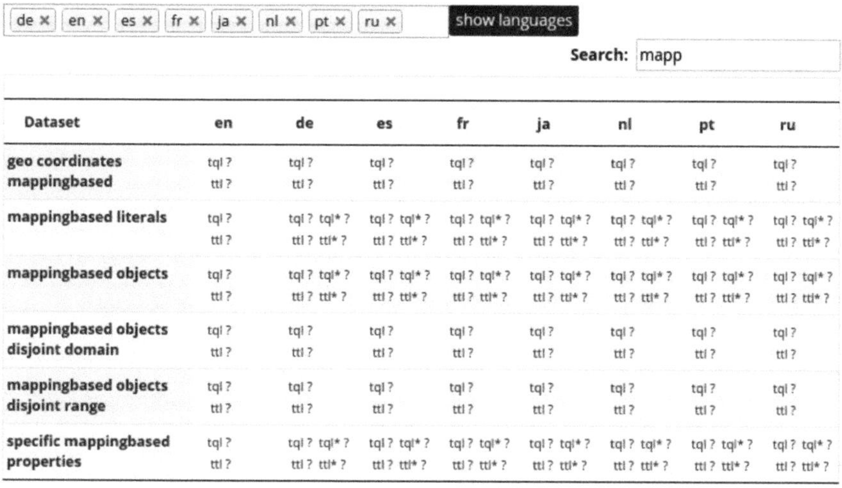

6.5 Data Validation at DBpedia

Figure 6.74 DBpedia download page through DataID.

Figure 6.75 Dockerised DBpedia.

of application distribution. DataID is used as the core metadata component to identify and filter DBpedia datasets.

6.5.6 Evaluation

6.5.6.1 Productivity

The basic unit of analysis for productivity is a comparison of time elapsing between two DBpedia releases. Typical tasks are code maintenance, release management, ontology editing, release documentation creation and dealing with user queries.

For a sufficient evaluation of productivity changes between two DBpedia releases, one has to consider the changes to data sources, ontology, mappings,and the code base. In addition, the number of published datasets tends to increase over time when incorporating new extraction methods and algorithms. Nonetheless, over the time of this project, DBpedia has managed to cut the time between releases in half (13 months to 6 months), while producing at least three times as many pieces of information (triples). Currently, DBpedia is pushing for regular updates for the 10 most widely used language editions on a bi-monthly basis (synchronising with the bi-monthly data releases by Wikimedia). Multiple efforts to increase productivity are closely related to Quality and Agility (see below).

A significant improvement of time spent on dissemination activities was achieved by introducing DataID as dataset metadata format. Extensive metadata descriptions of datasets allow for many automation tasks, such as automated downloading of relevant dataset files, generic implementations of dataset overviews, and download tables. In addition, extensive and high-quality metadata of datasets helped DBpedia to check 31 of the 35 Data on the Web Best Practices of the eponymous W3C working group.

6.5.6.2 Quality
Instance Validation

To create high-quality data, a validation method for DBpedia instance data has to provide sufficient metadata to distinguish between three different possible sources of a violation:

- The Wikipedia editor (entering erroneous values)
- Incorrect mappings, between source and DBpedia ontology
- A software issue in the DBpedia Extraction Framework

RDFUnit was created with these demands in mind, providing necessary metadata to any violation found and creating links between a software issue and the violating instance (see D5.8). The resulting violations and their pertaining metadata provide the exact coordinates of a violation, the grounds for this violation and the possible source. Thus, violations recorded in such a manner are used as feedback medium, relating possible mistakes to Wikipedia editors, the mapping community or software developers. DBpedia is running all published data through RDFUnit, validating it against an up-to-date version of the DBpedia ontology. The validated outputs generate consistent data that are termed DBpedia+, whereas the wider, more exhaustive data are published as the standard DBpedia datasets.

Mapping Validation

In addition to validating the resulting instance data, DBpedia started to validate the mappings between DBpedia ontology and the Wikimedia data sources on a nightly basis with RDFUnit. Thus, most of the mapping-related violations can be caught before ever starting the data extraction, preventing possible reruns of whole extraction steps and increasing productivity in turn.

Ontology Validation

The DBpedia ontology has been maintained by the DBpedia community in a crowdsourced manner at the mappings wiki. There is an ongoing effort to move ontology development onto GitHub for easier collaboration and for the sake of more control over the ontology structure.

At the time of writing, a set of constraints ensure that each DBpedia class and each DBpedia property conform to DBpedia community requirements. RDFUnit is used to perform the validation (using SHACL constraints) and to integrate with Travis CI and automate the checks on each commit and pull request.

Link Validation

The DBpedia-Links repository maintains linksets between DBpedia and other LOD datasets. A system for maintenance, update and quality checks, which validates various aspects of the link submission, is in place and is integrated with common continuous integration services, such as Travis CI. It offers a way to publish linksets between DBpedia and any given dataset, which are published alongside the DBpedia dataset files.

Quality checks include:

- The SUMMR Mapping validation tool
- RDFUnit for validating (using SHACL constraints) the link manifest (basic metadata providing a minimum of provenance)

Workflow Validation

To ensure quality regarding the extraction workflow, DBpedia extended the extraction framework to produce metadata for any extraction process, extensive logging of progress and exceptions, as well as high-level summaries of extractions. These efforts support extensive monitoring, metadata propagation and logging (on triple and dataset level) and the deployment of ETL frameworks and Workflow Management Systems to further decrease the time needed for extraction and to automate this process completely. Currently, a concerted effort to adapt the Unified Views Framework of SWC for this purpose is underway, which will continue until after this project has finished.

6.5.6.3 Agility

The greatest need for agility in DBpedia is the ability to rapidly respond to changes in source datasets like Wikipedia. This is the focus of the use case scenario Wikipedia/Wikidata change. Example Wikipedia changes that impact DBpedia are: the introduction of new pages that represent new concepts, the introduction of new infobox templates that represent additional instance data in DBpedia, and changes in infobox structures. Adapting to those changes in a (semi-) automated way will prevent the loss of data (due to changes to Wikipedia templates) and incorporate new instance data automatically.

As a prerequisite to automate mappings, DBpedia will switch its complete mapping infrastructure to an RML-based mapping approach in October 2017. This is a direct result of one of our Google Summer of Code projects of 2016. As a superset of the W3C recommended mapping language R2RML for relational databases to RDF, RML offers a way to completely represent all DBpedia mappings in RDF. It enables:

- Full support of RDFUnit mapping validation (no transformation necessary)
- The complete range of mapping possibilities of RML (incl. functions, conditions, etc.)
- Rule-based automation of mappings using all RML features
- Replacing the rigid wiki text mappings used by DBpedia until now

Concurrently, DBpedia helped to implement a taxonomy learning system based on Wikipedia categories. Set up as one of our annual participation with the Google Summer of Code program, this project realised the concept laid out in the "Unsupervised Learning of an Extensive and Usable Taxonomy for DBpedia". These automatically derived types are a reliable backbone for the automated mapping generation ahead.

7

Evaluation

Pieter Francois[1], Stephanie Grohmann[1], Katja Eck[2], Odhran Gavin[3], Andreas Koller[4], Helmut Nagy[4], Christian Dirschl[2], Peter Turchin[5] and Harvey Whitehouse[1]

[1]University of Oxford, UK
[2]Wolters Kluwer Germany, Germany
[3]Trinity College Dublin, Ireland
[4]Semantic Web Company, Austria
[5]University of Connecticut, USA

7.1 Key Metrics for Evaluation

The evaluation of productivity, quality and agility requires concrete metrics to be evaluated prior to the introduction of ALIGNED tools. This gives us a baseline measurement for gains in the three evaluation areas. Once ALIGNED tools and processes are then deployed, concrete comparisons can be made to assess the progress, which results from ALIGNED tools and processes. The units over which evaluation takes place, and the measures over these units must be designed such that they can be assessed both prior to, and after, the integration of ALIGNED tools and processes.

In order to evaluate the tools that we produced during the ALIGNED project, we took the following steps:

- Baseline studies: an initial estimate of how the use cases perform before the introduction of ALIGNED tools.
- Studies on initial prototypes: focussed initial prototypes will be developed for three ALIGNED use cases in phase 1 of the project (up to month 9) that only depend on the work of a single technical workpackage (WP3, WP4, WP5) and the tools can be evaluated in this initial phase to gain rapid user insight and feedback.
- Longer-term evaluations based on the empirical evidence collected from the four use cases for ALIGNED methods and tools developed during phase 2 and phase 3 of the project.

314 *Evaluation*

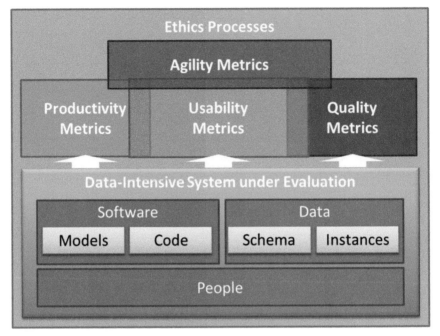

Figure 7.1 The ALIGNED Evaluation Framework.

There were three key target areas for the impact of ALIGNED methods and tools on the development and evolution of data-intensive systems: productivity, quality, and agility. Each of these is defined below to allow cross-tool and cross-use case comparisons to be made. In addition, each target area can be split into data and software aspects as well as system-wide measures, for example data management productivity, software development productivity and overall system productivity. For data management, it is often useful to split tasks into schema-oriented and dataset or instance-based measures since these often have different actors, timeframes and scopes. Figure 7.1 illustrates the ALIGNED evaluation framework, which is made up of the data-intensive system under study and the four evaluation aspects plus ethics processes covered by this handbook.

An important aspect of system evaluation that has cross-cutting impacts on quality, productivity, and agility is the well-developed concept of usability[1] and ALIGNED performed usability evaluations on all tools developed within the project.

---

[1] Ergonomics of Human System Interaction ISO 9241, in particular part 11 – Human-Computer Interaction, 1998.

In general, ALIGNED stressed quantitative evaluation over qualitative measures (information or data based on quantities obtained using a quantifiable measurement process) as befits automated systems such as model-driven software tools. However, the nature of systems development and maintenance (evolution) are that of a socio-technical system and as such qualitative evaluation (qualities that are descriptive, subjective or difficult to measure) based on user feedback were used to supplement quantitative evaluations. This is especially true in cases where informal or semi-automated human-based systems are either currently deployed (for baseline studies) or are necessary to produce the best outcomes (e.g., domain expert-based data curation).

7.1.1 Productivity

For evaluation purposes, we understand productivity as being a measure of the amount of human effort required to produce some unit of software, schema or dataset change for a given use case scenario. This effort may be measured in person-hours, but other measures are possible such as task completion time, task completion rate, or task error rate. For largely user-interface-driven processes, there are a number of popular keystrokes[2] or click-based models[3] for estimating productivity. For software engineering, there is prior work on evaluating the productivity of new engineering processes that should be considered.[4] In the first instance, it is possible to find a number of proxies which, when taken together, may act as a crude guide to measuring software size. Lines of Code,[5] Control-flow or Cyclomatic complexity,[6] and various feature counts[7] have traditionally acted as primitive metrics for software scale and complexity.

---

[2] The Keystroke Level Model for User Performance Time with Interactive Systems S. Crad, T. Moran, A. Newell, CACM, v23 n7, July 1978.

[3] Project Ernestine: Validating a GOMS Analysis for Predicting and Explaining Real-World Task Performance W. Gray, B. John, M. Atwood, Human-Computer Interaction, Vol. 8, Issue 3.

[4] Measuring and predicting software productivity: A systematic map and review K. Petersen, Information and Software Technology, Vol. 53, Issue 4, pp. 317–343.

[5] A Survey on Impact of Lines of Code On Software Complexity S. Bhatia, J. Malhotra, ACM SIGSOFT Software Engineering Notes, Vol. 39, pp. 1–6.

[6] Cyclomatic Complexity Metric for Component Based Software S. Chidamber, C. Kemerer, International Conference on Advances in Engineering and Technology Research (ICAETR), pp. 1–4, 2014.

[7] A metrics suite for object oriented design U. Tiwari, S. Kumar, IEEE Transactions on Software Engineering, Vol.20, No. 6, pp. 476–493.

316 *Evaluation*

There are also several cases in which cross-cutting productivity concerns are of importance, the one most particularly relevant to ALIGNED being the productivity costs of parallel development of software, schema, and datasets changes. In this case, productivity measures should look at the cost of changes from one area to the others in terms of productivity.

7.1.2 Quality

Quality is generally taken as the assessment of "fitness for purpose"[8] of the output of a given tool, process, or method. The measurement of quality is generally more context-dependent, and different measures are used in the areas of software, schema, and data.

For software quality, evaluation of software generation tools is difficult, especially as ideal tools produce no defects, and validating the absence of something is hard. It is possible to measure "churn" of software development or counts of bugs found and that can act as metrics for software quality and reliability.[9]

For data, we assess the ability of the data to satisfy properties, which are either desirable or required by consumers of the data. In particular, we will reuse the methods of assessment of Linked Data Quality defined by Zaveri et al.[10] This gives us 27 separate dimensions on which to evaluate data quality and specifies multiple metrics for all of them.

7.1.3 Agility

We define agility as the speed at which the ALIGNED tools can be adapted and reconfigured in the face of ongoing changes in requirements. It is often measured in terms of the human effort required to enact the change and so is closely related to productivity measures. When software or data management task sizes are combined with measurements of man-hours spent on development, some approximations can be made for notions of agility.[11]

---

[8]The Quality Control Handbook J. Juran, McGraw-Hill, New York, 1974.

[9]Evaluating Complexity, Code Churn, and Developer Activity Metrics as Indicators of Software Vulnerabilities S. Yonghee, A. Meneely, L. Williams, J. Osborne, IEEE Transactions on Software Engineering, Vol. 37, No. 6, pp. 772–787.

[10]Quality Assessment Methodologies for Linked Open Data A. Zaveri, A. Rula, A. Maurino, R. Pietrobon, J. Lehmann, S. Auer, Semantic Web Journal

[11]Survey on agile metrics and their inter-relationship with other traditional development metrics S. Misra, M. Omorodion, ACM SIGSOFT Software Engineering Notes. Vol. 36, Issue 6, pp. 1–3.

Agility for our use cases will often be measured with respect to parallel co-development of software, schema and datasets as agility is a cross-cutting concern. For instance, a change to a schema or ontology will generally require both migration of datasets, as well as changes to the programme interface to consumption of the data.

7.1.4 Usability

ISO 9241[12] on human computer interaction defines usability as "The extent to which a product can be used by specified users to achieve specified goals with effectiveness, efficiency, and satisfaction in a specified context of use". Effectiveness and efficiency can be measured through productivity-style measurements of task outputs and work rates. However, it is also considered valuable to analyse the user error rates generated and the quality of work produced (linking to our quality measures). Satisfaction is probably the hardest aspect to accurately measure but we will deploy System Usability Scale (SUS)[13] user questionnaires as a baseline. It is simple, ten-item attitude Likert scale giving a global view of subjective assessments of usability. Despite its simplicity, SUS is well-understood and widely deployed, and this gives us access to decades of comparative usability studies and analysis to interpret SUS results.[14] In addition, it is easy to augment SUS with additional questions that are specialised to the system under study or which follow recent best practice in user surveys such as Perlmans Practical Usability Evaluation questionnaire.[15]

In addition to questionnaire-based approaches to usability, we deployed, where appropriate, the "thinking-aloud" protocol where participants are asked to verbalise their thinking while performing a task.[16] Other techniques deployed are "co-discovery", where participants are asked to verbalise their thinking while performing a task and "retrospective testing" or "coaching".[17]

---

[12] Ergonomics of Human System Interaction ISO 9241, in particular part 11 – Human-Computer Interaction, 1998.

[13] SUS: a "quick and dirty" usability scale J. Brooke, Usability Evaluation in Industry. London: Taylor and Francis, 1986.

[14] An empirical evaluation of the system usability scale A. Bangor, P.p T Kortum, and J. T. Miller, Intl. Journal of Human-Computer Interaction, Vol. 24, Issue 6, pp. 574–594, 2008.

[15] Practical usability evaluation G. Perlman, CHI'97 Extended Abstracts on Human Factors in Computing Systems. pp. 168–169, ACM, 1997.

[16] Protocol analysis: verbal reports as data, revised edition K. A. Ericsson, H.A. Simon MIT Press, Cambridge, MA, 1993.

[17] Usability Engineering 2nd edition J. Neilsen, Morgan Kaufmann, San Francisco, 1994.

318   *Evaluation*

7.2 ALIGNED Ethics Processes

This section provides a set of guidelines followed by the coordinators of ALIGNED pilot studies and trials. Specific instructions are provided for each step in the life cycle of these pilot studies that involves ethical considerations. Taken together, these guidelines provide ALIGNED collaborators with detail on when and how to engage with the Ethics and Society sub-committee of the ALIGNED project and on how to ensure the pilot studies and trial confirm to both relevant national and EU regulation.

Over the life cycle of a pilot study, coordinators need to engage with ten sets of action points.

- BEFORE THE START OF THE PILOT STUDY: Coordinators need to familiarise themselves thoroughly with the Ethics section of the contract signed between the ALIGNED project and the EC. This is an important first step to understand the full range of potential ethical issues at stake when setting up a pilot study.
- BEFORE THE START OF THE PILOT STUDY: Coordinators need to obtain the appropriate internal institutional ethical approval. The bodies responsible for internal institutional approval are your first port of call to ensure that the pilot study respects institutional, national and European regulation. This is especially important for any pilot study that involves the storage of personal data as some categories of these data are classed as 'sensitive' (e.g., health, sexual lifestyle, ethnicity, political opinion, religious or philosophical conviction), and this data may only be processed according to specific rules. The ALIGNED Ethics and Society sub-committee has the details of the relevant institutional bodies for each partner.
- BEFORE THE START OF THE PILOT STUDY: Coordinators need to forward the institutional ethical approval obtained to the ALIGNED Ethics and Society sub-committee. This step is crucial as it is vital that the ALIGNED project forwards all ethical approvals to the EC. Furthermore, this will allow the Ethics committee to double check that all necessary steps have been taken and that the pilot study fulfils all necessary requirements.
- BEFORE THE START OF THE PILOT STUDY: As most pilot studies will involve voluntary participants, the coordinator must submit the consent form to be used to the ALIGNED Ethics and Society sub-committee. This consent form must be modelled on the template attached in appendix A and any change to the template must be approved by the ALIGNED Ethics and Society sub-committee.

- BEFORE THE START OF THE PILOT STUDY: As most pilot studies will involve voluntary participants, the coordinator needs to ensure that all staff associated with the pilot study fully understand the ethical considerations when handling voluntary participants. For this, all staff need to familiarise themselves with the relevant Ethics sections of the contract signed between ALIGNED and the EC. Special attention must be paid to those sections dealing with the recruitment of voluntary participants, the control of their personal data, the nature of their participation, the right of participants to cancel their involvement at any time in the process, the rights of voluntary participants to privacy and appropriate treatment, and the definition of informed consent. It is absolutely vital that no participation can take in any form without informed consent.
- AT THE START OF THE PILOT STUDY: the coordinator will ensure that sufficient measures are in place to store all personnel data password protected and all 'sensitive' personnel data encrypted.
- AT THE START OF THE PILOT STUDY: the coordinator, in collaboration with the ALIGNED Ethics and Society sub-committee, will prepare and share with the voluntary participants detailed information on the procedures that will be implemented for data collection, storage, protection, retention, and destruction. The ALIGNED Ethics and Society sub-committee will provide the coordinator with input to ensure that this information conforms to national and European legislation.
- THROUGHOUT THE ENTIRE LIFESPAN OF THE PILOT STUDY: the coordinator needs to assess on a continuous basis whether any of the ALIGNED methodologies result in discriminatory practices or unfair treatment. The pilot study coordinator needs to inform the ALIGNED Ethics and Society sub-committee even in case of the slightest doubt that the pilot study results in discriminatory practices or unfair treatment.
- DURING AND AFTER THE PILOT STUDY: In the case of incidental findings of value arising from research activities (e.g., psychological trauma arising from productivity-related questions), the coordinator needs to inform participants when such results will be disseminated. Participants will be given the right to withdraw their information.
- AFTER THE PILOT STUDY: as personnel data can only be archived during the lifespan of the ALIGNED project and thus needs to be deleted at the end of the project, the coordinator will work together with the ALIGNED Ethics and Society sub-committee to ensure the deletion of all personal data.

7.3 Common Evaluation Framework

Productivity, quality, and agility are the three dimensions that are most usefully measured in order to practically evaluate data-intensive systems. However, they are not separate dimensions but in fact have close semantic connections between them.

7.3.1 Productivity

Productivity is the overarching dimension used to measure the performance of all work systems – the ratio of the value provided by a service to the cost of delivering the service. If we were to implement two alternative systems in parallel and maintain them over time so that they provided exactly the same service, the relative cost would provide us with an unambiguous guide as to which system had performed better. Similarly, if we were to spend exactly the same time and money on delivering the same service over a period of time, through two alternative systems, the relative value provided by each would again tell us which system had performed better.

However, while costs are normally reasonably easy to measure, the value provided by a system can be more difficult as systems can be embedded within larger systems and provide value that cannot easily be distilled into economic units.

7.3.2 Quality

In the context of information systems, quality is a proxy measure for value. The better the Quality of Service (QoS), the greater the value provided by the system. If this is not the case, then the QoS has not been well defined. In general, therefore, if two systems provide the same QoS, we can compare them directly in terms of costs. In data-intensive systems, we are primarily focussed on the data quality because much of the behaviour of the system is driven by data. However, data quality only has meaning in the context of the services that are based on the data. We care about the overall service quality, and data quality is only interesting to the extent that it affects the business value provided by the system.

In any given system, it should be the case that improving quality increases the value provided by the system and vice versa. Quality is a multi-dimensional concept,[18] often with complex non-linear interactions

---

[18] A metrics suite for object oriented design U. Tiwari, S. Kumar, IEEE Transactions on Software Engineering, Vol. 20, No. 6, Pages 476–493.

between variables in different dimensions. For any given system, we can imagine a function Qual(sys) → $n which generates the value provided by a given system. In practice, we normally really want to know Qual(sys') ≥ Qual(sys), the effects of a given change in a system. We need a function which, for any given change to a system, will tell us what the change to business value will be. Our quality model defines the variables that will be passed to this function, and the function's implementation defines how changes to the values of variables impact service value.

7.3.3 Agility

We would like to be able to forecast the performance of systems and not just compare them in retrospect. Agility is essentially a measure of future productivity which attempts to capture such a forecast. How much future value will this system provide and at what cost? The trouble with this measure, of course, is that we do not know what opportunities for value the future holds. For any given system, agility to make changes that we never end up wanting to make have essentially no value. This means that agility, like quality, is very domain and context-dependent. We therefore need to know which types of changes are likely to be important in a given system before we can assess its agility. Because this is a prediction about the future, it can never be more than probabilistic, but previous behaviour is normally a good guide to future behaviour, so we can normally extract at least some characteristics of the types of changes that are important in a particular domain by observing existing systems.

In data-intensive systems, scale – considered as the volume, velocity and complexity of the data – tends to have significant influence upon the system's agility and tends to increase over time. As a general rule of thumb, service value and cost both increase with scale. Therefore, one of the most important aspects of understanding a data-intensive system's agility is understanding the interaction between these two variables and the different components of scale – in the context of the likely evolution of the system over the course of its operation.

Ultimately, the value of any work-system can be characterised by its productivity curve over time. The more agile the system, the more this curve will tend to rise in the future; the less agile the system, the quicker it will fall. This is because the more agile the system, the quicker and more effectively changes can be tested to meet emerging requirements. The most important way to compare systems is the net value that they deliver over their lifetime. We cannot know this in advance, but we can normally make reasonable predictions based on proxies for agility in any given context and use them to predict the likely future productivity trajectory.

Because the dimensions and metrics used in any given data-intensive system are heavily dependent on the specific context of the service, they cannot be directly compared. In one context, better accuracy and precision of data might be considered to have a uniformly positive effect on Quality of Service. In another context, it might cause the system to crash (e.g., because it causes the program to trigger a bug in a floating point operation that was not used when the data had lower precision).

Rather than comparing data quality directly, we can compare it indirectly through the cost of providing a given quality of service. There are several aspects that must be considered in this comparison:

Data Curation Cost: The cost of maintaining the data at a given quality level (to provide constant Quality of Service) over a period of time, given changes in scale. There are two particularly important data quality levels that are worth focussing on here. DQmin is the minimum level of data quality required in order for the service to work. The threshold is multi-dimensional and complex and includes, among other things, all the database conditions which cause the software service to crash. The second quality threshold worth considering is DQmax – the maximum level of data quality that the service can exploit. Examples of data quality that exceeds DQmax: data stored as floating points with high precision that are then cast to integers by a program, metadata about data semantics that is not used by programs. As a general rule, there is no return on investment for exceeding DQmax. Between these two thresholds, quality can vary in any number of dimensions. If the dimensions used are well chosen for the service, then increases in quality will translate into an increase in the overall quality of service provided, and if the service is well aligned with the business needs, this translates directly into increased business value.

Data Agility Cost: the cost of increasing the overall value provided by the system by using existing data in a new way – for example, how much time and money is required to make a slice of the data in a database available for use by a new program (with whatever data-formatting requirements it has). The cost includes any changes to the code of programs that consume the data, everything that is required to produce and deliver the new service.

Model Agility Cost: the cost to change the overall behaviour of the system in situations which require changes to the structure of the data. This includes the costs of changing the structure of the data, changing the software to encode the new behaviour, and returning the QoS to the level that it had before the change. The last part is important, because, for example, when changing the structure of a SQL database, all the existing programs that use the changed part of the database normally stop functioning. The measure of agility therefore includes all of the effort required to return the existing system to the QoS level that it had before the structure was changed, as well as achieving the required QoS for the new behaviour.

Based on this interpretation of productivity, agility and quality, a general methodology for evaluating and comparing data-intensive systems can be derived.

- Start by defining the overall value provided by the system and identifying proxies where possible.
- Define the data quality dimensions that are most important for the domain and how they translate into changes in quality of service and value.
- Define the data quality metrics and thresholds that are most important for the context.
- Take a given quality threshold and estimate the data maintenance, data agility, and model agility costs of maintaining that threshold over time.
- Forecast the evolution of the system and how the value it provides will depend on data and model agility and the characteristic requirements of the domain.
- The total cost of providing a service can be compared as the cost of maintaining a given quality of service over the lifetime of the system, plus the cost of model and data agility to support the required changes to the service, multiplied by their frequency.

It is important to emphasise that these comparisons are only valid at a particular quality of service level and should be made at the broadest possible level, where for example, manual processes are included where they are required in a given approach to achieve a given quality level.

7.4 ALIGNED Evaluation Ontology

In parallel with the development of the common methodological framework described above, ALIGNED has developed an ontology for the description of evaluation results (Figure 7.2). It contains classes and properties designed to capture the most important types of evaluation metrics and related concepts. The ontology is available at: https://github.com/nimonika/ALIGNED_Ontologies/blob/master/evaluate.owl

324 Evaluation

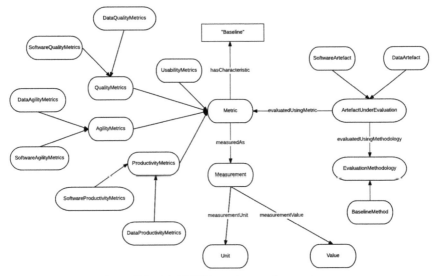

Figure 7.2 ALIGNED metrics ontology – classes.

At the core of the ontology is the concept of a metric. This is any property of the system that can be measured or analysed, such as the error rate of historical data variables in the Seshat: Global History Databank, or the number of data constraint violations on import in PoolParty. Metrics can be related to quality, agility, productivity, or agility and can be further subdivided into data and software metrics. A metric also includes information about the baseline of the metric (its initial value before any changes are made, used as a comparison to show change) and how it is measured.

These metrics are used to analyse an artefact. An artefact is any system or subsystem that is being evaluated for data and software quality analysis purposes. This also contains information about how the artefact is being analysed. The collection of metrics and the evaluation methodology provide a description of how the system in question is being analysed.

Appendix A – Requirements

JURION

Scenario 1

XSLT uplift of XML sources to RDF metadata. In area 1 (Figure 7.1) of the content pipeline, metadata is extracted from the proprietary WKD XML schema and transformed in RDF. Due to regular changes in the XML format, the correct transformation process based on existing XSLT scripts must be secured, so that no inconsistent data are fuelled into the metadata database in area 3.

Scenario 2

PoolParty quality checks in thesaurus management system. In area 2 (Figure 7.1) of the content pipeline, controlled vocabularies and domain models are created, maintained, and delivered for further usage, based on SKOS standard. The integrity of the knowledge management system as a whole needs to be ensured. Therefore, regular local and global quality checks need to be executed, so that, e.g., inconsistencies across different controlled vocabularies can be detected and resumed.

Scenario 3

Verifying available instance data for linked external metadata (area 2 in Figure 7.1). The Linked Data paradigm aims at linking information across sources, in contrast to simply link local copies of datasets. The technologies for linking are available, but there is not yet a proper support to maintain these links, especially when it comes to changes of the source – both on schema as well as on instance data level, e.g., when geo data of a linked organisation is updated or deleted on the source side.

Scenario 4

Verification of availability of appropriate metadata for Solr Indexer (areas 3 and 4 in Figure 7.1). Data-intensive systems like JURION are highly dependent on the metadata that steers many of their core functionalities. The indexing process of a search engine includes more additional information on top of the pure text. Extensive and smart quality checks on content based on the requirements derived from this process could strengthen the stability of the overall system.

Scenario 5

Cross-check of CMS instances and references in metadata database (area 3 in Figure 7.1). The metadata database is storing relationships between concepts maintained in PoolParty and documents maintained in the CMS as well as semantic relationships between documents in the CMS; e.g., document A overrules document B.

JS1 – XSLT Uplift of XML Sources to RDF Metadata

Description

A new XML document is available. A CMS expert checks for its conformity and then feeds it into the transformation process by executing XLST scripts on it. The resulting RDF data are stored in the metadata database, and a log file is created. We need to better control the transformation process of our XML metadata to RDF from a semantic and schema point of view.

Approach

Based on the schema, test cases are automatically created, which are run on a regular basis against the data that need to be transformed. The errors detected lead to refinements and changes of the XLST scripts and sometimes also to schema changes, which impose again new automatically created test cases.

Goals

- Better control over RDF metadata
- Streamlined transformation process from XML to RDF

- Early detection of errors in RDF metadata, since the resulting RDF metadata are a core ingredient for many subsequent process steps in production and application usage
- More flexibility in RDF metadata creation

JS1.1 – XSLT Uplift of XML Sources to RDF Metadata

User Story

We need to have better control over RDF metadata. As a legal editor being responsible for a legal domain within JURION, I must be sure that the metadata used in the application is correct.

Detailed Description

The resulting RDF data coming from the transformation process are stored in the metadata database, and a log file is created. An SME is checking the metadata database on a random basis, and if data errors are detected, it resolves the errors on a manual basis. Any systematic error is reported back to the CMS team.

JS1.2 – Streamlined Transformation Process from XML to RDF

User Story

We need a streamlined transformation process from XML to RDF, so that changes in any step of the process can easily be made, without adding complexity or increasing error rate.

Anybody working in the process or with the respective data want a transparent and lean process, which is avoiding any unnecessary errors.

Detailed Description

Changes to the schema can be initiated from anybody involved in the process or the data. A CMS expert can initiate it in order to reduce complexity; a product owner can initiate it in order to support new functionality in the customer application; a legal domain expert can initiate it in order to be able to support new legal concepts, and so on.

These schema changes are collected by the content architect and then it is evaluated, what the exact impact is on the existing processes.

Then the changes are confirmed and executed by a schema expert. Afterwards, all other depending processes are adapted accordingly if necessary by the responsible teams.

Then the whole pipeline is tested and evaluated by the responsible teams as well as by the product owner for the desired results.

Errors are reported and collected by the content architect, in case they cannot be solved locally within the responsible team.

Then the new schema and process are approved and ready for daily production.

JS1.3 – Early Detection of Errors in RDF Metadata

User Story

Errors in RDF metadata based on the transformation process from XML data need to be discovered as early as possible because many subsequent processes rely on the correctness of this data.

As a product owner, I rely on the correctness of the metadata in my applications.

Detailed Description

The CMS expert runs tests on a regular basis, especially if new versions of a schema are involved. A log file is created, and reported errors or unusual results are detected and further analysed by the CMS expert. The errors are grouped according to impact factors, and then the necessary experts like legal domain experts are approached for giving advice on error correction.

Afterwards, the required bug-fixing takes place and the tests are re-run until the results are satisfactory.

Important parts of error detection and recovery are documented, mainly in lessons-learned documents.

JS1.4 – More Flexibility in RDF Metadata Creation

User Story

It is gaining importance to be as flexible as possible in creating new metadata that are materialised as RDF. This metadata originates from very different sources and is collected through many channels.

As a product owner, I want to use as many types and instances of metadata as necessary for being able to offer smart and innovative applications to customers.

Detailed Description

The need of a product owner for more and more metadata being used in smart applications also requires that the metadata transformation process is as automised as possible, so that frequent changes and extensions are still safe and do not require a lot of intellectual resources. So new metadata should be added to the process, where it can be gained from most efficiently and with a high quality and not where it is easiest to be integrated from a process point of view. Since a lot of metadata will remain being descriptive for domain specific texts written by authors, one major source for generating smart metadata will also be the author, even if he is supported by semi-automatic tools like tag recommenders. This metadata could then be stored within the XML file in order to preserve, e.g., granularity issues like what part of a document this piece of metadata is actually referring to. Therefore, frequent schema changes could also be derived from data assignment and governance reasons.

JS2 – PoolParty Quality Checks in Thesaurus Management

Description

We want to check the data quality in the thesaurus management system.

Approach

Wolters Kluwer is already using the thesaurus management system PoolParty for several vocabularies. With the growing operational use and number of contents and the extended functionality to define customised schemas, we encounter various pre-existing and new challenges:

- Transparency of vocabulary dependencies
- Resulting from this also the consistency of vocabulary dependencies
- Versioning issues due to model changes, deletions, and so on.
- As different users with different expertise are using the tool, there are subsequent changes executed that need to be tracked
- Process definition for the maintenance of vocabularies.
- Usability related to the understanding of the data models

- Ambiguities
- Doublets

The approach is to build partly on existing approaches for quality management that were developed in LOD2. These works concern the challenges of the prototypical thesaurus management tool.

By using the tool in an operational environment, we gained new insights. With additional features and complexity of functionalities on the one hand and growing contents and operational use on the other hand, we will have to extend the requirements and related to that the approach for gaining appropriate data quality.

Goals

The goal of this epic is to deploy existing prototypical approaches in the operational system of WKD and to investigate additional approaches to ensure data quality.

- Enhance the transparency and consistency of dependencies
- Resolve versioning issues
- Deploy tracking functionalities
- Deploy a maintenance process
- Identify and encounter ambiguities and deploy a solution for dealing with doublets

JS2.1 – Enhance the Transparency and Consistency of Dependencies

User Story

As more people are working with the thesaurus management tool, we encounter problems of transparency and consistency.

Detailed Description

In order to address consistency issues that arise with a broader usage of PoolParty and with a more tightened integration in an overall content development process, the following challenges need to be addressed:

- A query/functionality that shows which vocabularies use specific classes, attributes and relations
- Statistical information (e.g., number of links between a dataset and a certain PoolParty project)

- Enhanced usability (e.g., The overall content is hidden for a non-expert user who is not able to understand RDF data)
- User data (show responsibilities of users, vocabularies they worked with, schemas they created, etc.)
- Possibility to adapt datasets with regard to changing kinds of relations, directions of relations, URIs and so on.
- Extended descriptions of properties, functionalities, and so on.

JS2.2 – Deploy Tracking Functionalities

User Story

Tracking and propagating changes of information into the correct channels in a personalised way is an efficient way of keeping the whole knowledge development and maintenance process accurate.

Detailed Description

Within the LOD2 project, we developed a prototypical subscription and notification service for specific scenarios of data changes. We would like to deploy these notifications in a realistic environment and to extend the areas of notification. Therefore, we need to execute the following tasks:

- Evaluate existing prototypical notifications with regard to current data processing
- Deploy notifications in an operational environment
- Extend notifications with regard to custom scheme creation, edition, and so on.

JS2.3 – Resolve Versioning Issues

User Story

Versioning issues have to be resolved so that subsequent processes are aware of model changes; especially deletions, extensions, and so on.

Detailed Description

The idea is to define a process for data versioning that is aligned with the software development life cycle. We will collect strategies based on existing processes, as for example the DTD versioning or software versioning that will also work for data and schema versioning. Currently, we have no established process in place to manage versioning of vocabularies and schemas.

JS2.4 – Improve Maintenance Processes

User Story

As the complexity of metadata and metadata structures is evolving, we want to ensure that all vocabularies and respective schemata are developed in an efficient and effective way.

Detailed Description

As the number of vocabularies and changes in the thesaurus management system is growing, the issue about who of the existing users (legal domain expert, CMS expert, etc.) should execute changes that are demanded by external departments gains importance. Sometimes, the person who gets demands for changes is not the person who created the content or not the only person who uses it for specific functionalities. Especially, core vocabularies serve several users as a base for further development. The maintenance of the vocabulary must therefore be managed with clear roles and responsibilities.

JS2.5 – Disambiguation and Doublets

User Story

Ambiguities and doublets are known issues that come with large vocabularies. With regard to classification and entity extraction, their handling is also on the agenda of unsolved issues.

Detailed Description

Basic prototypes for doublet handling exist from the LOD2 project. In this prototype, concepts with the same label (persons) were compared based on related data. In case it was obvious that the same person was meant, there was a functionality for concept merging in place. This approach could be deployed. For disambiguation, there has been a small initiative that could be revived.

JS3 – Verifying Available Instance Data for Linked External Metadata

Description

WKDinternal information both in PoolParty as well as in the metadata database is linked to external data sources like DBpedia. Changes within the external sources need to be recognised.

Approach

There is a strong governance of external sources that are used for information ingestion in place. A database has all information about the source, which data are used for what purpose in store. On a regular basis (based on the importance of the external source), the gathered information is collected and checked for changes. Therefore, a local copy of the data needs to be stored. In case the external source offers alerting mechanisms or at least logging mechanisms for changes, this is used for quality control as well.

Goals

- Changes of data in external sources are recognised
- Changes of data initiate different internal processes within WKD
- Some types of changes are processed automatically within the WKD ecosystem
- Statistics about the reliability of external sources are available

JS3.1 – Changes of Data in External Sources are Recognised

User Story

WK datasets are in several cases linked to external data, e.g., definitions, synonyms, images, and so on. This can cause problems when the external data are changed (quality issues) or deleted (functional or graphical issues). In existing prototypical installations, e.g., error messages were shown instead of linked images.

Detailed Description

Information from external sources is included via linking. When changes in the external sources appear, the responsible internal stakeholder is informed, so that he can change the link if necessary. The more Information about these changes is available, the more sophisticated the reaction can be – from a purely manual look-up to a completely automatic processing.

JS3.2 – Changes of Data Initiate Different Internal Processes within WKD

User Story

Once changes in the external data are recognised, there should be a mechanism that initiates different processes depending on the changes and the kind of instance data.

Detailed Description

Based on the type of change (deletion, change, new suggestion), the type of process (approval, quality check, editing) as well as the type of responsible user role and the severity of the change/changed content, different processes need to be initiated. These processes can be purely manual, semi-automatic and automatic. Based on previous experiences, deletions and structural changes need to be specifically addressed.

JS3.3 – Some Types of Changes are Processed Automatically within the WKD Ecosystem

User Story

Specific changes can be processed automatically depending on their severity and automatic quality control.

Detailed Description

Based on the analysis of changes, a certain subset like minor deletions can be processed completely automatic, which leads to a performance improvement. These automatic processes require a QA step afterwards, which will normally be purely technical, e.g., if the process has terminated successfully. Some automatic changes could however also include intellectual checks at least on a random basis. It is desirable to extend the number of use cases for automatic processing over time.

JS3.4 – Statistics about the Reliability of External Sources are Available

User Story

Statistics about the reliability of external sources should be available based on the executed change processes.

Detailed Description

We want a detailed monitoring of the change processes including the approved changes, the rejected changes, the kind of change, the external source, and so on. Based on this, automatic processes could be initiated when the quality of external changes is satisfying.

JS4 – Verification of Availability of Appropriate Metadata for Solr Indexer

Description

The Solr indexing engine is using more metadata for enhancing search functionalities. The appropriateness of the available metadata needs to be ensured.

Approach

Information about the required metadata is currently hidden in the code of the indexing procedure. This information needs to be extracted and test procedures need to be executed, which verify the existence as well as the sufficient quality of the required metadata. This is especially important when it comes to personalised data and applications, where scenarios get so specific, that a comprehensive testing is hard to achieve.

Goals

- Support of successful indexing process
- More transparency to metadata usage

JS4.1 – Support of Successful Indexing Process

User Story

We want to support the indexing by better understanding and documenting the processes. We want to support these processes with the extraction of required metadata.

Detailed Description

To support a successful indexing process, we need to identify the required metadata. Once they are identified, they need to be extracted. Test procedures need to be executed to verify the existence and quality of the required metadata.

JS4.2 – More Transparency to Metadata Usage

User Story

By investigating the required metadata for the indexing process, we want to gain more transparency for the usage of metadata – this does not only apply for indexing processes but also for other processes. This way we can also easier track dependencies in case of errors or any other complications, and so on.

Detailed Description

Metadata is relevant for the production process, but also for supporting features in digital applications like JURION. A direct feedback loop from the application to the data life cycle enables a better performance and sustainability of metadata management. So any request from the software side, which touches metadata (such as false metadata, missing metadata, inconsistent metadata) can and must improve the data life cycle directly. By prioritising these requests, reasonable KPIs for current and future improvements can be determined.

JS5 – Cross-Check of CMS Instances and References in Metadata Database

Description

Metadata Database, PoolParty, and CMS form an information ecosystem. We need to ensure that coherence is guaranteed.

Approach

References in the metadata database that cover document instances in the CMS are checked for existence and for change (e.g., new versions) on a regular basis. These checks lead to either automatically updates of the relationships or manual tasks for human intervention.

Goals

- There are no dead links to documents in the metadata database
- Certain types of document changes automatically lead to reference updates
- Certain types of document changes automatically lead to initiation of editorial QA checks

JS5.1 – There are no Dead Links to Documents in the Metadata Database

User Story

Dead links should be recognised automatically. In some cases, the physical deletion of documents should even be delayed until potential metadata risks coming with it are sorted out.

Detailed Description

We want a mechanism that recognises dead links. Depending on the kind of link, kind of deletion, data severity, and so on, there will be initiated an automatic deletion. This could also traverse towards a whole sub-network of metadata that is directly or even indirectly connected to this document link.

JS5.2 – Certain Types of Document Changes Automatically Lead to Reference Updates

User Story

Based on the type of document and data that are changed, this will lead to automatic updates.

Detailed Description

There are different kinds of metadata, which form references between documents in the CMS and the information stored in the metadata database. Some pieces of information changes on the document lead automatically to updates in the metadata database. This is true, e.g., for a new legal reference added to the document or when the impact factor is changed.

JS5.3 – Certain Types of Document Changes Automatically Lead to the Initiation of Editorial QA Checks

User Story

Based on the type of document and data that are changed, this leads to manual quality checks and editorial work.

Detailed Description

First, we will analyse which kind of data is appropriate for this purpose and then we will implement a prototypical application that recognises these changes. This will then trigger alerting mechanisms within the editorial Interface of the metadata database, where the required editorial tasks are finally executed.

JS 6 – Synchronisation of Relational Data and RDF

Description

Although there is transformation towards semantic technologies, most data are still stored in relational databases. There exist well-rehearsed processes concerning maintenance and even more importantly usage of this data from this source, which are often hard to change on a short-term basis. Therefore, a transformation support of relational database data to RDF format and even vice versa would enhance many internal processes and improve the acceptance of the semantic data format, since a slow shift can be initiated instead of a big bang.

Approach

As a pre-processing step, we initiate a mapping between the relational database schema and the desired RDF output data using SHACL as the basic model. If available, we use tools that support this process semi-automatically as much as possible. We make Semantic Booster sit on top of the existing relational database that is currently used, so that it understands the existing relational database schema. We use R2RML to generate the desired RDF data. Then we go into a quality assurance round for the resulting data – also with the help of semantic Web tools until the quality is fine. Then we use this final SHACL model, so that Semantic Booster can create its internal model and we compare the result with the relational database model in the database. When these models are the same, we have a complete roundtrip scenario. Once changes in the RDF requirements or in the relational database occur – which happens in any operational system – we can run hopefully minor modifications in this roundtrip and make sure that no information gets lost.

Goals

- Supported transformation process from relational database data to RDF
- Supported transformation process from RDF to relational database data
- Implementation of a synchronisation process
- Simple yet efficient and complete maintenance processes in place

JS 6.1 – Supported Transformation Process from Relational Database Data to RDF

User Story

An easy and reliable transformation process of relational database data to RDF data would improve productivity, use, and acceptance of RDF data.

Detailed Description

Initially, the relational database schema is mapped to the RDF database schema. Additional business rules for adding required information for the RDF are defined and integrated in the overall process. Any change of the relational database schema requires only an adjusted mapping. The transformation process itself is performant and transparent.

JS 6.2 – Supported Transformation Process from RDF to Relational Database Data

User Story

In order to ensure data consistency throughout the processes, we also need to implement a transformation process from RDF to relational data. This does not mean that all data are stored twice, but that all data are accessible via the source it is best suited for a distinct purpose.

Detailed Description

Although it is very clear that all relevant data should be available in RDF, it can still be the case that data originating from sources in RDF also need to be available in systems and processes that access relational databases for that purpose. Therefore, we need to put a process in place, where we can also transform RDF data to relational data. This should be an exception, but it is definitely an existing requirement in existing operational systems.

JS 6.3 – Implementation of a Synchronisation Process

User Story

It is not enough to have a transformation process in place. In addition, a controlled and transparent overall process helps to make consistent data sources part of the operational environment in a trusted and reliable way.

Detailed Description

This requirement is more dedicated towards operational excellence than towards pure availability. Since we assume that both technologies and data sources will remain within the overall ecosystem for good reasons for a long time, the costs for having information in two different places need to be minimised. Therefore, a lean and efficient synchronisation process is important. This does not only cover the transformation as such, but also the documentation on errors and changes, an easy-to-use interface and a clear definition of roles and responsibilities, e.g., in case something goes wrong.

JS 6.4 – Simple Yet Efficient and Complete Maintenance Processes in Place

User Story

Once the transformation is in place and a reliable process is built around it, it is finally important to introduce a maintenance process, which is efficient and complete.

Detailed Description

There is always a trade-off between an optimised process and a process which is open for frequent changes. This challenge needs to be addressed when looking at an efficient synchronisation process on the one hand and a proper maintenance process on the other hand. In order to meet this challenge, it is important that the maintenance process transparent, so that every actor knows what to do and why. It should also be as lean as possible, so that few components are affected as needed. Finally, regular review cycles on potential improvement areas both for the synchronisation as well as for the maintenance process need to be initiated by the process owners.

JS 7 – Schema Change

Description

When a schema is changing in the content life cycle, this has measurable impact on both content and software development life cycle and the communication between them. Therefore, we use this UCS for showing what a better integration of content and software development life cycle would require. In order to build upon the story of JS 3, explained in D2.1, we assume that adding external data to the respective metadata model triggers this schema change.

Approach

Currently, the information about schema changes mainly remains within the content team until test data are forwarded to the software development team, so that they can evaluate the impact on their tasks. There is some general information to all stakeholders about schema changes as such available, but since there is no detailed knowledge about what parts of the schema influence which processes, this general information is most of the times more or less ignored. Therefore, more fine grained information about processes and information flows needs to be available across teams and cultures in order to make things more effective and in order to reduce errors and misunderstandings.

Goals

- Support of early and consistent knowledge flow between life cycles
- Clarity about impact of schema change
- Enabling iterative changes based on interests of different stakeholders
- Streamlining change process over time

JS 7.1 – Support of Early and Consistent Knowledge Flow Between Life Cycles

User Story

A better and more consistent and controlled knowledge flow about schema changes will help us to get teams work more closely together and foster a mutual understanding about the different needs and approaches.

Detailed Description

To support a streamlined process, we need to establish a proper information flow. This covers not only information about the schema change as such, but also about how this schema and its respective instance data are used in both life cycles. Feedback and a proper documentation of decisions add to the required transparency. The more direct and to the point of use this communication takes place, the more efficient the process gets.

JS 7.2 – Clarity about Impact of Schema Change

User Story

In order to make the schema changes as lean as possible, knowledge about the impact of the changes is key. This does not only cover how these changes enable new functionality, but also at what cost, e.g., from a complexity point of view.

Detailed Description

The lack of knowledge on overall impact is the main blocker for efficient progress. As long as the respective teams work in isolated silos, a common understanding will not happen, and therefore a lot of friction on delays will be a daily business – without being able to discover the real causes. In order to address this, communication needs to take place, but also tools and processes need to be available, so that everybody knows what a schema change really means for everyone else. This touches aspects like different "jargons" as well as having tools in place, where all usages of a certain schema or instance data can easily be documented and followed.

JS 7.3 – Enabling Iterative Changes Based on Interests of Different Stakeholders

User Story

The target process aims at more timely and more direct communication within and between teams. This will ensure to easily introduce short iterations as a means to accelerate things between people within the overall process.

Detailed Description

One of the advantages of SCRUM methodology in software development is the high flexibility in adapting to new and changing requirements. This includes also that the necessary information for doing things is available on a timely basis. Therefore, direct communication introducing iterations of the schema change are required. This means not only that people talk more directly and more frequently, but also that their work environment is prepared to leverage the outcome of this communication. A proper tool support, e.g., about decisions and why these were made is helpful to make the exercise sustainable in the end.

JS 7.4 – Streamlining Change Process Over Time

User Story

By introducing integration points between content and software development life cycles, a learning curve is initiated, which will finally lead to a streamlined and lean process, whenever new schema changes occur.

Detailed Description

A higher degree of integration comes with an additional cost of complexity in the early stages of process implementation. This additional cost should more or less disappear over time, when an optimum of efficiency is achieved. Therefore, a control and governance procedure needs to take care of this and needs to ensure that this optimum is achieved in the end.

JS 8 – Bug Reporting Governance

Description

Bug reporting is often quite isolated within specific processes. Problems can arise when the cause of a bug comes from another process. In many cases, communication needs also to take place between processes like software and data development. Therefore, we want to enable a bug reporting that connects both life cycles to improve processes and productivity of error resolutions.

Approach

In case data errors cause bugs in the software, there is not yet an adequate service that supports the bug reporting over/between both life cycles of data and software development. Since there is a tendency at Wolters Kluwer to move towards more and more data-intensive systems, the probability of application errors based on wrong or incomplete or simply misinterpreted data is growing exponentially. Therefore, a better integration and communication is key, and within the JURION team, we regard this use case as being of very high relevance for our business.

Goals

- Enable processes for the interaction between software and data development
- Define error resolution strategies/processes based on the process status
- Improve transparency over bug fixing processes
- Improve productivity of bug fixing processes

JS 8.1 – Enable Processes for the Interaction Between Software and Data Development

User Story

An integration of bug fixing processes for the software and data development is essential to support an alignment of both engineering cycles. Especially with a growing number of involved parties, there is a need for clear and simple interfaces with low information loss.

Detailed Description

With help of the meta model, we want to express specific processes, user roles, and actions with regard to specific bugs. These specifications need to be integrated into the running bug fixing systems, so that users have a plan at hand how to proceed in the bug fixing process in a comprehensible and reproducible way.

JS 8.2 – Define Error Resolution Strategies/Processes Based on the Process Status

User Story

For a productive bug reporting strategy, we need to have a comprehensible and reproducible way in place how to proceed with bugs. Therefore, we want to develop (as far as possible) a generic strategy to handle specific bugs based on data issues.

Detailed Description

Based on the bug fixing status and other pieces of information like the error code, error description and other more technical metadata we will analyse required actions and see how these can be integrated into the meta model and used for an enhanced communication between both development live cycles. We are envisioning a standard clarification and resolution process as being part of the tools that both life cycles are using for their respective tasks.

JS 8.3 – Improve Transparency Over Bug Fixing Processes

User Story

One of the major gaps in the bug fixing processes between data and software development is transparency over processes, responsibilities, status of works, and so on. The described scenarios JS 8.1 and 8.2 will already support this transparency.

Detailed Description

To enable an overview over integrated bug fixing processes, there needs to be some overview to track the status of processes including additional information. We will investigate how this could be integrated into the running bug fixing processes, but we envision some sort of dashboard, where all stakeholders can easily detect the information they need for fulfilling their tasks. Also, a messaging system that acts as a push service could be helpful, but only if the information flow is restricted to relevant information – so some sort of automatic filtering needs to be in place.

JS 8.4 – Improve Productivity of Bug Fixing Processes

User Story

The scenario of improving productivity builds on the previous scenarios. An increased productivity is in the end the overall goal of the improvement of the bug fixing process.

Detailed Process

The productivity will be increased by better interfaces for communication and bug resolution and supported bug-fixing processes. This improvement in productivity can be measured by objective means like a reduced number of bugs in total as well as a faster time-to-market for new or modified software features.

JS 9 – Dynamic Relevance Ranking

Description

To improve relevance ranking of documents on Wolters Kluwer JURION platform, we want to implement a dynamic relevance ranking that adapts to contextual information. Documents will get a relevance score by additional related data like time, area of law, and so on; but also by data from user profile and user search history. So this information is directly gathered both from the information repositories and software functionalities and driven by application requirements.

Approach

Currently, the relevance or general importance of a document is either defined as editorial metadata in the document once and does not change afterwards, but influences relevance ranking as long as the document exists. Or relevance is calculated during indexing using classical statistical methods. With the help of entity extraction, links and even inclusion of external data on the one hand and local events like previous searches, we want to enable a dynamic relevance ranking. A good application field within the legal domain could be a better ranking of court decisions.

Since this use case covers all three areas of ALIGNED core interests – content life cycle, software development life cycle and finally JURION application itself – a more detailed description is currently not yet possible due to complexity issues. Once e.g., integration points are defined and a more sophisticated domain model is available, it will be worth to revisit this use case in phase 3 of ALIGNED.

Goals

- Automatic dynamic and contextual relevance calculation
- Integration of generated relevance score in search result list

Requirements

XSLT uplift of XML sources to RDF metadata

In area 1 of the content pipeline shown in Figure 6.1, metadata is extracted from the proprietary WKD XML schema and transformed in RDF. Due to regular changes in the XML format, the correct transformation process based on existing XSLT scripts must be secured, so that no inconsistent data are fuelled into the metadata database in area 3.

PoolParty quality checks in thesaurus management

In area 2 of the content pipeline shown in Figure 6.1, controlled vocabularies and domain models are created, maintained, and delivered for further usage, based on SKOS standard. The integrity of the knowledge management system as a whole needs to be ensured. Therefore, regular local and global quality checks need to be executed, so that e.g., inconsistencies across different controlled vocabularies can be detected and resumed.

Verifying available instance data for linked external metadata

In area 2 (Figure 6.1), the Linked Data paradigm aims at linking information across sources, in contrast to simply link local copies of datasets. The technologies for linking are available, but there is not yet a proper support to maintain these links, especially when it comes to changes of the source – both on schema as well as on instance data level, e.g., when geo data of a linked organisation is updated or deleted on the source side.

Seshat

Goals

Data Validation – a harvester enters a value or set of values into the dataset which is syntactically, semantically, or factually invalid. Currently, there is very limited support for validation of these values, and it is thus easy for incorrect or invalid entries to be added to the dataset.

Capture of Data Complexity – the Seshat architects desire that the harvesters express the full complexity of the data – and capture where values are uncertain or disputed. However, the harvesters tend to prioritise speed over complexity for a variety of reasons and will often neglect to express the full complexity of the evidence. Another problem is that, when dealing with pre-historical societies, the data required by the schema require significant interpretation which is often beyond the competence or confidence of data harvesters.

Schema Evolution – the Seshat schema has been developed iteratively and continues to evolve. Data that have been collected with earlier versions of the schema currently need to be manually updated to make it consistent with schema updates.

Dataset Evolution – the Seshat dataset has been in rapid evolution since its inception and is expanding at an increasing rate. The Seshat researchers would like to continue to increase the rate at which high-quality data are added to the system. They would also like to gain greater understanding of how the dataset has evolved: in what context was a given variable added? Why was a value changed? How do the overall characteristics of the dataset change over time?

Expert Interpretation – the Seshat dataset entries capture information about uncertain and disputed values and, in some cases, includes a range of different opinions drawn from a variety of sources. Expert interpretation is required in order to turn this raw information into time-series datasets that can be statistically analysed. This is currently a manual process.

RA-based Data Collection – Seshat data are currently collected and stored using a wiki-based manual system. This currently requires extraction and conversion in order to be converted into computer-readable form. Validation of input is not an automatic process.

Multi-format Data Publication – Seshat is used by four different user roles such as editors, contributors, knowledge engineers, and data analysts. These four roles have different requirements and require different views on Seshat data.

DBpedia-based Candidate Generation – Manual generation of candidates requires significant effort from RAs and experts. If users were able to quickly view and assess potentially relevant information, this would provide a significant saving in time and effort.

User management and dataset productivity – Seshat administrators need a means by which they can control the scope of the work which RAs are doing, as well as assess their productivity. Dataset managers need to monitor progress of data collection.

SS1 – Enter Invalid Data

Description

We want to be able to identify and rectify whenever invalid data are entered into the system.

S1.1 – Syntactic Error

User Story

Goal: Prevent syntactic errors from entering the dataset

As a harvester, I want to ensure that the system identifies any syntactic mistakes on my part and, where possible, suggests how to correct it. As an editor, I want to be able to analyse the dataset to identify any syntactic mistakes, which have entered the dataset and efficiently correct them.

Detailed Description

A harvester enters a set of variable values in the Seshat data input form, some of which have syntactic mistakes. Each variable is analysed in real time (when the data are changed and the variable loses focus) and a warning, with a suggested remedy if possible, is indicated whenever the value does not parse correctly. When the harvester attempts to save the form, an error message is produced if there are any outstanding syntactic errors, highlighting the outstanding errors and suggesting corrections – the harvester is presented from saving the page until the error is corrected.

An editor selects a subset of the Seshat dataset and launches an analysis of it. The results return a list of all syntactic errors in the selected dataset and provides a form which allows the editor to correct the errors. The form presents a detailed description of the error and suggestions to the editor for ways in which it can be corrected. The editor updates a subset of the errors and saves them. The system prevents the editor from saving any updated variables that contain errors – non-updated errors are saved.

S1.2 – Typo Detection

User Story

Goal: ensure that only variables specified in the schema enter the dataset.

As a harvester, I want to ensure that any mistakes I make in variable names are identified by the system and I am given a chance to correct them. As an editor, I want to be able to view all variables in the dataset that are not in the schema and correct them.

Detailed Description

A harvester enters a set of variables in the Seshat data input form, some of which have typos or are variables that are not defined in the schema. Each variable is analysed in real time (when the data are changed and the variable loses focus) and a warning is indicated whenever the variable is not recognised. When the harvester attempts to save the form, an error message is produced if there are any outstanding variable name errors, highlighting the outstanding errors – the harvester is presented from saving the page until the error is corrected.

An editor selects a subset of the Seshat dataset and launches an analysis of it. The results return a list of all variable name errors in the selected dataset and provides a form which allows the editor to correct the errors. The form presents a detailed description of the error and suggestions to the editor for ways in which it can be corrected. The editor updates a subset of the errors and saves them. The system prevents the editor from saving any updated variables that are not in the code book – non-updated errors are saved.

We would also like to detect general cases where misspellings are made.

S1.3 – Semantic Error

User Story

Goal: Ensure that the dataset is semantically consistent

As a harvester, I want to ensure that the system identifies any semantic mistakes on my part and, where possible, suggests how to correct them. As an editor, I want to be able to analyse the dataset to identify any semantic inconsistencies that have entered the dataset and efficiently correct them.

Detailed Description

A harvester enters a set of variable values in the Seshat data input form, representing the opinions of a single source (e.g., a book, an expert, a primary source). When the harvester has completed the inputting and clicks 'save' the system runs a series of semantic checks on the data entered, which will identify inconsistencies such as multiple overlapping values for a single variable that is specified as having a single value (e.g., two different population figures for the same date). The system will highlight these errors and, where possible suggest ways in which the inconsistency can be resolved. The system will give the harvester the option of correcting these inconsistencies before proceeding. The harvester can choose to do so or can choose to keep the inconsistencies (indicating that they were present in the source).

An editor selects a subset of the Seshat dataset and launches an analysis of it. The results return a list of all semantic inconsistencies across the dataset and launch the expert interpretation interface to allow the editor to create a consistent dataset.

SS2 – Capture of Data Complexity

Description

The Seshat architects desire that the harvesters express the full complexity of the data – and capture where values are uncertain or disputed.

S2.1 – Facilitating Harvesters' Input Complexity

User Story

The Seshat harvesters can under-report the complexity of data, particularly by not reporting the time boundaries of variable values. One possible cause of this is the difficulty of specifying complexity in the Seshat variable syntax. By providing a simple user interface, the harvesters will find it easy to specify the full complexity of the underlying data.

Detailed Description

When a harvester decides to enter a variable value, a user interface is displayed which allows them to easily specify time boundaries and uncertainty.

S2.2 – Comparing Harvester Complexity Capturing

User Story

Harvesters code Seshat variables over time and have different characteristics in terms of their propensity to enter the full complexity of the data. Editors would be like to be able to track this propensity between harvesters and over time.

Detailed Description

A Seshat editor can view the harvesting history of each user and compare them to one another with their relative propensity towards encoding complexity highlighted.

SS3 – Schema Evolution

Description

Managing changes to schema in such a way that instance data and schema evolve in synch.

S3.1 – Architect Breaks Schema

User Story

Architects should be prevented from making changes to the schema that damage the integrity of the schema.

Detailed Description

When the architect changes the dataset schema, the system will

- Prevent structural changes that damage the dataset's integrity
- Version/publication naming scheme.
- Indication of instance data that need updating
- Queuing that workload and continuing with existing data alongside partially updated dataset.
- A harvester can go through the queue of pending updates with a user interface which helps him or her in making sure it is valid.

S3.2 – Indicating Required Instance Data Updates

User Story

The architect makes a change to the schema that requires instance data to be updated. The system indicates to the architect what changes need to be made.

Detailed Description

When an architect updates the schema:

- The system indicates to the user what instance data need to be updated to reflect the schema change.
- The system indicates which data can be automatically updated and which needs manual updating
- If the architect chooses to carry out the update, the relevant instance data are either manually updated or sent into the queue for processing by editors/harvesters

SS4 – Dataset Evolution

Description

The Seshat dataset has been in rapid evolution since its inception and is expanding at an increasing rate. We want to improve their ability to manage the evolution of instance data and understand what changes have happened over time.

S4.1 – Track Dataset History

User Story

Seshat editors have the ability to select slices of the dataset (sliced according to harvester, entities, time (both entry time and entity lifespan).

Detailed Description

Seshat editors have a page available to them which allows them to select slices of data using a number of criteria and analysing those slices for:
- Completeness
- Complexity and other measures of data quality

SS5 – Expert Interpretation

Description

The Seshat dataset entries capture information about uncertain and disputed values and, in some cases, includes a range of different opinions drawn from a variety of sources. Expert interpretation is required in order to turn this raw information into time-series datasets that can be statistically analysed.

S5.1 – Interpreting Data

User Story

The Seshat data have many uncertainties, disagreements, missing portions of data, and so on. The system will identify these situations and help experts to create an ambiguity free dataset.

Detailed Description

- The system will identify variables and time-slices where expert interpretation is required to resolve ambiguity.
- The system will provide some support for mapping variables into different theoretical frameworks.
- The system will provide supports for mapping input variables from external databases

SS6 – RA-based Data Collection

Description

Seshat data are currently collected and stored using a wiki-based manual system. This currently requires extraction and conversion in order to be converted into computer-readable form. Validation of input is not an automatic process.

S6.1 – Data Collection

User Story

The RA can enter data into the system via a Wiki page. These data are not machine-readable. The goal here is to make data collection of RA data as easy as possible while preparing it for conversion into a structured, machine-readable form.

Detailed Description

When an RA decides to enter new data, they should be provided a user interface to allow them to easily input and structure their knowledge.

SS7 – Multi-format Data Publication

Description

Seshat is used by four different user roles such as editors, contributors, knowledge engineers, and data analysts. These four roles have different requirements and require different views on Seshat data.

S7.1 – Data Publication

User Story

Publication in different data formats for editors, contributors, knowledge engineers, and data analysts. Implementation of different views to represent data.

Detailed Description

Presentation of data in different formats such as RDF (N3, Turtle, RDF/XML, etc.), JSON, XML, and so on.

SS8 – DBpedia-Based Candidate Generation

Description

Manual generation of candidates requires significant effort from RAs and experts. If users were able to quickly view and assess potentially relevant information, this would provide a significant saving in time and effort.

S8.1 – Candidate Generation

User Story

Manual generation of candidates requires significant effort from RAs and experts. The goal is to make the process more efficient by making viewing and assessing potentially relevant information faster.

Detailed Description

Automatically provide relevant candidates for assessment by Seshat RAs. This removes an entirely manual step in the current process and allows Linked Data sources to be harnessed as a basis for dataset bootstrapping.

SS9 – User Management and Dataset Productivity

Description

Seshat administrators need a means by which they can control the scope of the work which RAs are doing, as well as assess their productivity. Dataset managers need to monitor progress of data collection.

S9.1 – User Management

User Story

Seshat administrators need a means by which they can control the scope of the work which RAs are doing, as well as assess their productivity.

Detailed Description

The productivity of individual workers or roles can be evaluated through a convenient user interface.

S9.2 – Productivity Assessment
User Story

Dataset managers need to monitor progress of data collection.

Detailed Description

Dataset work rates will be made available to enable prediction of resources required to complete/curate the dataset.

SS10 – Wiki Import
Description

Applications using the data need a means to import wiki data and convert it to RDF. Application users need an overview of the dataset, and a simple and easy to use method of manipulating the dataset.

S10.1 – Import Data
User Story

Applications using the data need a means to import wiki data and convert it to RDF. Application users need an overview of the dataset and a simple and easy to use method of manipulating the dataset.

Detailed Description

Provide integrators with tools to import wiki data from Seshat and use it in their applications. Efficiency and usefulness to users for visualisation and manipulation of RDF graphs. Evaluation of existing data through graph analysis methods and import checks.

S10.2 – Validate Wiki Page
Goal

A user wants to validate a semi-structured wiki page to check if it is ready for import and fixes any detected errors in the wiki. No actual import takes place.

Detailed Description
- User navigates to wiki page
- User invokes the import validation tool
- A list of import errors are displayed
- User edits wiki to resolve errors and validates again.

358 *Appendix A – Requirements*

Requirements

The high-level goal of ALIGNED in this use case is to produce tools for the Seshat researchers which will increase productivity and data quality and improve the availability of data for analysis. These tools will be largely generated from OWL models. In terms of combined software and data engineering, therefore, the major requirements are concerned with developing models, which are sufficiently rich to generate software tools to support:

- Data Validation: Automatic generation of software tools which will ensure that all data produced will be valid.
- Capture of Data Complexity: Automatic generation of Web-based User Interfaces which allow data harvesters to efficiently enter complex data.
- Schema Evolution: Automatically generate software tools which will automate the evolution of instance data given complex changes to schemas.

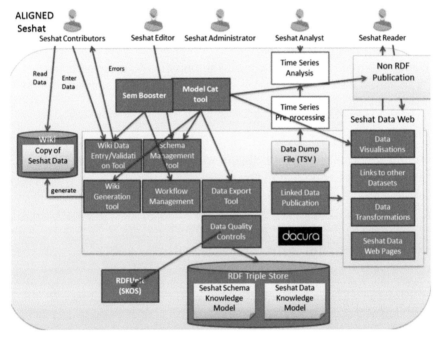

Figure A.1   Seshat Architecture for Month 19 Demo.

Figure A.1 shows the interaction points in the technologies used to support the Seshat use case.

The Dacura platform will support the system with the following tools:

- Wiki-data validation tool: for extracting data from the mediawiki site where they have been collected and validating them interactively – supporting Seshat contributors
- User-interface elements: generated from the schema to help users in more easily specifying complexity – supporting Seshat contributors
- Schema-management tool: for updating, validating, migrating and deploying changes to the Seshat code book – supporting Seshat editors
- A workflow-management tool: for managing the flow of information through the system – supporting Seshat administrators
- A data-publishing tool for exporting of data to a range of formats

The system will leverage ALIGNED partners technology as follows

- RDFUnit will be used for implementing extra quality controls on imported data (e.g., SKOS validation)
- Semantic Booster will be used for workflow automation and data validation
- Model Catalogue tool will be used for import/export to various non-Linked Data formats.
- Dacura will be used to generate user interface elements, manage schemas, and publish data

PoolParty

Scenario 1

Data Consistency Constraints. PoolParty components impose constraints on data they process (see figure below). Users of the software often wish to import arbitrary datasets, vocabularies, or ontologies which do not always meet these constraints. Currently, when users attempt to import data that violate the constraints, the data will simply fail to display, or in the worst case, cause unexpected behaviour and lead to/reflect errors in the application. SWC would like to be able to tell the user why the import has failed, suggest ways in which the user can fix the problem and also identify potential new constraints that could be applied to the structure. Apart from this import functionality, different software components (e.g., taxonomy editor, reasoning engine) drive RDF data constraints and vice versa.

Scenario 2

Schema Generation and Validation. Linked Data resources are often published without schemata or with a minimally specified schema. Users of PoolParty would like to be able to import Linked Datasets but in order for the system to be able to manage the data, a schema describing its structure is required. However, end users typically find schema specification to be extremely difficult to do correctly. SWC would like to be able to offer users as much support as possible in specifying schemas to describe both schema-free datasets that they are importing and to correctly specify the constraints that they wish to include in the schemata that they develop through the system. Specifically, they would like to be able to do the following:

- Generate or partially generate schemata, or suggestions for schema structures from instance data
- Check a taxonomy as a whole for being conform with a specific data schema

The system currently supports both creation of custom data schemas as well as automated quality control of the predefined SKOS data schema. The latter comprises eight quality controls, some of which can be selectively turned off, are run with a rule engine which has a ruleset describing schema-consistency constraints. However, these are relatively computationally expensive to run and are thus deployed only at 'save' time – when the user has decided that they have completed their modifications and wishes them to be deployed.

Scenario 3

Constraint Resolution. In the semantic space, constraint violations can be extremely difficult to understand as they may come from a number of different sources: inherited classes or properties. This is particularly difficult when dealing with multiple ontologies – for example when merging two ontologies or datasets. Users of PoolParty would greatly benefit from tools which could help them to understand more easily what the cause of constraint conflicts is and what pathways are available to resolve them. For this situation, we need a formal method for formulating actions for responding to constraint violations and a generic way to generate user interfaces for applying these actions.

Scenario 4

Development Process Integration. SWC uses a range of tools to manage the information flows that are relevant to their development processes: Atlassian Confluence Enterprise Wiki for project management and documentation; Atlassian JIRA Issue Management solution for continuous issue management; Q&A System for customer FAQ management; Sonr WebMining for market monitoring. Particular information events that occur in any of these systems have implications across the development process and across these systems. For example, a bug notification that arrives in the JIRA issue queue might require an update to a FAQ entry and an update to project management plans and project documentation. Currently, the information produced by these systems is largely document orientated and focussed on human consumption, and integration of this multi-source information is a manual process.

Scenario 5

Code and Data Co-evolution. Upgrades to the PoolParty software can inadvertently break data models that had been constructed with older versions. This is currently very difficult to predict, particularly as the models become more complex. Fixes require manual modification of the data model to make it consistent with the new software requirements. SWC would like to be able to automate the identification of such model-software mismatches, suggest remedies, and even automate the fix.

PS1 – Data Consistency Constraints

Description

We want to be able to manage and express constraints for data consistency in a standard way throughout the PoolParty Thesaurus Server (PPT) application. Data constraints are required in various components of the application:

- PPT-Internal data processing requirements: mandatory, built-in rules
- Reasoning: built-in as well as custom rules (in future) that can be deactivated
- General Schema-related constraints: bound to specific usage scenarios
    - SKOS quality: based on qSKOS quality issues, can be deactivated
    - Custom constraints (user-defined), constraints inferred from OWL (from ontology evaluation approaches like, e.g., RDFUnit)
- Data Import

In order to improve maintenance and handling of the constraints for these components, we seek to establish a unified method to consistently support the use cases that are covered by these components.

Mandatory and quality constraints

Within PoolParty, we define two categories (severities) for the consistency constraints. The mandatory constraints include constraints that have to be satisfied for PoolParty to function normally. Usually, constraint violations in this category prevent the project data to be used by the application because of occurring errors caused by the inconsistencies. The quality constraints include constraints that show modelling problems in the imported data (checked in conjunction with the already existing data) and provide useful information on how to improve the data quality. Quality constraint violations do not cause application errors but may cause other inconveniences because of inconsistent, redundant, or otherwise disadvantageous information. Violations of the SKOS standard, which is a fundamental part of PoolParty Thesauri, can be found in both categories. For example, the disjointness of SKOS classes has to be guaranteed for the instance data. In contrast, the pairwise disjointness of prefLabel, altLabel and hiddenLabel is also stated as part of the SKOS semantics, but does not need to be enforced in PoolParty. This is provided as a quality check that can be deactivated by the user on demand.

Approach

PPT uses a triple-store (Sesame by default) to persist taxonomy information. Changes to these datasets are currently performed by executing atomic "actions", which encapsulate triple changes and removals. Currently, the checks, if an action can be executed, are scattered in the code and sometimes performed multiple times, making them hard to maintain. The challenge is to specify constraints (preconditions) for each action in a concise and easily understandable way and which are checked upon action invocation. We furthermore experienced that consistency checking and reasoning can be expensive in terms of runtime and CPU/memory utilisation. Therefore, another challenge is to, whenever possible, (re-)formulate constraints to establish good performance. These may involve different technologies for constraint/rule evaluation (e.g., Jena rules vs. SPARQL/SPIN-based inferencing, SHACL) and it must be possible to utilise the technology whichever works best for the application scenario.

Summarising, the underlying assumption of our approach is that if it is only possible to interact with the triple store using exactly specified atomic actions in combination with constraints (pre- and postconditions), we will be able to avoid inconsistent or erroneous data and, as a consequence, avoid bugs in the application. Constraint metadata, such as version information, will also help in integrating the approach with process management tools such as JIRA or Confluence, which we cover in PS4 and PS5.

The approach in PoolParty is to only check data for consistency if it is necessary because of the performance implications described above. That means that we define different entry points and scenarios where data need to be checked when entering a PoolParty project.

Import

PoolParty features an RDF import and an Excel import to introduce arbitrary data into a PoolParty project. While the Excel import needs some conformance because of the tabular format which is checked separately, a user can introduce any RDF data with the RDF import. Depending on the proficiency of the user with the PoolParty data model, such an import is likely to introduce inconsistencies into the project data. This is the most important part of PoolParty where mandatory constraint checks have to be applied to ensure normal operations of the application. For convenience, also quality constraints can be checked on import to provide a unified data maintenance feature.

Quality

PoolParty also features a quality-checking component that can be used to run a complete constraint validation on the project data and present a detailed report with information that can be used to resolve the constraint violations found. In contrast to the mandatory constraint checks, where violations will break the application, the quality constraint checks can be run after an import and also just on an existing project to find flaws in data modelling and to improve data quality. These quality constraint checks can be configured in the project settings. They can be activated and deactivated on demand and will report only the activated checks while ignoring the other checks. With this feature, a project-specific quality control can be established.

Actions

Atomic actions for data modification are used by PoolParty to add and remove statements in the triple store. These actions may include reasoning consequences that can further alter the data. These actions are part of the PoolParty code base and are not provided by actors outside of the application. They are guaranteed to not produce inconsistent data and are not validated on execution. However, there can still be programming errors that may cause inconsistent data being produced by even these actions. Therefore, we propose constraint violation checks being performed as part of the testing process in PoolParty development to detect errors that cause violations.

Methodology

The goals of Phase 2 were:

- Deciding on a methodology for defining constraints for RDF datasets
- Deciding on a framework that checks these constraints and can be integrated with PoolParty

Constraint definition language

For defining the constraints, we decided to use the SHACL Shapes Constraint Language, a language for validating RDF graphs against a set of conditions. These conditions are provided as shapes and other constructs expressed in the form of an RDF graph. This approach synergises well with the requirement of many data constraints for PoolParty to have existing data of a specific form. These shapes can then be used for data validation instead of multiple independent constraint checks to provide a more integrated view on data consistency. Still, the option of having multiple constraint checks using SPARQL to define them is a valid option.

Framework

We decided to use RDFUnit for checking the constraints. RDFUnit is a test-driven framework for running test cases on RDF data. The test cases are executed as SPARQL queries using a pattern-based transformation approach. RDFUnit also supports defining test cases using SHACL. The provided implementation uses Java and can be easily integrated into PoolParty. However, the current supported SHACL features are not sufficient to express all mandatory constraints needed by PoolParty. Therefore, we decided to use a SPARQL-based approach for the current implementation.

Integration

Integration in PoolParty is based on independent RDFUnit test cases that can be used by PoolParty as part of the RDF Validation. These represent the constraints and can be used to find violations in the data. The approach is intended for high maintainability and reusability of constraints. RDFUnit test cases are formulated using RDF and are based on SPARQL. They can be executed independent of PoolParty and can be used wherever RDFUnit can be integrated. Also, the test case definitions can be modified without modifying the PoolParty code base and therefore provide high maintainability. Finally, the RDF definitions can be published as Linked Data for further use. PoolParty extends the RDFUnit test cases with its own test case descriptions, which are also defined using RDF with all the advantages described above. These descriptions add additional information about how the RDFUnit test case is used as part of the import validation. For example, the applicable repair strategies for the constraint are defined.

The constraint checks are currently used for validating the import of data into PoolParty projects. Each constraint check includes one or more repair strategies that can be used to correct the problem by modifying the data to satisfy the constraint. Some of the repair strategies feature bulk repair options to automatically correct all violations of a specific constraint.

Goals

As goals of this epic, we focus on:
- Evaluating the RDF Validation implementation in PoolParty to ensure normal operations by importing various datasets into projects and repairing them
- Further evaluation for options to replace the SPARQL-based approach with a SHACL implementation
- Adding quality constraint checks

S1.1 – Constraints for Internal Actions

User Story

We want to secure all changes to the triple store by easily understandable constraints so that we can ensure that the data are consistent before and after performing the change. The constraints are assigned to actions and must be checked before and after each action invocation and, in case of failure, a meaningful error message should be provided.

Detailed Description

We currently use atomic "actions", i.e., triple changesets, to modify data in the underlying triple store. Compilation of these changesets can be a complex task, which, in some cases, also involves changesets from other actions. Furthermore, the decision, if an action can be performed, is currently involved from various places in the source code, which sometimes is redundant. We want a unified way to express conditions when an action can be performed (all preconditions are met) and what the result of the action should be. This way, we can be sure that the action is correctly implemented and the application functionality that relies on the changed data is not affected. Constraints should also carry version information to find out when they have been introduced which is necessary to detect cases in which the triple store content does not match the installed version of the application (see PS5). We see the process of defining these constraints at an early development stage, similar to creating constraints in an object-relational mapping for applications that rely on data from relational databases. Therefore, constraint checking in our approach should also be done by utilising a closed-world view. The constraints should be formulated using RDF and standard ontologies, allowing them to be reused and linked by utilities that support the development process, e.g., JIRA or Confluence (see PS4).

S1.2 – Rules for Reasoning and Inferencing

User Story

We want a way to express all current as well as custom user-defined rules in a standardised, easily maintainable form. Performance should be at least equal to the existing solution.

Detailed Description

We currently use Jena rules (https://jena.apache.org/documentation/inference/) for inferring additional facts in a taxonomy that is bound to a user-specified schema. For some actions (e.g., type removal), we experience performance issues. We will therefore investigate if it is possible to replace the Jena rules with an SPARQL/SPIN-based approach, which we also proposed in S1.1 for internal actions. Depending on what implementation delivers the best performance (Jena, SPARQL or others), it must be possible to use either of these implementation for rule evaluation and inferencing. Therefore, the challenge is to provide a way of formalising rules using the various evaluation and inferencing technologies.

S1.3 – Constraints for Specific Schemas

User Story

For each used data schema, we want the possibility to define constraints that the instance data must comply to.

Detailed Description

Taxonomy developers use PPT to create taxonomies that use classes and properties defined in schemas such as SKOS, FOAF, GeoNames, and others. For each of these schemas, taxonomy creators often have a specific policy how to use them. For example, for SKOS, a large number of constraints exist that are defined in both the schema itself (as described in the SKOS reference) as well as by publicly available tools such as qSKOS (which derives a catalogue of quality issues from existing literature and practical experience). Based on qSKOS, PPT currently supports eight quality checks that alert users if the current taxonomy contains potential quality problems. We plan to extend this approach for multiple schemas (which can partly be automatically derived as shown in UL's RDFUnit contributions) and by custom constraints. These custom constraints are either PoolParty-specific (e.g., domain/range assertions are interpreted as data constraints) or user-definable (e.g., a thesaurus manager does not allow polyhierarchies). The rules must be easily understandable and editable, and we therefore seek to provide tools to easily formalise constraints and schemas. As with S1.1 and S1.2, these constraints must be easy to understand and maintain by both PPT users and developers.

S1.4 – Import of Third-Party Datasets

User Story

We want an import functionality that ensures the imported data are consistent with the constraints defined in S1.1–S1.3. Data that are not used by PPT should not be imported and in case of conflicts, unfulfilled constraints or missing information, it should be possible to provide easy ways to repair the issues (see PS3).

Detailed Description

PPT currently supports importing RDF datasets that are added to the taxonomy knowledge graph. As this method bypasses the action framework, we mentioned in S1.1, it has several drawbacks. For example, it is possible to import data that PPT currently cannot use, e.g., labels in languages that are not configured in the project settings. A more severe problem is the import of incomplete data that can cause data integrity violations, leading to unexpected behaviour of the application.

Therefore, the approach we follow to tackle this story involves multiple steps:

- Extraction of actions from the imported dataset
- Execution of these actions
- Collect of potential constraint violations
- Providing (batch) resolution strategies for the occurred constraint violations

PS2 – Schema Generation and Validation

Description

We want a way to describe how to infer facts from other sources (e.g., DBpedia) and to convert them to facts in the local taxonomy. It should also be possible to find contradictions and constraint violations when applying a schema to a taxonomy.

Approach

Similar to the approach of PS1, S1.2, we want a solution to specify rules which allow the deduction of a data schema from a given Linked Data source and make it available for application to a PPT taxonomy project. There will be a default ruleset that cannot be changed by the user and as well as a user utility to edit and provide new custom rules. On the other hand, we want to exploit existing schemas that are already created for a PPT taxonomy to support and improve the taxonomy creation process. Therefore, we identify the following key challenges:

- Derive schema information from thesaurus structure using available data (such as mappings to DBpedia). Using a rule-based approach, schema classes and relations should be imported automatically from a given set of Linked Data sources.

- Leverage existing schemas which are developed by taxonomists in PPT for a certain project, to improve term suggestions, automatic population and thesaurus mapping with external Linked Data sources.
- Check the whole thesaurus for contradictions and inconsistencies (e.g., owl:disjoint, interpretation of domain and range properties as constraints, custom rules).

Goals

The goals are to develop automated support for assisting users to more easily describe the structure of the datasets that they wish to manage and lighter-weight support for tools which help them to develop consistent rules and constraints. We can break them down into the following steps:

- Derive schema from thesaurus
- Exploit schema for Linked Data lookup
- Validate taxonomy as a whole against a schema

S2.1 – Derive Schema from Thesaurus

User Story

We want a method to automatically suggest a data schema for an existing PPT project based on a provided set of Linked Data sources.

Detailed Description

In PPT, taxonomy developers can impose a data schema to their projects. For example, it is possible to assert class membership to concepts or specify property domain and ranges which are used in the application to constrain the resources they can be applied upon. In order to create such schemas, taxonomy developers currently have to either start from scratch and define custom classes and properties, or they can import resources from existing ontologies, such as FOAF or GeoNames which they can reuse in their projects. However, currently, we have no way of automatically suggesting classes and relations from these external sources. Such a functionality would greatly increase the value of the developed taxonomies, as it improves the coherence of the various Linked Data resources on the Web. We see the key challenges of this story in:

- Establishing or adopting an existing a rule-based framework for schema deduction

- Provide an easy way for reviewing and editing the involved rules
- Integrate the deduced schema in the PPT workflow

S2.2 – Exploit Schema for Linked Data Lookup

User Story

We want to improve term suggestions and taxonomy mappings, based on the existing schemas of a taxonomy.

Detailed Description

Currently, the schema of a taxonomy is not taken into account when additional terms are retrieved from external Linked Data sources such as DBpedia. The schema is also ignored when suggesting mappings to other projects or Linked Datasets, as it currently utilises label string comparison algorithms. Leveraging schema information, we expect to improve the accuracy and relevance of mappings and term suggestions and can build on an extensive amount of literature in the field of ontology mapping.

S2.3 – Validate Thesaurus Against Schema

User Story

We want a way to validate a thesaurus against a provided data schema and create a report on detected conflicts and inconsistencies.

Detailed Description

As users should have the possibility to automatically deduce schemata, it will also become necessary to evaluate if data in the taxonomy are contradicting against certain rules. This kind of evaluation has already been implemented using the qSKOS library of quality issues; however, in order to support custom schemas, a more generic approach is needed that is also compatible to the real-time action-based checks as described in PS1. The key challenges are to

- provide an implementation that performs the taxonomy check against the set of PPT-specific schema constraints (e.g., domain/range or disjointness contradictions),

- display a concise report of constraint violations, and
- integrate with work done in PS3

PS3 – Constraint Resolution

Description

We want a software component that helps users to resolve violated constraints or contradictions in the thesaurus.

Approach

As outlined in S1.1–S1.3 and S2.3, constraint violations can occur in various areas of taxonomy development (e.g., basic editing actions, schema definition, and reasoning). A single change of the taxonomy can involve multiple constraint violations that can potentially be resolved in various ways. We want to find a unified way to deal with this problem which covers:

- supporting both action-based as well as complete taxonomy checking and the results of the reasoning step,
- automatic generation of meaningful reports,
- generic definition of resolution strategies,
- (batch) execution of these strategies

Goals

Our goal is to develop tools and methods which enable users to more easily identify and correct the sources of constraint violations in complex ontologies. We want to evaluate

- ways of expressing resolution strategies for constraint violations,
- the possibility of creating curation interfaces from constraint violation reports,
- how to integrate resolution strategies and documentation with constraint formulation,
- how to embed the approach into the PPT workflow.

S3.1 – Definition of Constraint Violations

User Story

As a PoolParty admin, I want to be able to specify constraint violations for issues that I want to have checked when introducing data into a PoolParty project.

These constraint violations should be defined declaratively using standards. They should be human-readably and machine-readable and support taxonomy developers to conveniently manage (a large number of) constraint violations.

Detailed Description

In this story, we want to provide a method to define constraint violations in a generic, machine-readable way for allowing the taxonomy developer to easily introduce additional constraint checks into the application.

S3.2 – Application of Constraint Violation Repair Strategies

User Story

As a PPT taxonomy developer, I want to be able to apply resolution strategies for each type of constraint violation. These strategies should be machine-readable and support taxonomy developers to conveniently resolve (a large number of) constraint violations.

Detailed Description

We encountering a constraint violation, multiple ways for resolution exist. In some cases, a violation must be treated individually, while in others, a set of similar violations can be treated identically. In this story, we want to provide a method to formulate resolution strategies in a generic, machine-readable way for allowing the taxonomy developer to easily select an appropriate strategy from a list. When needed, taxonomy developers must be prompted for input of missing data, required to execute the resolution strategy. Even if a violation has to be treated individually, we still want to have an option for automatic resolution of all violations of a specific type.

S3.3 – Formulation of Constraint Violation Repair Strategies

User Story

As a PPT admin, I want to be able to specify resolution strategies for each type of constraint violation. These strategies should be machine-readable and support taxonomy developers to easily resolve (a large number of) constraint violations.

Detailed Description

We encountering a constraint violation, in most cases, multiple ways for resolution exist. In some cases, a violation must be treated individually, while in others, a set of similar violations can be treated identically. In this story, we want to provide a method to formulate resolution strategies in a generic, machine-readable way for allowing the taxonomy developer to easily select an appropriate strategy from a list. When needed, taxonomy developers must be prompted for input of missing data, required to execute the resolution strategy.

S3.4 – Creation of Repair User Interfaces

User Story

Based on detected constraint violations, we want a way to automatically create interfaces that help users in repairing the violations.

Detailed Description

Users need to effectively deal with fixing constraint violations in the taxonomies they create. We will therefore research ways for automated creation of user interfaces and how they can be used together with defined resolution strategies (see S3.2).

PS4 – Development Process Integration

Description

We want a way to express development process data that are currently maintained by using, e.g., JIRA, Confluence, or a Q&A application, as Linked Data, utilising the ALIGNED ontology. This enables us to improve coherence of the tools and build the foundation to link process-relevant information with data-model constraints. By (internally) publishing this data, we will be able to query it in a standardised way (SPARQL) and therefore detect interrelations and dependencies of, e.g., feature requests and software defects.

Approach

In the requirements engineering phase, we collect information on the affected parts of the software and the data constraints that must hold. Currently, this is mainly done in an unstructured form, and the data reside in relational databases of the respective tools. In our approach, we will

- Improve the structure of process-related documents such as, e.g., requirements and defect descriptions,
- Generate Linked Data from the existing process-related data,
- Provide a way to link process-related data to integrity constraints (which can be seen as the data model consistency specification).

Goals

We want to integrate the information flows from across the most relevant process-management application in use by SWC in order to produce a unified, machine-readable view on the development process data and how process-specific documents (e.g., defect tickets, new requirements) relate to one another. Our goals are to

- Structure process-relevant data in more detail,
- Publish data from development process tools as Linked Data for internal use and reporting purposes,
- Detect related or conflicting development process situations

S4.1 – Extract Data From Confluence and JIRA

User Story

We want a way to (periodically) extract process-related data in a structured form from the JIRA and Confluence instances.

Detailed Description

We will establish a way to extract process-relevant data from the used tools, JIRA, Confluence. We plan to use and extend UnifiedViews for these tasks, setting up a UnifiedViews data extraction pipeline. Data extraction can either occur on a timely basis (e.g., each day at 1 am), event-triggered (e.g., during system idle times, by updates in the system), or user-triggered (e.g., an administrator is responsible to start the extraction process). It is currently unknown how to deal with data updates. Options are to

- Always fetch and convert all data from the process tools, overwriting the existing data. Despite easy to implement, this option may be slow but serves as a proof-of-concept.

- Fetch only changed data and persist them as, e.g., changesets. This should be faster and provides the possibility to build up a history of changes. However, it is unknown if APIs support it and it should be more complex to implement.

Depending on the interfaces provided by JIRA and Confluence, we furthermore have to decide if we can and want to use dedicated API calls to extract the data of interest or if we can directly access the underlying persistence layer, i.e., the SQL databases. So far, the requirements are entered as unstructured text, it will be necessary to create Confluence Templates or Blueprints so that the data are collected in a structured way to support exploitation of the data as a next step.

S4.2 – Create RDF Data From the Extracted Data

User Story

We want to express process-related data as Linked Data.

Detailed Description

SWC will provide input for the ALIGNED metamodel in the form of informally describing the software development process and the workflow within the development team. The Design Intent Ontology (DIO) should be able to capture this information so that it can be used to express the structured data extracted in S4.1 in RDF. We will identify a mapping between the data available from JIRA and Confluence and the DIO. This is a basic requirement to produce a valid and reusable RDF dataset that can be processed by several (also third-party) tools. We may also additionally use the SLO for formalising development process-related data (such as, e.g., planned features and bugfixes) as well as the SIP, which are both developed within the ALIGNED project.

S4.3 – Internally Published RDF Data

Detailed Description

In the final stage, we want to have all development process-relevant data in their most current version in a triple store, providing the possibilities to

- Run SPARQL queries against the data: find, e.g., what tickets are based on what requirements and when have they been solved in what time

- Utilise PowerTagging on the data extracted from JIRA and Confluence to build up a combined PoolParty application and development ontology that helps to find duplicate or conflicting requirements and tickets
- Feedback data into JIRA or Confluence

S4.4 – Link Development Process Data with Data Model Integrity Information

User Story

We want a way for linking process-related information with data-related information (consistency constraints) in order to better align the software development process with the data model development.

Detailed Description

To get a comprehensive dataset of process-related information linked to data (and consistency constraints) objects, there are several possibilities. For example,

- Requirements engineers and consultants should have a way of linking process-related information (e.g., defect tickets, requirements) with data consistency constraint definitions. This can be done by, e.g., providing a plugin for the development management application that offers a pick-list of the available constraint and their associated version and description.
- A JIRA plugin exists that automatically links git commits to the processed tickets that are in turn linked to the requirements they originate from (see S4.1)

We will use the DLO to track and formalise changes of the PoolParty data model and consistency constraints.

PS5 – Code and Data Co-evolution

Description

We want to be able to identify mismatches between the deployed version of the application and the status of the data in the triple store.

Approach

As the code evolves and the data model is changed, existing installations must also be updated. If this is not or only partially done, the new version of the software would operate on an outdated data model which causes failures. SWC currently tackles this problem by using migration scripts which must be run manually after each deployment of a new software (PoolParty Thesaurus Server) version. However, it is currently not easily possible to get an overview which feature requires which migration script to be executed or to automatically find and run all needed scripts to prepare the triple store data in order to match the current software version. We also do not cover conflicts between the migration scripts, i.e., one script converting data to a format that is incompatible with the format expected by another script. We will address this issue by linking integrating specification of data constraints into the requirements documentation for component changes. This way it would be possible to find out the set of data constraints (based on the unified ALIGNED data model) and alert developers on potential conflicts. Our underlying assumption is that we can express data evolution in a constraint (rule) language that can be represented in a standardised, machine-readable way (as described in detail in PS1).

Goals

Increase the system's ability to automatically identify model-code mismatches suggest remedies to such mismatches and automate the execution of fixes.

Related Work

Results of PS1 and PS4

Examples

A practical example for usage of a migration script is as follows: earlier versions of PoolParty Thesaurus Server used the property dcterms:identifier to control URI creation of new concepts. As this was found to cause problems with some customer datasets, we needed to express the identifier of a concept by a property from the SWC ontology an, as a consequence, had to change the existing data in all customer projects. In future our goal is to specify data constraints as preconditions for all actions that rely on (in this case) identifier information and, on deployment, check for violation of these conditions, which can be combined with information from development process-planning applications.

S5.1 – Develop Plugins for Confluence and JIRA

Actors
- JIRA
- Confluence
- Developers
- Requirements engineers

Detailed Description
Having the extraction of relevant process-related data as RDF in place, this data should be leveraged to support collecting and editing of requirements. SWC will provide plugins for JIRA and Confluence that give users (requirements engineers, developers) UI support for answering some of the competency questions that the developed system should be able to cover (see S5.2).

Affected Components
- JIRA
- Confluence

Acceptance Criteria and Test Scenarios
Creating/Editing/Finding requirements and issues take less effort than without plugin support

S5.2 – Make Use of Collected Process-Related Data

Actors
- Requirements engineer
- Developer

Detailed Description
Based on the extracted RDF data, we define a set of exemplary competency questions, i.e., queries the system should be able to answer:
- Identify duplicate requirements and issues.
- Which issue has not been updated in the last defined time range and who did it?
- Which issues have similar characteristics?

Appendix A – Requirements 379

- What part of the software has the most "major" issue
- Which issue has the most watchers (e.g., to get an additional measure of issue importance)

Affected Components

- JIRA, Confluence Plugins
- RDF store

Acceptance Criteria and Test Scenarios

- Coverage of a subset of the identified competency questions
- Responses to competency question lie within an acceptable threshold of accuracy

S5.3 – Integrate Data Constraints Information with PPT Data Migration and Deployment Strategy

User Story

We want a way to combine information from the "data model" (i.e., integrity constraints) with the data migration tools used by SWC for new or updated PPT installations.

Actors

SWC system administrator

Detailed Description

By integrating data constraints with tools that are currently in use for migrating existing datasets to be compliant with new versions of PPT, we expect an easier and more effective deployment process. We expect a couple of advantages that will optimise SWCs deployment process:

- Automatically detect incompatibilities between data and software versions in advance and not only by user testing,
- Have better overview on the status of the customer data (which is beneficial if the development process shifts from version-based deployment to single feature deployment)
- Detection of missing migration scripts

Affected Components
PPT migration framework

Acceptance Criteria and Test Scenarios
Construct exemplary cases of incompatibility between data constraints and available customer data and show how combining data from constraint mismatches with the existing development process data can retrieve additional information on this incompatibility.

Requirements
The developed platform has been used for the following use cases:
- **RDF Validation:** Importing RDF data in PoolParty and using the integrated validation checks to identify problems, which are reported to the user as constraint checks. The user is then given options to repair the data consistency. After fixing the inconsistencies, the user can then import the data without the risk of application failure.
- **Issue Integration:** In the Semantic Middleware Configurator of PoolParty, the user can add a JIRA instance. When the user tries to import RDF into PoolParty that creates an inconsistency that is not covered by the RDF validation or that is caused by a software defect, it can cause an application failure. The user can then automatically report this issue to PoolParty Support.
- **Graph Search:** The faceted search application can be used to analyse the data. It is used to detect duplicate bugs and identify requirements or stories which are related to a specific bug.
- **RDF Validation:** repairing reported problems using the integrated repair strategies that are presented to the user for each constraint violation. Using the combined identification and repair of the violations, the user is able to fix import problems without help from consultants. The user does not have to know about PoolParty data conformance because the list of violations is presented. The problem does not have to be fixed in the original RDF data and can be repaired in the application itself. Therefore, the solution is more stable because the repair is controlled by the RDF Validation. Even if a repair introduces a new constraint violation, it will not be undetected, in contrast to repairs done on the RDF level. It will appear as a new constraint violation in the list and can then be repaired as usual. We will show for RDF data that were reported by customers to break the project on import

that it can be repaired using the RDF Validation by a user not knowing the conformance requirements of PoolParty.
- **Unified Governance:** The development artefacts are imported into the triple store using a UnifiedViews pipeline. This pipeline runs daily to keep the data up to the date. The pipeline also calculates similarities between the issues and requirements.
- **Unified Governance Search:** search for the RDF data representation of the development artefacts will be provided by a SPARQL endpoint for querying and by a Web application providing search features like facetted search. We will show advantages in terms of precision, recall and time consumption for retrieving specific requirements. Evaluation is done using an experiment where users search for issues based on predefined requirements texts. The results will be analysed regarding precision and recall. Also, the time needed to retrieve the correct results will be measured and it will be shown that the retrieval of issues can be done faster.
- **Unified Governance Similarity:** the similarity computation can be applied to find related development artefacts. It can be used to find corresponding requirements for JIRA issues for an automatic proposal of links. It can also be used to detect duplicate JIRA issues, which would also be linked as part of the software development process if already existent, but can be prevented in practice if the duplicate detection is applied before creation of an issue. Both use cases will be evaluated by comparing the results of the similarity computation to a gold standard created by consultants and developers of the PoolParty team. It will be shown that the similarity computation provides an advantage in issue management by reducing the management overhead and therefore saving time.
- **Unified Governance Statistics:** Visualisation of statistical values based on the RDF data. The information will be presented in form of diagrams as part of the search application. Evaluation will be based on a gold standard created by developers and will show tendencies in the development timeline that can be used for development process improvements.

DBpedia

Scenario 1: Wikipedia/Wikidata Evolution

DBpedia uses a very big code-base to handle an – as correct and as complete as possible – parsing of a Wikipedia language edition. This code base comprises a framework written in Java/Scala, a crowd-sourced ontology and crowd-sourced mappings between the ontology and the Wikipedia templates. As a Wikipedia evolves, the ontology, the mappings and the code must be adapted to match the current state of each distinct language edition.

Challenge

Decrease the effort to identify new errors or data loss due to the source evolution.

Scenario 2: Schema Evolution

The DBpedia ontology is most times adapted to match the Wikipedia contents; however, this may lead to fragmentation and require a refactoring of the DBpedia ontology. Any changes in the ontology, however, must be reflected in the crowd-sourced mappings.

Challenge

Decrease the effort to identify schema alignment errors.

Scenario 3: Validation

DBpedia is based on a crowd-sourced and semi-structured content. Data extracted from DBpedia must be checked for validity.

Challenge

Decrease the probability that the extracted dataset introduces invalid or erroneous values while minimising productivity overheads.

Scenario 4: Data Dissemination Life cycle

DBpedia will provide two static releases per year, following the release of new data, one has to adapt to the new datasets.

Challenge

Decrease the data adaptation effort.

Scenario 5: Feedback

Along with the data validation, the identified errors must be analysed and provide appropriate feedback to the error sources. Possible sources of an error can be (1) the data in Wikipedia, (2) the extraction framework, or (3) the data and schema mappings.

Challenge

Decrease the effort of associating errors to a specific source and providing feedback to each community (data editors, mappers, and developers).

DS1 – Wikipedia/Wikidata Evolution

The main goal of this epic to minimise the effort to align the DBpedia tool and data stack with regard to the Wikipedia/Wikidata Evolution.

Description of Epic

DBpedia uses a very big code base to handle the parsing of a Wikipedia language edition (as correct and as complete as possible). This code base contains for a framework (DBpedia Information Extraction Framework – DIEF) written in Java/Scala, a crowd-sourced ontology and a crowd-sourced mappings between the ontology and Wikipedia templates. As Wikipedia evolves the ontology, mappings and code base have to be adapted to match the current state of each distinct language edition.

Approach

For every release, the release manager and the extractors analyse the latest Wikipedia community trends and try to adapt the DBpedia mappings and the Information Extraction framework to the latest version. Different tools have been developed that facilitate change detection but require manual steps to complete. One approach that can make this process more efficient is the comparison of different Wikipedia language editions. Having complete coverage of all the Wikimedia projects will allow us to do a cross-project comparison and easier identify data and schema trends. There are mainly two Wikimedia projects left for DBpedia to incorporate in the DBpedia data stack: Wikimedia Commons and Wikidata.

Goals

Decrease the effort to identify new errors or data loss due to source evolution.

DS1.1 – Wikimedia Commons Integration as a DBpedia Dataset

User Story

As a DBpedia user, I want to be able to query multimedia content from Wikimedia Commons and get metadata as well as license information.

Detailed Description

Wikimedia Commons is a wiki describing multimedia content with different licenses that is linked through different Wikipedia language editions. We want to integrate Wikimedia Commons in the DBpedia dataset ecosystem to provide queryable metadata.

DS1.2 – Wikidata Integration as a DBpedia Dataset

User Story

As a user, I want to be able to query Wikidata using the stable DBpedia ontology.

Detailed Description

This case is similar to DS1.1 but the difference with Wikidata is that the data they store are not in WikiText like all other Wikimedia wikis but in JSON which requires changes in the DBpedia Information Extraction Framework architecture. However, this case is essential for DS1.3 to be able to have complete coverage of all Wikimedia information. The goal is to integrate Wikidata in the DBpedia dataset ecosystem.

DS1.3 – Evaluate Fact Overlap and Conflicts

User Story

We want to find ways to evaluate the overlapping and conflicting facts between different Wikipedia language editions, Wikimedia Commons and Wikidata.

Detailed Description

As a Release Manager, I want to know the overlapping and conflicting facts between different DBpedia datasets originating from Wikipedia language editions, Wikimedia Commons, and Wikidata. This report can provide an overview of quality for each dataset and can be used to build a new fused knowledge base.

DS2 – Schema Evolution

The DBpedia ontology is most times adapted to match the Wikipedia contents; however, this may lead to fragmentation and require a refactoring of the DBpedia ontology. Any changes in the ontology however must be reflected in the crowd-sourced mappings.

Description of Epic

The DBpedia ontology is a crowd-sourced collaborative project that evolves together with the DBpedia Infobox to ontology mappings. As Wikipedia infoboxes change over time, ontology adjustments are required to facilitate these changes. However, this sometimes leads to ontology fragmentation, especially when there is not adequate coordination between the mapping and ontology editors. In turn, when ontology changes are attempted, they must be made in care to not break any existing mappings

Approach

The approach to tackle this problem is to develop or deploy tools that can validate both the ontology and the mappings. The tools will feed reports to the ontology and the mapping editors and allow them to faster identify alignment errors.

DS2.1 – Mapping Validation for Conformance to DBpedia Ontology

Goals

The main goal of this epic is to decrease the effort for the mapping and ontology community to identify alignment errors.

User Story

As a mapping editor, I want to be able to determine if a mapping I create is in conformance to the DBpedia ontology.

Detailed Description

Writing Infobox to ontology mappings is a manual process. The mapping editor cannot easily get feedback for the mappings he/she created in respect to conformance to the DBpedia ontology. For example, mapping an infobox to class 'Place' and an infobox property to 'isbn' is a mapping that produces data but 'isbn' should be used with the class 'Book' (the rdfs:domain of isbn is class Book). If such an error passes unnoticed, we can only identify it after a release. The idea is to validate directly the mapping without requiring an actual extraction and provide direct feedback to the mapping editors. For this case, we export the mappings from WikiText that they are currently defined in the mappings wiki to RML, a Relational Mapping Language and treat the mappings as an RDF dataset. Goal: Validate the Infobox to ontology mappings.

DS2.2 – DBpedia Ontology Validation

User Story

Goal: Validate the DBpedia ontology

As an Ontology Editor, I want an integrated tool that can analyse the DBpedia ontology and identify various errors such as inconsistencies and unsatisfiable classes.

As a Release Manager, I would like to ensure that the ontology is consistent before it is published in a release.

Detailed Description

- Examine what is the best way to integrate ORE in the ontology editing workflow
- Make it easy for ontology editors to run ORE on the current version of the DBpedia ontology
- Check ORE report for errors
- If errors exists, try to resolve them

DS3 – Validation

DBpedia is based on a crowd-sourced and semi-structured content. Data extracted from DBpedia must be checked for validity.

Description of Epic

We want to ensure that all DBpedia ontologies are well-formed that instance data conform to the appropriate ontology and that interlinks to other datasets contain no bad links to missing or inappropriate targets.

Approach

Currently, there is very limited validation of new releases of DBpedia, and most checks are manual, depending on crowd-sourced feedback on the datasets. The DBpedia release cycle consists of a large number of steps performed by independent tools. This release process needs to be formalised and structured to include iteration and a validation step that provides feedback in the form of error/validation reports to the other stages.

Goals

Decrease the probability that a DBpedia release candidate dataset includes invalid or erroneous values while minimising productivity overheads. Add feedback to other stages of release cycle so that confidence in the outputs of those stages can be increased and detected errors can be fixed.

DS3.1 – Instance Validation

User Story

As a Release Manager, I want to be able to have an overview of the quality of the release before it is made available.

Detailed Description

Before a DBpedia release is announced, the Release Manager needs to have a unified quality overview of the datasets which are going to be released. If the quality is adequate, the release is marked for announcement; otherwise, the datasets are further inspected and the release is postponed until all major errors are resolved. The validation will be performed with RDFUnit for instance validation and the report will include both the instance validation and high-quality metadata.

DS3.2 – Interlink Validation

User Story

Goal: Identify bad outlinks (e.g., owl:sameAs) between a DBpedia release candidate and an external dataset, e.g., Freebase

As a Release Manager, I want to check a release candidate for bad links to other datasets and validate that the links included in the release are correct. I need a report describing the checks performed, any outlink triples that have problems and a characterisation of the problem to help me or another member of the team to fix it.

Detailed Description

- Identify release candidate to be validated.
- Identify target external dataset to be validated.
- Identify type of outlink predicates/properties to check.
- Run command line script to validate links, generates report.
- Check report for errors.
- If errors, fix errors in release candidate and goto 4, else finished.

DS4 – Data Dissemination Life cycle

DBpedia provides two static releases per year. Disseminating the data of the new releases to the community or to testers requires adaptation effort that we need to decrease.

Description of Epic

This Epic targets the consumption of a new release, which involves a lot of manual steps.

- Visit the release download page
- Select part or all the datasets and download them
- Setup and build a local Virtuoso Server (or another Triple-Store)
- Load the downloaded datasets in the Triple-Store
- Query the data

Approach

We will try to use the expressive power of DataID (dataid.dbpedia.org). Representing the release data in DataID provides a machine-readable format of the release that different tools can harvest. Another technology we will try to exploit is Docker (*www.docker.com*) that provides prepackaged software that can run on different operating systems.

Goals

Automate as many steps as possible and minimise human interaction

DS4.1 – Generate a DataID Based on the New DBpedia Release

User Story

As a complex, semantically rich metadata format for describing datasets, DataID is a central part of the disseminated metadata about a new DBpedia release. Thus, generating a DataID containing all information about datasets, licenses, contact information, and so on is a necessary step in the publishing process of DBpedia.

Detailed Description

Manual creation: use the Web interface on dataid.dbpedia.org, follow all steps

Automated creation: use the DataID-Hub API

DS4.2 – Create DBpedia Docker Image

User Story

Running a SPARQL endpoint of DBpedia is currently a complex task that requires a lot of effort and technical support for domain experts. To simplify the task, we want to provide dockerised version of a triple store loaded with all necessary data for a DBpedia endpoint. Part of this task, as well as a common use case for DBpedia users, is the upload of multiple DBpedia files into a triple store. We want to automate this operation to avoid general mistakes and expedite the process.

Detailed Description

- Create a list of all files needed for the SPARQL endpoint (e.g., use distribution list of DataID) (plain text, dcat, DataID).

- Provide a dld.yml bootstrap script, which is the basis for the consequent Docker Compose file. The script also downloads or copies DBpedia files into a given or default working directory and let the docker compose file know where your data are.
- Execute the dld script in conjunction with the file list. Part of this process is to automatically import the necessary files into the triple store (e.g., Virtuoso).
- Commit the triple store to the new docker image. Add additional components if needed to the image.
- Export and upload the docker image.
- Test the image by installing the new docker image on a target machine.

DS4.3 – Deploy a Dockerised DBpedia Release

User Story

DBpedia users should be able to deploy a dockerised version of a DBpedia endpoint as easy as possible.

Detailed Description

- Download a prepared Docker image archive containing the Triple Store database files for the endpoint dataset collection
- Import the image into the Docker Engine of the desired host machine
- Use DLD command line script with appropriate provided configuration file to start a container with the triple store

DS4.4 – DBpedia Release Download Page Autogeneration

User Story

Based on the newly extracted DBpedia files, a download Web page is created before the final announcement. Besides a detailed list of the downloadable content, statistics about the release as well as additional information (e.g., license, contact, etc.) pertaining to the described datasets are displayed. An automatic generation of this Web page based on the DataID (S4.4) is paramount to a decrease in time needed to publish a new DBpedia release.

Detailed Description
- Create a script which generates the download Web page in the expected format based solely on a DataID of the new DBpedia release
- Run the script and deploy the resulting html

DS5 – Feedback

Explore ways of providing validation feedback to the error source.

Description of Epic

Along with the data validation, the identified errors must be analysed and provide appropriate feedback to the error sources. Possible sources of an error can be (1) the data in Wikipedia, (2) the extraction framework, or (3) the data and schema mappings.

Approach

Identifying value errors in a single Wikipedia edition is not easily feasible unless we can compare the values to an authoritative source. However, authoritative sources are specialised in certain domains and need a lot of manual configuration to set up properly. The main idea to automate this step is to use data from all different Wikipedia language editions, Wikimedia Commons, and Wikidata. If we align the articles using the interlanguage links, we are able to compare values. When the values have minor disagreements, we send them to the respected Wikipedia communities and let them fix the errors in the source.

Another approach for error feedback to the mapping community is to move the validation step closer to the actual data (the mappings in this case). If we are able to emulate the mapping application without having to perform a full extraction, we provide a tighter and quicker feedback to the mapping editors.

Goals

Decrease the effort of associating errors to a specific source and providing feedback to each community (data editors, mappers, and developers).

DS5.1 – Automated Validation Reports for Infobox Mapping Definitions

User Story

Goal: Mapping Editors should get feedback on invalid mapping definitions. As a Mapping Editor, I want to get better feedback on the infobox mapping definitions. This feedback will allow me to revise wrong mappings and in turn improve the quality of DBpedia.

DS5.2 – Report Erroneous Facts to Wikipedia and Wikidata

User Story

Goal: Identify erroneous facts in a Wikipedia language edition or Wikidata and report them to the respecting communities. As a Release Manager, I want to fix errors in the data sources I use. Identifying these errors and reporting them back to the source maintainers for fixing, allows future extractions to contain less errors and provide a better experience to the DBpedia users and an easier task for the Extractors.

Requirements

The DBpedia release process requires a number of processing stages in order to validate and prepare each dataset for publication. The platform has been constructed to provide automation support for the following DBpedia processing stages:

- Validation of a DBpedia release to identify instance violations.
- Validation of the DBpedia infobox-to-ontology mappings to prevent mapping errors from propagating on the instance data.
- Validate interlinks from DBpedia to other datasets and remove links to targets that no longer exist.
- Automatically generate a download page for a DBpedia release
- Automatically deploy a triple store with a DBpedia release in a docker container.

The requirements for this are:

- Wikipedia/Wikidata Evolution: Decrease the effort to identify new errors or data loss due to the source evolution.
- Schema Evolution: Decrease the effort to identify schema alignment errors.

- Validation: Decrease the probability that the extracted dataset introduces invalid or erroneous values while minimising productivity overheads.
- Data Dissemination Life cycle: Decrease the data adaptation effort.
- Feedback: Decrease the effort of associating errors to a specific source and providing feedback to each community (data editors, mappers and developers).

Index

A
ALIGNED metamodels 79, 113
ALIGNED tools 19, 40, 215, 254
API (Application Programming Interface) 176
artefacts 3, 79, 127
automation 32, 138, 261

B
Booster 34, 73, 142, 220

C
catalogue 9, 34, 71, 127
catalogue contents 129, 134, 136, 137
classes 86, 95, 103, 129
classification schemes 116, 117
code book 24, 130, 193, 236
conceptual 51, 83, 115, 161
consistency constraints 19, 35, 176, 277
crowd-sourced public datasets 97, 103, 106, 185
custom schemas 214, 282, 393, 370

D
Dacura 17, 33, 99, 172
Dacura Quality Service 89, 170, 177, 198

data agility 40, 219, 233, 255
data and software engineering 23, 54, 60, 76
data consistency violations 215, 276, 279
data elements 71, 138, 216, 252
data engineers 21, 53, 147, 299
data modelling 51, 59, 75, 116
database schemas 18, 112, 251
data-intensive software systems 3, 59, 75
Data-Intensive System Evolution 1
data-intensive systems 8, 13, 53, 75
DataID 17, 22, 38, 68
dataset curators 164, 165, 167, 172
dataset metadata 87, 109, 268, 306
datasets 18, 30, 42, 88
DBpedia 36, 69, 86, 103
DBpedia ontology 27, 107, 310, 382
DBpedia release 105, 109, 302, 388
demonstrator system 23, 26, 241, 226
design 188, 197, 235, 262
Design Intent Ontology 81, 83, 274, 375

design intents 8, 12, 112
design rationale 81, 83
development artefacts 38, 278, 293, 381
development data 293, 295, 301
development process 22, 57, 110, 138
documentation 28, 52, 89, 112

E

editors 27, 130, 134, 150
ELV (External Link Validation) 22, 216, 219
engineering activities 79, 82
engineering domain 3, 71
engineering life cycles 17, 43, 58, 76
engineering phases 55, 62
engineering processes 8, 39, 54, 315
engineering projects 50, 86
engineering systems 13, 37
engineering tools 7, 72, 108, 263
entities 2, 81, 203, 354
evaluation 5, 67, 218, 381
experts 11, 165, 260, 389
external dataset 30, 107, 223, 388
geographic information system (GISs) 25, 237
graphs 26, 122, 206, 364

I

imported data 240, 279, 299, 368
information systems 1, 15, 139, 320
instance graph 170, 180, 183, 198
integration points 40, 53, 220, 347
interlink validation 108, 184, 388

Interlink Validation Tool 105, 184, 188, 388
interlinks 29, 71, 189, 392
iterative approach 42, 46, 61, 65
iterative process 32, 46, 66, 260

J

JIRA 22, 111, 278, 381
JSON (JavaScript Object Notation) 101, 197, 235, 384
JURION 29, 66, 158, 217, 347
JURION IPG 23, 39, 215, 229

L

life cycles 5, 43, 79, 345
link validation 22, 303, 311
Linked Data 1, 79, 168, 375
linked data quality 12, 59, 182, 316
links 27, 48, 212, 392
Linksets 28, 311
log files 185, 190, 301

M

mappings 3, 152, 239, 392
MDE (model-driven engineering) 8, 49, 121, 267
metadata catalogue 18, 263, 266
metamodels 8, 55, 112, 294
metrics 4, 95, 262, 323
model agility 40, 232, 301, 323
Model Catalogue 17, 34, 112, 272
model catalogue tool 34, 70, 227, 359
model components 128, 137, 143
modelling 2, 58, 139

modelling languages 22, 49, 139, 162
modelling tools 193, 255
models 1, 60, 136, 361

N
named graphs 170, 206, 293
notification messages 207, 293
ontologies 5, 82, 113, 387

P
PoolParty 16, 98, 211, 361
Process Model 167, 171

Q
Quality of Service 218, 322

R
RDF Validation 35, 96, 155, 286
RDFunit 17, 35, 155, 387
Reasoning Violation Ontology 89, 90
Resource Description Framework 2

S
schema graph 158, 170, 180
Semantic Booster 17, 72, 143, 359
semantic models 2, 18, 165
semantic web 89, 116, 153, 339
Semantic Web Company 16, 125, 201, 313
Seshat 18, 101, 228, 350
SHACL (Shapes Constraint Language) 19, 156, 278, 365

SIP (Software Implementation Process) 80, 86, 375
SLO (Software Life cycle Ontology) 17, 104, 375
software and data engineering 79, 184, 358
software and data engineering life cycles 28, 73, 76
software artefacts 5, 248, 261
software development 27, 109, 203, 381
software development processes 40, 51, 55
Software Engineering 43, 82, 249, 316
software engineers 37, 197, 246, 268
software life cycle 36, 85, 113, 202
software systems 3, 75, 125
software tools 17, 81, 165, 358
source dataset 29, 184, 312
SWC (Semantic Web Company) 16, 112, 274, 379

T
test cases 19, 156, 277, 365
triple-store 168, 190, 368, 388

U
UML (Unified Modeling Language) 2, 71, 166, 263
URIs 81, 178, 280, 331

X
XML schema 32, 158, 259, 347

About the Editors

Dr. Kevin Feeney was a senior research fellow in the School of Computer Science and Statistics, Trinity College Dublin, the Information Technology Editor of Seshat Global History Databank, and leader of the ALIGNED research project. He has more than 20 years of experience in developing innovative socio-technical systems in both research and industry, including pioneering distribution of prepaid mobile phone credit through credit card networks and terminals, and developing one of the first citizen journalism plaforms. He is now CEO of DataChemist.

Prof. Jim Davies is Professor of Software Engineering and the director of the Software Engineering Programme in the Department of Computer Science, University of Oxford. He is a Fellow of Kellogg College. His research interests include the development of automatic generation of systems from re-usable models of structure and functionality, and he is the Principal Investigator on CancerGrid, a consortium to develop open standards for clinical cancer informatics.

James Welch is a researcher in the Department of Computer Science, University of Oxford. His research interests are the development of enterprise software, "model-driven" technologies, the Unified Modelling Language, model transformations and metamodelling. He is also interested in formal methods, refinement and in particular the languages of Z and B. These interests are combined in a current project on the Booster language and Toolkit, and put into practice on a number of development projects.

Dr.-Ing. Sebastian Hellmann is the head of the Knowledge Integration and Linked Data Technologies group in University of Leipzig's Agile Knowledge Engineering and Semantic Web Group. He is also the executive director and a board member of the non-profit DBpedia Association. He focusses on semantic technology research – often in combination with other areas such as machine learning, databases, and natural language processing.

About the Editors

Christian Dirschl is Chief Content Architect and Head of Content Strategy and Architecture for Wolters Kluwer Germany. Here, he is responsible for Wolters Kluwers' taxonomies, ontologies, content structures and metadata, and managing automatic classification and text mining projects. He has represented Wolters Kluwer Germany in international research projects such as LOD2, WDAqua, and ALIGNED.

Andreas Koller is the Chief Information Officer and co-founder of the Semantic Web Company. He is responsible for the IT infrastructure of the company and PoolParty cloud services. One focus of his activity is on the evaluation of tools for the Semantic Web. In this role, he has contributed to the technical implementation of semantic technologies in numerous projects, with the design of system architectures and concepts being a key aspect where customer added-value can be generated through the use of semantic technologies.

Dr. Pieter Francois is an Associate Professor in Cultural Evolution at the University of Oxford. Since 2017 he is the Dean of St. Benet's Hall. In 2011 he founded the Seshat: Global History Databank which attracted several large grants from the ESRC, ERC, Horizon2020 and the John Templeton Foundation. He published extensively on evolutionary anthropology, digital humanities and the history of nineteenth century travel and migration. He is the author of 'A little Britain on the Continent'. British perceptions of Belgium, 1830–1870 (Pisa University Press, 2011).

Prof. Dr Hab. Arkadiuz Marciniak is a professor at the Institute of Archaeology at Adam Mickiewicz University in Poznań. His research interests include the Neolithic period. He is the regional editor for Mesopotamia in the Seshat Global History Databank.